Theory and Practice of
Foundation Design

Theory and Practice of
Foundation Design

N.N. SOM
Professor of Civil Engineering
Jadavpur University, Kolkata

S.C. DAS
Professor of Civil Engineering
Jadavpur University, Kolkata

Prentice-Hall of India Private Limited
New Delhi - 110001
2003

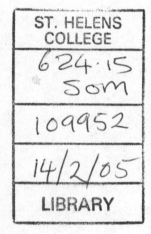
Rs. 250.00

THEORY AND PRACTICE OF FOUNDATION DESIGN
by N.N. Som and S.C. Das

ISBN-81-203-2190-1

The export rights of this book are vested solely with the publisher.

Second Printing ... ... **August, 2003**

Published by Asoke K. Ghosh, Prentice-Hall of India Private Limited, M-97, Connaught Circus, New Delhi-110001 and Printed by Jay Print Pack Private Limited, New Delhi-110015.

To
Professor NOEL E. SIMONS
and
Indian Geotechnical Society

Contents

Preface

Over the long years of our association with the profession of teaching and research in civil and geotechnical engineering, our experiences suggest that foundations are often designed without taking all the essential parameters into consideration particularly those with regard to geology of the site, soil data, ground conditions, type of structure, and land use pattern in the vicinity of the construction area. Though the theoretical aspects of foundation design are well understood, the field situations are usually not given the due importance. As a consequence, the implementation of the proposed design runs into problems necessitating changes in construction methodology or in some cases, even the design.

The book covers the essential features of foundation design through fourteen chapters. The foundation design requires understanding of the soil type, its strength and deformation characteristics, ground water table, and the other details of the site and the structures. We have striven to incorporate all these criteria so as to acquaint the reader with all necessary aspects of foundation design. This treatment begins with Chapter 1 that details the engineering properties of soil as required for the design of sound foundations. Chapter 2 discusses site investigation as the next step towards foundation design. This chapter elaborates on the various methods of soil exploration and testing. The design parameters and the importance of proper interpretation of the data collected from soil investigation are described in Chapter 3 while Chapter 4 introduces different types of foundations and their characteristics.

Chapters 5 through 7 offer detailed study of the mechanical properties of the soil including the stress distribution, the bearing capacity of foundations, and settlement. The effect of non-homogeneity and nonlinearity of the stress–strain relationships on the stress distribution in soils is elucidated and the concept of stress-path method of settlement analysis is introduced in these chapters. The important considerations for obtaining the design parameters from a large amount of soil test data are highlighted. The design procedures for shallow and deep foundations, as also for well foundations are presented in Chapters 8 through 10. The special requirements of expansive soils are covered and the earthquake response of soils and foundations are emphasized in Chapters 11 and 13 respectively. The ground improvement techniques commonly used in practice and the construction problems generally encountered at site are adequately dealt with in Chapter 12. The reader should also find in the book a comprehensive treatment of the design procedures vis-a-vis the construction problems and practices (as discussed in the final chapter). The chapters on analytical aspects are followed by worked-out examples taken from real-life problems which make the reading both topical and interesting. Besides, numerous references have been made to actual cases of foundations for better clarity and understanding of the topics covered.

We have drawn upon our experiences in teaching and consultancy in the same discipline to compile this volume. Almost forty years of teaching and research at Jadavpur University have given us the opportunity to face many field situations which had to be treated in unconventional ways. The interaction with our students has also been very helpful during all these years. This book is designed to serve undergraduate and postgraduate students of civil engineering with interest in foundations—their design, development, and maintenance. It will be equally useful for practising civil and structural engineers who have to design foundations of structures in difficult subsoil conditions.

We acknowledge the immense help derived out of our association with colleagues in the civil engineering department at Jadavpur University, Kolkata. We are thankful to Prof. R.D. Purkayastha, Prof. P. Bhattacharya, Prof. S.C. Chakraborty, Prof. S.P. Mukherjee, Dr. S. Ghosh, and Dr. R.B. Sahu who participated in many fruitful discussions on the subject and gave their invaluable suggestions. We are also grateful to our staff at the soil mechanics division, particularly Mr. Robin Pal, Ms. Apurba Mukherjee, Mr. Sisir Mondal, Mr. Rantu Jana, and Mr. Bhupesh Ghosh for their skilled help in many ways. The painstaking task of compiling the manuscript was undertaken by Mr. Bivas Roy, Mr. Subhasish Ghosh, and Mr. Hrishikesh Nayak with great interest and patience. We express our sincere thanks to them all.

Last but not the least, we express our sincere gratitude to our spouses Smt. Rita Som and Smt. Sikha Das for bearing with the demand of time that was needed to complete the task often under trying circumstances.

N.N. SOM
S.C. DAS

Soil as an Engineering Material

1.1 INTRODUCTION

From an engineering viewpoint, soils and soft rocks comprise all the loose and fragmented materials that are found in the earth's crust. They are distinguished from solid rock, i.e., the hard and compact mass in the earth's body, which cannot normally be excavated by manual means. Soils are formed by disintegration or decomposition of a parent rock by weathering or they may be deposited by transportation from some other source.

Soils which are formed by the disintegration of a parent rock and remain at their place of formation are known as *residual* soils. The disintegration of the parent rock is caused by physical agents such as temperature changes, freezing, thawing etc. or by chemical agents like oxidation, hydration etc. When the soil is transported from its original bed rock by forces of gravity, wind, water or ice and re-deposited at another location it is known as *transported soil*. Transported soils are generally sorted out according to their grain size as the velocity of the transporting medium gets reduced away from the source. After deposition at a new place, these soils may be subjected to further weathering with the passage of time.

Transported soils are classified into different types according to their mode of transportation. Deposits of soil that are formed by wind are called *Aeolian deposits*. Sand dunes and loess are examples of these deposits. Loose sand is generally swept by wind and transported close to the surface. If the motion is stopped, it is deposited in the form of sand dunes. The common transported soils are, however, those which have been carried by water or ancient glaciers. *Marine soils* which have been carried by sea water and *Alluvial soils* which have been carried by rivers and streams constitute probably the largest group of transported soils on earth. These deposits may also be called *sedimentary* deposits as they have been formed by deposition from either standing or moving water. The deposition is primarily caused by the gradual decrease in velocity of river carrying the sediments. A larger part of the great Indian plains is made up of alluvial deposits. *Glacial* deposits are remnants of the ice age that were carried along by the moving ice. They are generally found as big boulders at places away from their parent rock and are heterogeneous in nature with little or no stratification. Other sedimentary deposits are the *Lacustrine* soils which are deposited on a lake bed and the *Estuarine* soils which are deposited at the mouth of an estuary.

1.2 NATURE OF SOIL

Properties of soil are complex and variable, being primarily influenced by the geological environment under which they have been deposited. An understanding of soil composition is important in appreciating the mechanical behaviour of the soil. Natural soil consists essentially of discrete solid particles which are held together by water and/or gas filling the pore space. These particles are, however, not bonded as strongly as the crystals of a metal are and can, therefore, move freely with respect to one another. The size and shape of grains and the mineralogical composition of soil particles varies widely in nature. However, in coarse-grained soils the most important properties do not depend on the constituent minerals although locally, the minerals may control the frictional characteristics of the individual grains. In these soils, the particles are so large that the forces between the grains other than those due to externally applied forces and gravity are small. The non-clay minerals such as mica, feldspar, and quartz which constitute sand and silt do not render any plasticity and cohesion to the soil. Thus, the influence of the constituent minerals becomes appreciable with the decrease in size.

Clay minerals are hydrated aluminium silicates in crystalline form. These are generally of three different types (Scott 1965):

$$
\begin{array}{ll}
\text{Kaolinite} & : \text{Al}_4\text{SiO}_4\text{O}_{10}(\text{OH})_8 \\
\text{Montmorillonite} & : (\text{Al}_{1.67}\text{MgNa}_{0.33})\text{Si}_4\text{O}_{10}(\text{OH})_2.\text{H}_2\text{O} \\
\text{Illite} & : \text{K}_y(\text{Al}_4\text{Fe}_4\text{Mg}_{4.6})(\text{Si}_{8-y}\text{Al}_y)\text{O}_{20}(\text{OH})_4
\end{array}
$$

Kaolinite has a very stable structure. It generally resists the ingress of water and consequently, undergoes little volume change when in contact with water. On the other hand, Montmorillonite attracts water and undergoes large swelling and expansion when saturated with water. Most of the black cotton soils of India contain clay minerals of this variety. Illite is less expandable than montmorillonite.

A single particle of clay consists of many sheets of clay minerals piled one on another, as shown in Fig. 1.1. As each sheet has a definite thickness but is large at right angles to its thickness, clay particles are believed to be plate-shaped. The flat surfaces carry residual negative charges, but the edges may carry either positive or negative charges depending upon the environment.

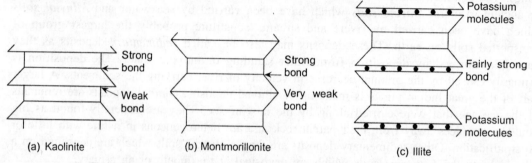

Fig. 1.1 Structure of clay particles.

Particle orientation has an important bearing on the engineering properties of a soil. Spacing and orientation of particles influence the development of interparticle bonds. For cohesionless soils, individual grains may be approximated as spheres with loose, dense, or honeycombed structure, as shown in Fig. 1.2. A dense structure is more stable than the loose or honeycombed structure. Figure 1.3 shows some simplified structures found in clay. The development of structure is influenced by the origin and nature of deposition of the soil. Thus, the flocculated structure is typical of clay deposits in salt water. This structure may change due to leaching or by external influences, such as loading, drying, freezing, electro–osmositic processes, and so on.

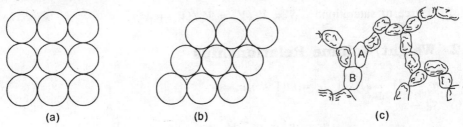

(a)   (b)   (c)

Fig. 1.2 Structure of cohesionless soils.

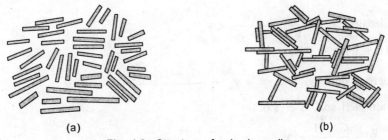

(a)   (b)

Fig. 1.3 Structure of cohesive soils.

1.3 THREE-PHASE SYSTEM

Figure 1.4 shows a typical soil skeleton consisting of three distinct phases—solid(mineral grains), liquid(usually water), and gas(usually air). These phases have been separated to facilitate the quantitative study of the proportional distribution of different constituents.

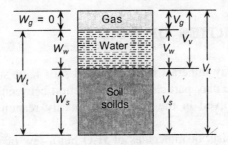

Fig. 1.4 Three-phase system.

1.3.1 Definitions

(a) Unit weight, Water   : $\gamma_w = W_w/V_w$
             Solid particles : $\gamma_s = W_s/V_s$
             Specific gravity: $G_s = \gamma_s/\gamma_w = W_s/V_s\gamma_w$

(b) Water content, $W = (W_w/W_s) \times 100\%$

(c) Void ratio, $e = V_v/V_s = V_w + V_g/V_s$

(d) Porosity, $n = V_v/V = V_v/(V_v + V_s)$

(e) Degree of saturation, $S = V_w/V_v = V_w/(V_w + V_g)$

1.3.2 Weight/Volume Relationships

(a) $n = e/(1 + e)$ or $e = n/(1 - n)$

(b) $Se = wG_s$

(c) Bulk density, $\gamma = \dfrac{W}{V} = \dfrac{Se + G_s}{1 + e} \times \gamma_w$

(d) Dry density, $\gamma_d = \dfrac{W_s}{V} = \dfrac{G_s}{1 + e_o} \times \gamma_w$

(e) $\gamma = (1 + w)\,\gamma_d$

(f) Submerged density, $\gamma' = \gamma - \gamma_w = \dfrac{(G_s - 1) - (1 - S)e}{1 + e} \times \gamma_w$

For saturated soil, $S = 1.0$,

$$\therefore \quad \gamma' = \frac{G_s - 1}{1 + e} \times \gamma_w$$

The range over which the typical values of the above parameters vary are as follows:

(i) $G_s = 2.60–2.75$

(ii) $\gamma = 1.60–2.25$ g/cc

(iii) $\gamma_s = 1.30–2.00$ g/cc

(iv) $n = 0.25–0.45$ (for sand)

(v) $S = 0$ (for dry soil)–100% (for fully saturated)

1.4 INDEX PROPERTIES OF SOIL

As already mentioned, the clay minerals in fine-grained soils have sufficient surface forces to attract water molecules to the clay particles. The interaction between the clay minerals, water, and various chemicals dissolved in the water is primarily responsible for developing the consistency of these particles.

Pure water mainly consists of molecules of H_2O but a few of them get dissociated into hydrogen ions, H^+ and Hydroxyl ions, OH^-. If impurities such as acids and bases are present, they also dissociate into cations and anions. Salt, for example, breaks up into Na^+ and Cl^-. Since plane surfaces of the clay minerals carry negative charges, cations including H^+ from water are attracted towards the surface of these particles. The water molecules closest to the clay particles, called the adsorbed water, are tightly held to the clay and exhibit properties which are somewhat different from those of ordinary water. This adsorbed water is believed to lead cohesive and plastic properties to clayey soils. It is obvious, therefore, that the amount of water present in a clay will determine its plasticity characteristics and, in turn, its engineering properties. The Atterberg limits are designed to serve as an index of the plasticity for clayey soils, refer Fig. 1.5.

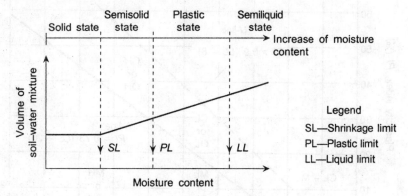

Fig. 1.5 Limits of consistency.

Starting with a low water content a clayey soil first appears to be a solid and moves to the plastic state with increasing water content. The word *plastic* here refers to the ability of a soil to be moulded into different shapes without breaking up. At even a higher water content, the soil begins to flow as a viscous fluid. The Atterberg limits, that is, the liquid limit (LL), the plastic limit (PL), and the shrinkage limit (SL) indicate the limits of water content at which the consistency of clayey soil changes from one state to another.

The Atterberg limits along with the natural water content give useful indication of the nature of the clayey soil. A natural water content close to the liquid limit indicates a soft compressible soil while a natural water content close to the plastic limit is characteristic of a stiff and less compressible clay.

Plasticity index, *PI*

The range of water content over which a soil remains plastic is called the plastic limit.

i.e. $$PI = LL - PL(\%) \tag{1.1}$$

Liquidity index, *LI*

It is the ratio of natural water content, w of a soil in excess of its plastic limit to its plasticity index and is indicative of the state of the water content in relation to the liquid limit and the plastic limit of the soil.

$$LI = \frac{w - PL}{LL - PL} \times 100\% \tag{1.2}$$

1.4.1 Plasticity Chart

It has been observed that properties of clay and silt can be correlated at least qualitatively with the Atterberg limits by means of the *Plasticity Chart*, as shown in Fig.1.6. The liquid limit and the *plastic limit* of a soil are plotted on the plasticity chart and the soil is classified according to the region in which it falls, the A-line being an arbitrary boundary between inorganic clays and inorganic silt/organic clays. Table 1.1 gives the liquid limit of some cohesive soils.

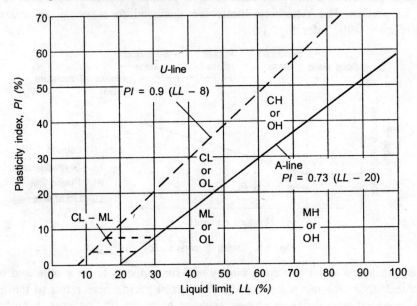

Fig. 1.6 Plasticity chart.

Table 1.1 Consistency limits of some soils

Soil	Liquid Limit (%)	Plastic Limit (%)
Alluvial Deposits		
Boston Blue Clay	41	20
Chicago Clay	58	21
Normal Calcutta Soil	55	28
Marine/Estuarine		
London Clay	75	29
Norwegian Quick Clay	40	17
Bombay Marine Clay	90	40
Cochin Marine Clay	90	45
Shellhaven Clay	97	32
Clay Minerals		
Illite	100	45
Kaolinite	50	25
Montmorillonite	500	50

1.5 SOIL CLASSIFICATION

Density, void ratio, and water content are fundamental soil parameters which help to identify and assess—at least qualitatively, the nature of the soil deposit. For example, a high void ratio of a sandy soil would indicate a loose state of compaction while a clayey soil with high water content is likely to be more compressible than one with low water content. Table 1.2 gives the density, void ratio, and the water content of some common soil deposits.

Table 1.2 Density, void ratio, and water content of some soils

Soil	Geologic type	Void ratio	Bulk density (g/cc)	Water content (%)
Mexico City	Volcanic	9.0	1.0	350
Shellhaven clay, England	Estuarine	1.6	1.8	60
London Clay, Selset	Marine	0.7	1.9	35
Boulder Clay, England	Glacial	0.4	2.0	20
Normal Calcutta Clay (Upper)	Alluvial	1.3	1.7	50
Normal Calcutta Clay (Lower)	Alluvial	0.8	2.0	30
Bangkok Clay (Soft)	Marine	2.1	1.5	80
Bangkok Clay (Stiff)	Marine	0.8	2.0	30
Norwegian Quick Clay	Marine	1.0	1.9	38

Soil consists of solid grains that have various sizes ranging from coarse grained particles such as boulder, gravel and sand down to the fine grained particles like, silt and clay. The grains are classified according to their sizes. The most common system of classification is the M.I.T. system as illustrated in Fig. 1.7.

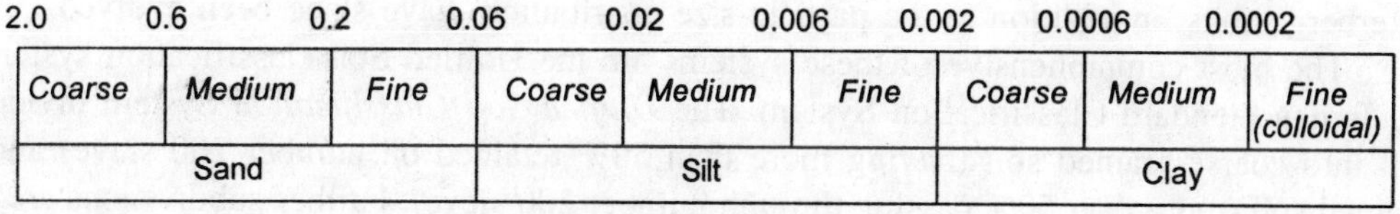

2.0	0.6	0.2	0.06	0.02	0.006	0.002	0.0006	0.0002
Coarse	Medium	Fine	Coarse	Medium	Fine	Coarse	Medium	Fine (colloidal)
	Sand			Silt			Clay	

Fig. 1.7 M.I.T. classification system.

Natural soil generally consists of mixture of several groups and the soil, in such cases, is named after the principal constituent present. For example, a soil that is predominantly clay but also contains some silt is called silty clay.

The grain-size distribution of a soil is best represented by the grain-size distribution curves, refer Fig. 1.8. The shape of the curve indicates whether the soil is uniform or poorly graded, or well-graded.

The uniformity coefficient of the soil is defined as

$$c_u = \frac{D_{60}}{D_{10}} \tag{1.3}$$

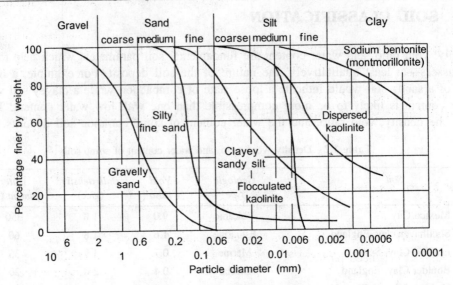

Fig. 1.8 Grain-size distribution curve.

where D_{60} is the diameter of particles corresponding to 60% finer and D_{10} is the diameter of the particle corresponding to 10% finer. The gradation of soil is determined by the following criteria:

$$\text{Uniform soil} \qquad : c_u = 1$$
$$\text{Poorly-graded soil} : 1 < c_u < 4$$
$$\text{Well-graded soil} \quad : c_u > 4$$

It must be considered, however, that the particle size alone is not an adequate criterion for the classification of a soil, as the shape of grains and clay fraction may vary widely depending upon the constituent minerals. More elaborate soil classification systems, making use of the Atterberg limits, in addition to the particle size distribution, have since been evolved.

The most comprehensive of these systems are the Unified Soil classification system and the Indian Standard Classification System. The *Unified Soil Classification* System divides the soil into coarse-grained soil (having more than 50% retained on number 200 sieve) and fine grained soil (more than 50% passing through number 200 sieve). Further subdivisions are made according to gradation for coarse-grained soils and plasticity for fine-grained soils and each soil type is given a group symbol (Table 1.3). The *Indian Standard Classification System* (IS 1498) is similar in some respects except that the fine-grained soils are divided into three ranges of liquid limit as opposed to only two in the unified soil classification system.

1.6 RELATIVE DENSITY OF GRANULAR SOIL

The engineering properties of granular soil primarily depend upon its relative density, grain-size distribution, and shape of grains. The relative density determines the compactness to which the solid grains are assembled in a soil skeleton and is expressed as

$$R_D\% = \frac{e_{\max} - e}{e_{\max} - e_{\min}} \times 100 \qquad (1.4)$$

where e_{max} = void ratio in loosest state
e_{min} = void ratio in densest state
e = in-situ void ratio

The properties of granular soil are also dependent on their particle size distribution, that is, whether the soil is well-graded or poorly-graded. In well-graded soils, the smaller grains tend to fill the voids between the larger grains and thus, make the soil more compact. The shape of grains (angular sub-rounded or rounded) may also have some effect on the properties of granular soil.

1.7 SOME SPECIAL SOIL TYPES

Apart from the common soil types that may be identified by the different soil classification systems, certain natural soils are characterized by the properties of their chemical and mineral constituents. Such soils exhibit characteristic features with regard to their strength and compressibility and need particular care when used to support a foundation.

Organic soils are those which contain large quantity of organic/vegetable matter in various stages of decomposition. Natural soils may contain varied percentages of organic matter and only a small percentage may be sufficient to affect its properties. In *organic clay*, vegetable matter is intermixed with the predominant clay mineral while *peat* consists almost wholly of vegetable matter. Peat is characterized by high liquid limit and sponzy structure with low specific gravity. The top 10–12 m of the normal Calcutta deposit is a common organic clay of the Bengal basin. A thin layer of peat is often encountered in this deposit at many locations as shown in Fig. 1.9.

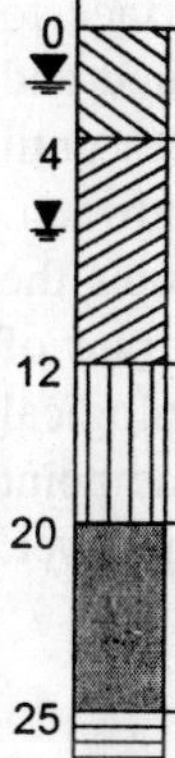

Description of strata	N blows /30cm	w (%)	LL (%)	PL (%)	c_u (t/m²)	m_v m²/t
I Light brown/brownish grey silty clay/clayey silt with occasional lenses of fine sand	5–8	30	54	24	4.0	0.002
II Grey/dark grey silty clay, clayey silt with semi-decomposed timber pieces	2–4	55	12	28	2.5	0.007
III Bluish grey silty clay with calcareous modules	10	30	60	25	6.0	0.002
IV Brown/yellowish brown sandy silt/silty fine sand with occasional lenses of brown and grey silty clay	20	38	20	30	4.5	—
V Mottled brown/grey silty clay with laminations often with rusty brown spots	20–30	25	65	25	10.0	0.001
VI Brown/light brown silty fine to medium sand	40	$\phi' >> 40°$				

Soil properties (average)

Fig. 1.9 Normal Calcutta deposit.

Expansive soils are found in many parts of India, Africa, and the middle-east. The black cotton soils of India and Africa are the most common types of expansive clays. These soils have high expansive potential because of the predominant presence of montmorillonite minerals. They are apparently stiff when dry but undergo swelling when saturated with water, e.g., due to seasonal fluctuation of ground water table. The Atterberg limits along with the percentage of solid particles less than 0.001 mm is taken as criteria for identification and classification of the expansive soils. Table 1.4 gives the classification of expansive clays.

Table 1.4 Classification of expansive clays

% of Particles finer than 0.001 mm	*Index Test Data* Plasticity index (%)	Shrinkage limit (%)	*Probable expansion under pressure of* 0.07 kg/cm² *(Dry to saturated condition)*	Degree of expansion
28	35	11	30	Very high
20–31	25–41	7–12	20–30	High
13–23	15–41	10–16	10–20	Low
15	18	15	10	Low

1.8 GROUNDWATER

The water which is available below the ground surface is termed as *groundwater* or *subsurface water*. Practically all ground water originates from the surface water. The process by which the surface water infiltrates into the ground surface and percolates deep into the ground is termed as *natural recharge and artificial recharge*. Main sources of natural recharge of groundwater include precipitation, rivers, lakes, and other natural water bodies. Artificial recharge of groundwater occurs from excess irrigation, seepage from canals, leakage from reservoirs or tanks, or from water purposely applied on the ground surface to augment groundwater storage. Water from any of these sources infiltrates into the ground and percolates downwards under the action of gravity through soil pores and, rock crevices until further movement is prevented by an impermeable stratum. It is then stored as groundwater. The groundwater exposed to atmospheric pressure beneath the ground surface constitutes the *water table*. Water table rises and falls based on the amount of precipitation, the rate of withdrawal or recharge, and climatic conditions. Groundwater held by the geological formation is, however, not static but moves slowly in the lateral direction towards some point of escape and appear as springs, infiltration galleries or wells, or reappears to join the river, or lake, or the sea.

1.8.1 Types of Water-bearing Formations

Groundwater occurs in most geological formations, of which the most important ones are aquifers. An *aquifer* is defined as a geological formation that permits storage as well as transmission of water through it. Thus, an aquifer contains saturated soil which yields significant quantity of water to wells and springs. Sands and gravels are typical examples of formations which serve as aquifers.

Other geological formations include aquiclude, aquitard, and aquifuge. An *aquiclude* may be defined as a geological formation of relatively impermeable material which permits storage of water but is not capable of transmitting it easily. Thus, an aquiclude contains saturated soil which does not yield appreciable quantities of water to wells. Clay is an example of such a formation.

An *aquitard* is defined as a geologic formation of poorly permeable or semipervious material which permits storage of water but does not yield water freely to wells. However, it may transmit appreciable quantity of water to or from adjacent aquifers. A sufficiently thick aquitard may constitute an important groundwater storage zone. A formation of sandy clay belongs to this category.

An *aquifuge* is a geological formation of relatively impermeable material which neither contains nor transmits water, for example, solid rocks.

1.8.2 Division of Subsurface Water

As shown in Fig. 1.10, subsurface water can be divided into the following zones:

(a) Soil–water zone
(b) Intermediate zone
(c) Capillary zone, and
(d) Zone of saturation

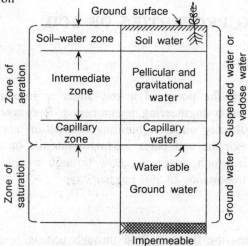

Fig. 1.10 Zones of subsurface water.

Soil–water zone

The soil–water zone extends from the ground surface to the major root zone. The soil in this zone becomes saturated either during irrigation or rainfall. The water in the soil–water zone is gradually depleted by evaporation from within the soil and by transpiration by vegetal growth on the ground surface and if it is not replenished, the water content may be reduced to such an extent that only thin film of moisture known as *hygroscopic water* remains adsorbed on the surface of the soil particles.

Intermediate zone

The intermediate zone occupies the space between the lower edge of soil–water zone and upper limit of the capillary zone. This zone usually contains static water which is held by molecular and surface tension forces in the form of hygroscopic and capillary water. Temporarily, though, this zone may also contain some excess water which moves downward as gravitational water. The thickness of this zone may vary from zero when the water table is high to more than 100 m under deep water table conditions.

Capillary zone

The capillary zone extends from the water table upto the limit of capillary rise of water. In this case, the pore space may be considered to represent a capillary and hence, just above the water table almost all pores contain capillary water.

Zone of saturation

In the zone of saturation, all the interstices are filled with water under hydrostatic pressure. The zone of saturation is bounded at the top either by the ground water table or an overlying impermeable stratum, and stretches upto underlying impermeable strata (or bed rock). Generally, all soils below ground water table are fully saturated.

1.9 ENGINEERING PROPERTIES OF SOIL

1.9.1 Permeability

The flow of water through the pores of a soil under a pressure gradient or under a differential head is a common engineering phenomenon. Seepage of water through earth dams and consolidation of clay under a building foundation are some instances where percolation of water through the soil plays an important role on the performance of the foundation. The ease with which water can flow through a soil, called *permeability*, is therefore, of fundamental importance in soil mechanics.

Darcy's law

Darcy's law which governs the flow of fluid through porous media is also found to be applicable to soils when flow is due to a combination of pressure and positional gradient. With the exception of flow through coarse gravels, the flow of water through soils is streamlined, and can be expressed as:

$$v = ki \qquad (1.5)$$

where,

v = average rate of flow of water in unit time.

i = hydraulic gradient, i.e., head loss per unit length of soil measured in the direction of flow, and

k = coefficient of permeability of the soil.

For any given soil, k depends on the porosity of the soil, the structural arrangement of particles, the size of particles, the properties of pore fluid (e.g. density and viscosity) etc. (Taylor 1948). The temperature of the pore fluid should theoretically have some effect but for practical purposes, the variation of k with the range of temperature normally encountered in the soil is small.

The coefficient of permeability of a soil can be measured from the laboratory constant head test, variable head test, or the field pumping test. In view of the hetrogeneity and non-homogeneity of natural soil, field pumping tests give a better measure of the permeability of the soil in the field. Typical values of permeability for different types of soil are given in Table 1.5.

Table 1.5 Permeability of different types of soil

Soil type	Coeff. of permeability	Drainage quality
Gravel	1	
Coarse sand	$1-10^{-1}$	
Medium sand	$10^{-1}-10^{-2}$	
·Fine sand	$10^{-2}-10^{-3}$	Good
Silty sand	$10^{-3}-10^{-4}$	Poor
Silt/weathered clays	$10^{-4}-10^{-7}$	
Intact clays	$10^{-7}-10^{-9}$	Very poor

Range of validity of Darcy's law

Darcy's law is valid only for laminar flow. Since Reynolds number serves as a criterion to distinguish between laminar and turbulent flow, the same may be employed to establish the limit upto which Darcy's law holds good. Reynolds number for this case is expressed as

$$R_e = \frac{pvd}{\mu} \tag{1.6}$$

where p is mass density of fluid,
      v is the discharge velocity, and
      μ is the dynamic viscosity of the fluid.

Most of the natural groundwater flow occurs with $R_e < 1$ and hence, Darcy's law is valid. However, Darcy's law is not applicable in aquifers containing coarse gravels, rockfills, and also in the immediate vicinity of wells where the flow may not be laminar due to steep hydraulic gradients.

1.9.2 The Principle of Effective Stress

In a multi-phase system composed of solids and voids, the behaviour of the material under applied stresses depends on how total stress is distributed amongst several components in the soil aggregate, namely the intergranular pressure that acts between the soil grains at their points of contact and the pore pressure which acts in the pore fluid. The normal stress on any plane is, in general, the sum of two components, namely the stress carried by the solid particles and the pressure in the fluid in the void space. The principle of effective stress provides a satisfactory basis for understanding the deformation and strength characteristics of a soil under an applied load. This can be simply be stated as:

(a) The volume change of a soil is controlled not by the total normal stress applied on the soil, but by the difference between the total normal stress and the pressure in the fluid in the void space, termed the pore pressure. For an all-round pressure increase, this can be expressed by the relationship.

$$-\frac{\Delta V}{V} = C_c(\Delta\sigma_n - \Delta u) \tag{1.7}$$

where,

$\Delta V/V$ = volume change per unit volume of soil
$\Delta\sigma_n$ = total normal stress/pressure
Δu = pore pressure
C_c = compressibility of soil skeleton

This relationship can be illustrated by a simple test where a saturated soil sample is subjected to undrained loading followed by drained compression as shown in Fig. 1.11. Not until there is a change of effective stress, is there a change of volume of the soil. This is the primary cause of long term consolidation settlement of foundation on clay. This also explains the settlement of an area due to ground water table lowering, either for construction work or for water supply.

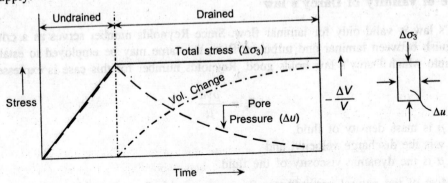

Fig. 1.11 Principle of effective stress.

(b) The shear strength of a soil is determined by the frictional forces between the solid particles. These are clearly a function of the component of normal stress that is carried by the solid grains rather than the total normal stress on the plane considered. This may be expressed by the equation.

$$\tau = c' + (\sigma_n - u)\tan\phi' \tag{1.8}$$

where,

c' = apparent cohesion
ϕ' = angle of shearing resistance
σ_n = total normal pressure
u = pore-pressure.

In Eqs. (1.7) and (1.8), the term $(\sigma_n - u)$ is termed the effective stress and is denoted by the symbol σ', that is,

$$\sigma' = \sigma - u \tag{1.9}$$

In most engineering problems, the magnitude of total normal stress can be estimated from considerations of statics while the magnitude of pore pressure depends on the hydraulic boundary conditions. Bishop (1955) has shown that the effective stress in a soil can be related to the intergranular pressure at the points of contact.

For a unit area perpendicular to the plane X–X through a soil mass, the total stress, σ acting on the plane of contact can be divided into two components, (refer Fig. 1.12)

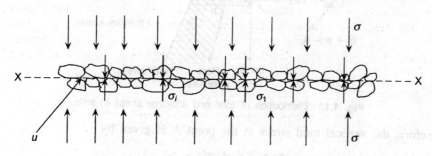

Fig. 1.12 Effective stress and intergranular pressure.

$$\sigma = \sigma' + u \tag{1.10}$$

where,

u = Pore water pressure and

σ' = Effective stress

If a = effective contact area per unit area of the plane and σ_i = average intergranular pressure, then, for a unit area

$$\sigma \times 1 = \sigma_i \times 1 + (1 - a)u$$

or

$$\sigma_i = (\sigma - u) + au \tag{1.11}$$

But a is small (though not equal to 0) and hence,

$$\sigma_i = (\sigma - u) \tag{1.12}$$

From Eqs. (1.11) and (1.12)

$$\sigma_i = \sigma' \tag{1.13}$$

Thus, for practical purposes, effective stress may be considered equal to the intergranular pressure, the average pressure between the solid gains.

In a natural soil deposit, (refer Fig. 1.13), the total stress at any depth is given by the overburden pressure at that depth while the pore pressure, in the absence of any artesian condition, is given by the hydrostatic head of water at that depth. The distribution of total and effective stress in the soil is shown in Fig. 1.13.

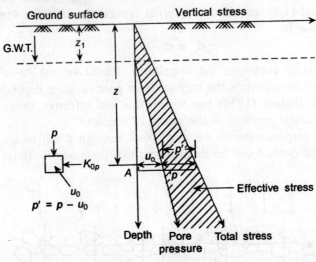

Fig. 1.13 Distribution of total and effective stress in soils.

Therefore, the vertical total stress at the point A is given by

$$\sigma_v = \gamma_1 z_1 + \gamma(z - z_1) \qquad (1.14)$$

where,

γ_1 = unit weight of soil within depth z_1

γ = unit weight of soil below depth z_1

The pore water pressure

$$u = \gamma_w(z - z_1) \qquad (1.15)$$

where γ_w = unit weight of water.

Hence, the effective stress at A is given by

$$\sigma_v' = \gamma_1 z_1 + \gamma(z - z_1) - \gamma_w(z - z_1)$$
$$= \gamma_1 z_1 + \gamma'(z - z_1) \qquad (1.16)$$

where γ' = submerged density of soil below water table

1.9.3 Pore-pressure in Soil due to Applied Load

The application of structural load causes an increase in the total stresses in the ground, the magnitudes of which can be determined from the theory of elasticity. If the subsoil consists of clay of low permeability and construction is sufficiently rapid, these changes in total stress occur under conditions of no volume change and are associated with simultaneous development of excess pore water pressure.

The concept of pore pressure coefficient is utilized to obtain a clear picture of how the pore-pressure in a soil responds to different combinations of applied stress (Skempton 1954). This concept not only explains the relationship between different types of triaxial test, but also provides a basis for estimating the magnitude of pore-pressure to be encountered in practical problems.

Skempton (1954) expressed the change of pore-pressure in a soil under axi-symmetric stress changes in terms of two empirical parameters A and B where

$$\Delta u = B[\Delta\sigma_3 + A(\Delta\sigma_1 - \Delta\sigma_3)] \tag{1.17}$$

where,

Δu = change in pore-pressure

$\Delta\sigma_1$ = change in total vertical pressure

$\Delta\sigma_3$ = change in total lateral pressure

A and B = pore-pressure parameters

By putting $\Delta\sigma_3$ and $(\Delta\sigma_1 - \Delta\sigma_3)$ equal to zero successively, it can be shown that the parameters B and A represent the effect of the allround stress increase and the deviatoric stress increase respectively on the pore-pressure developed in a soil element. Accordingly,

$$\Delta u = B\cdot\Delta\sigma_3 \qquad \text{when } \Delta\sigma_1 - \Delta\sigma_3 = 0$$

and

$$\Delta u = A[\Delta\sigma_1 - \Delta\sigma_3] \qquad \text{when } \Delta\sigma_3 = 0 \tag{1.18}$$

The parameter B depends on the degree of saturation of soil (for fully saturated soil $B = 1$). The parameter A, however, depends upon a number of factors such as stress history of the soil, stress level, strain level, and so on, the most influential being the stress history, that is, whether the soil is normally consolidated or overconsolidated (Lambe 1962). Table 1.6 gives typical values of pore-pressure parameter A for different stress history of the soil.

Table 1.6 Pore-pressure parameter A of different soil types

Soil type	Value of A at failure
Sensitive clay	1.2–2.5
Normally consolidated clay	0.7–1.2
Overconsolidated clay	0.3–0.7
Heavily overconsolidated clay	−0.5–0.0

Data from Winterkorn and Fang (1975).

1.9.4 Shear Strength of Soils

The shear strength of a soil under any given condition of drainage is defined as the maximum shear stress which the soil can withstand. When a structure is erected on a soil, the soil elements beneath the foundation are subjected to increased shear stresses. The capacity of the foundation to bear load is a function of the shear strength of the soil. The maximum shear stress a soil can withstand depends to an appreciable extent on the manner of loading and the boundary conditions. A soil specimen does not, therefore, have a unique shear strength and it depends on factors such as rate of strain, drainage condition, and size of sample.

The failure criterion most commonly used for defining the shear strength of a soil along a plane is expressed by the Mohr–Coulomb equation, as illustrated in Fig. 1.14,

$$\tau = c + \sigma_n \tan\phi \tag{1.19}$$

where

c is the cohesion intercept,

σ_n is the normal pressure on the plane considered, and

ϕ is the angle of shearing resistance

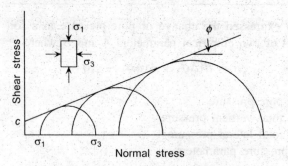

Fig. 1.14 Mohr–Coulomb failure criterion.

In Eq. (1.19), the shear strength parameters c and ϕ for any particular soil depend on several factors, the most important being the condition of drainage. Therefore, Eq. (1.19) expressed in terms of the total normal stress on the plane considered may be used to study the shear behaviour of soil under undrained condition. A more general expression may be written in terms of effective stress (Bishop 1955) as:

$$\tau = c' + (\sigma_n - u)\tan\phi'$$

or
$$\tau' = c' + \sigma'\tan\phi' \tag{1.20}$$

where c' and ϕ' are the shear strength parameters in terms of effective stress.

Measurement of shear strength

While the Vane shear test (Cadling and Odenstad 1950) or the Pressuremeter test (Menard 1956, 1969) may be used to determine the undrained shear strength of soils in the field, the shear behaviour of soils is best understood from the laboratory triaxial test. Figure 1.15

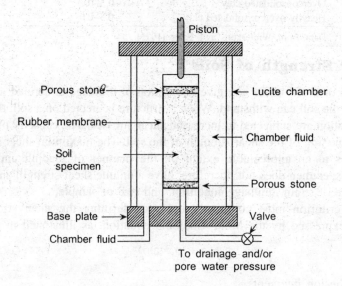

Fig. 1.15 Triaxial test set-up.

shows triaxial test set-up. Here, a cylindrical soil specimen is enclosed in a thin rubber membrane with rigid caps at top and bottom. The soil is placed inside a triaxial cell which is then filled with water. Pressure is applied to the specimen through water and deviator stress is applied through the end caps until the specimen fails in compression. Drainage of water from the pores of the soil may be controlled through suitable valves at the base of the triaxial cell. If required, volume change and/or pore water pressure can be measured. The details of the triaxial apparatus and various tests that can be performed with it have been described by Bishop and Henkel (1962).

Types of triaxial test

There are, in general, three conditions of drainage under which the triaxial test is performed, namely unconsolidated undrained (UU), consolidated undrained (CU), and consolidated drained (CD).

In the unconsolidated undrained (UU) test, the sample is first subjected to an all round pressure σ_3 and then to a deviator stress $(\sigma_1 - \sigma_3)$ under undrained condition. The deviator stress is applied rapidly—usually at a rate of strain of 1–2% per minute, and failure is achieved in 10–20 minutes. For saturated soil, the Mohr envelope remains horizontal giving $\phi_u = 0$. The shear strength obtained from the UU test is called the *undrained shear strength* of the soil c_u, see Fig. 1.16. The shear strength obtained from the UU test is used in the study of bearing capacity of foundations on clay or in the rapid construction of embankments on clay. The unconfined compression test is a special case of UU test, where no confining pressure is applied to the specimen prior to shear.

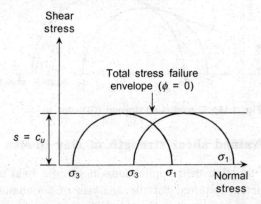

Fig. 1.16 Unconsolidated undrained (UU) test on saturated clay.

In the consolidated undrained (CU) test, the sample is allowed to consolidate under an all round pressure σ_3 but no drainage is allowed during shear. If pore pressure is measured during the test, both the pore pressure parameter A and B, and the shear strength parameters c' and ϕ' can be determined. Figure 1.17 depicts consolidated undrained (CU) test on saturated clay. The data from CU test may be used in the analysis of stability for stage construction of embankments on clay.

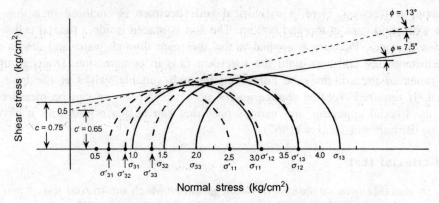

Fig. 1.17 Consolidated undrained (CU) test on saturated clay.

The consolidated drained (CD) test differs from the consolidated undrained test in the way that both, the initial all round pressure and the subsequent shear stresses are applied under fully drained condition. The test, therefore, gives the shear strength parameters of a soil in terms of effective stress, as given by Fig. 1.18. The results of this test can be used in the study of long term stability problems.

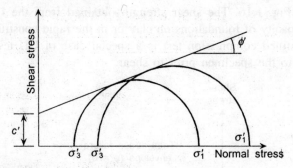

Fig. 1.18 Consolidated drained (CD) test results.

Factors affecting undrained shear strength of clay in-situ

Different types of triaxial test find their applications in specific field problems but, the most important strength parameters required for the analysis of foundations on clay are those obtained from the quick unconsolidated undrained (UU) triaxial test. For saturated clays, this shear strength is expressed by the undrained cohesion $c_u \cdot \phi_u$ should theoretically be zero, if the soil is fully saturated (Skempton 1948). Therefore, the shear strength of the clay in this condition may be expressed as

$$\overline{C} = c_u \qquad (1.21)$$

Although the UU test is easy to perform in the laboratory, application of the test data to field problem has to be done with care. Factors such as anisotropy, rate of shear, sample size, and so on affect the test results significantly (Skempton and L. Rochelle 1955). For stiff-fissured clays, in particular, the shear strength of the clay mass in the field may be considerably less than the strength obtained from conventional laboratory tests on small size samples (Bishop 1966).

Sensitivity

Most clays have been found to lose a part of their strength when remoulded. This remoulding may be caused by physical or mechanical means such as pile driving. However, with time, the clay may regain this strength either wholly or partly, by a phenomenon known as *thixotropic hardening* (Skempton and Northey 1952). From the point of view of sensitivity towards remoulding, clays may be classified as follows (Table 1.7):

Table 1.7 Sensitivity of clays
(Skempton and Northey 1952)

Sensitivity	Classification
Less than 2	Insensitive
2–4	Moderately sensitive
4–8	Sensitive
8–16	Very sensitive
16	Quick clays

Sensitivity, in this context, is defined as the ratio of the undrained shear strength of the undisturbed soil to that of the fully remoulded soil.

1.9.5 Consolidation

The gradual squeezing out of water from the pore space of a soil skeleton under the influence of externally applied load or gravity is called *consolidation*. The process results in a net change in volume of the soil and is time-dependent.

When an element of soil is subjected to an increase of total stress under undrained condition, the pressure is distributed among the solid grains and water depending on the relative compressibilities of the two phases and their boundary conditions. For a confined saturated clay–water system with no drainage, the compressibility of the mineral skeleton is so large compared to that of water alone that virtually all the applied pressure is transmitted as an excess pore water pressure. If drainage is now permitted, the resulting hydraulic gradient initiates a flow of water out of the clay and the soil consolidates. There is a consequent transfer of the applied load from the water to the mineral skeleton.

The mechanism of consolidation and the factors that govern the process of consolidation of clayey soils are studied experimentally in the laboratory in the consolidation test or the oedometer test. The arrangement for oedometer test is shown in Fig. 1.19. A sample, usually 76 mm dia × 20 mm thick, is enclosed in a metal ring and sandwiched between two porous stones placed at top and bottom. A load is applied to the sample through the porous stones using a lever arrangement. As the sample consolidates, thickness of the sample decreases. Being laterally confined within the metal ring, the sample is prevented from expanding laterally and the entire volume change takes place in the vertical direction. The flow is, therefore, one dimensional and the rate of consolidation is governed by the permeability of the soil in the vertical direction only. However, consolidation is not always one-dimensional in the field and the consolidation test as performed in the laboratory represents the field condition only under certain boundary conditions.

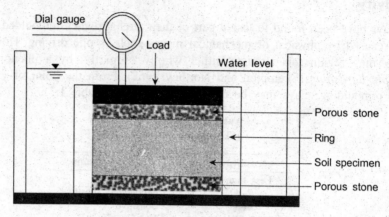

Fig. 1.19 Oedometer test apparatus.

Pressure–void ratio relationships

The change of volume of a sample as measured in the consolidation test is a function of increment in applied stress and is generally expressed in terms of the pressure–void ratio relationship of the type shown in Fig. 1.20. For a normally consolidated soil, this relationship is found to be linear on a semi-log plot.

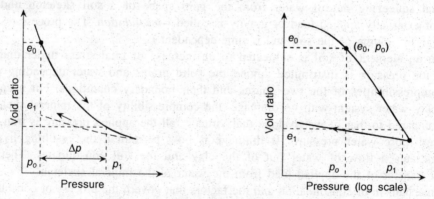

Fig. 1.20 Pressure–void ratio relationship.

The coefficient of volume decrease is defined as the volumetric strain per unit increase of effective pressure and this can be expressed, in terms of void ratio, as

$$m_v = -\frac{de}{dp}\frac{1}{1+e_0} \tag{1.22}$$

where,

e_0 = initial void ratio for the stress increment considered.

For the linear void ratio versus effective stress relationship on the semi-log plot, the compressibility of the soil is given by the compression index, C_c defined as,

$$C_c = \frac{de}{d(\log p)} \tag{1.23}$$

and

$$\frac{\Delta V}{V} = \frac{C_c}{1 + e_0} \log(p_1 p_0) \tag{1.24}$$

The compressibility of a soil is a measure of its consistency. Greater the compressibility, softer is the soil and vice versa. Again, compressibility of a soil is not a constant property. It decreases with increasing effective stress, as is evident from the decreasing slope of the pressure–void ratio relationship with increasing pressure. But for the range of stresses usually encountered in practice, Table 1.8 gives one dimensional compressibility of some representative clays.

Table 1.8 One-dimensional compressibility of some clays

Stress History	*Clay*	m_v (cm²/kg)	$C_c/(1 + e_0)$
Normally Consolidated	Gosport Clay, England	0.15	—
	Shellhaven Clay	0.25	—
	Normal Calcutta Clay	0.05	0.15–0.20
	Cochin Marine Clay	0.06	0.20
Over Consolidated	Brown London Clay	0.017	—
	Blue London Clay	0.010	—
	Normal Calcutta Clay (desiccated)	—	0.05

(Data from Skempton and Bishop (1954), Chummer (1976) and Author's files)

Normally consolidated and overconsolidated clays

Let us consider an element of soil during deposition under water, Fig. 1.21. As more and more soil is deposited on the element, the overburden pressure on the element increases and its void ratio decreases along the curve AB. When the maximum height of deposition, H_c is reached, the pressure increases to P_c and the void ratio decreases to e_c. The soil anywhere on the curve AB is called *normally consolidated*, to indicate that it has never in its past, been subjected to a pressure greater than that corresponding to the curve AB. Marine clays and alluvial soils of India are typical examples of normally consolidated clays.

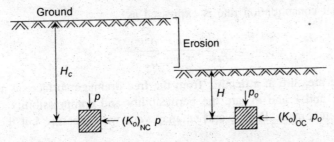

Fig. 1.21 Stresses in an element soil below ground surface.

Now, if some part of the overburden is removed by, say, erosion and the remaining height of deposition is only H, corresponding to which the pressure is p_0, the element of soil will undergo swelling and the void ratio will increase to e. This means that the soil in its past has been subjected to a pressure greater than that exists now. Many clays and clay shales, for example, London clay and Bearpaw shale, are heavily overconsolidated in nature. The strength and deformation characteristics of a soil depend to a large extent on it being normally consolidated or overconsolidated. Soft clays are normally consolidated and their behaviour differs from overconsolidated clays and clay shales. Structures founded on normally consolidated clays, in general, experience much more settlement than those founded on overconsolidated clays.

Although erosion of overburden has been identified as one of the causes of overconsolidation, many residual and alluvial soils near the ground surface are rendered over-consolidated by desiccation. Alternate wetting and drying due to seasonal fluctuation of water table and changes of temperature introduce capillary forces in the soil and the latter develops a pseudo overconsolidation effect. This results in increased strength and decreased compressibility of the soil. Depending on the severity of changes during desiccation, the over consolidation effect may be quite appreciable. Most residual soils of India and occasionally the alluvial soils near the ground surface appear overconsolidated due to desiccation.

Rate of consolidation

The rate of consolidation of clayey soils is governed by the theory of one-dimensional consolidation (Terzaghi 1923), as shown in Fig. 1.22:

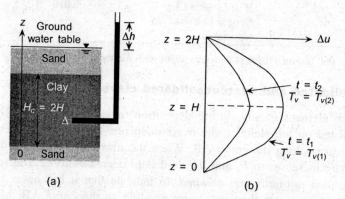

Fig. 1.22 One-dimensional consolidation.

The one-dimensional consolidation rate is expressed as

$$\frac{\delta u}{\delta t} = \frac{k}{\gamma_w m_v} \frac{\delta^2 u}{\delta z^2} \qquad (1.25)$$

where u is the pore pressure at a depth z from the free drainage surface, at a time t after the pressure increment; and k and m_v are the permeability and compressibility of the soil for a particular pressure increment. Both k and m_v may vary with pressure but their ratio remains approximately constant (Skempton and Bishop 1954). Consequently, Eq. (1.25) may be written as,

$$\frac{\delta u}{\delta t} = C_v \frac{\delta^2 u}{\delta z^2} \tag{1.26}$$

where C_v is defined as the coefficient of consolidation.

Solving Eq. (1.26) for the appropriate boundary conditions, we get the distribution of excess pore pressure with depth at a given time t as shown in Fig. 1.22(b). Then, integrating the area of the pore pressure dissipation diagram at a given time and expressing it as a ratio of the initial pore pressure diagram, we get the average degree of consolidation, U of the soil as a function of the time factor T_v. Thus, the degree of consolidation, U can be conveniently expressed as,

$$U = f(T_v) \tag{1.27}$$

where,

$$T_v = \frac{C_v t}{H^2} \tag{1.28}$$

and H = length of the drainage path, to be taken as full depth of clay when drainage is from one end and half the depth of clay when drainage is from both ends.

The coefficient of consolidation is governed primarily by the size and nature of particles as reflected by the water content or whether the soil is normally consolidated or overconsolidated. C_v is determined from laboratory consolidation test by curve fitting methods (Taylor 1948). The relationship between the U and T_v for the most common boundary conditions of single or two-way drainage is plotted in Fig. 1.23.

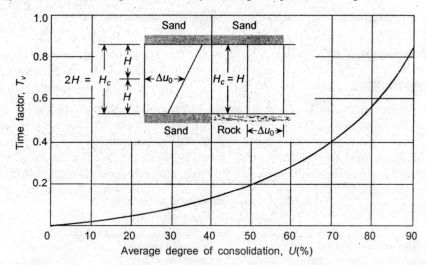

Fig. 1.23 Degree of consolidation versus time factor plot for one-dimensional consolidation.

This gives the C_v value for one-dimensional consolidation of small size specimens. U.S. Department of Navy (1971) proposed an empirical relationship between C_v and liquid limit for determining the field C_v for practical use and this is shown in Fig. 1.24.

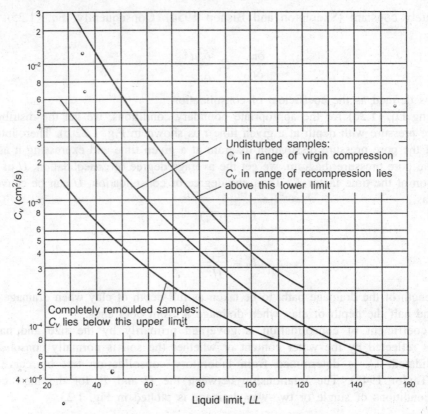

Fig. 1.24 Variation of C_v with liquid limit (after U.S. department of Navy (1971)).

1.9.6 Properties of Soil

Coarse-grained soils

The relative density of granular soils is measured from in-situ *standard penetration test*. This test involves counting the number of blows required to drive a standard split-spoon sampler to a depth of 30 cm by means of a 65 kg hammer falling from a height of 75 cm. An empirical correlation between standard penetration resistance, N (blows per 30 cm), the relative density, and shear strength of granular soils is shown in Table 1.9 (Terzaghi and Peck 1967).

Table 1.9 Relative density of granular soils
(Terzaghi and Peck 1967)

Relative density (%)	N (Blows/30 cm)	Angle of shearing resistance (φ)	Degree of compactness
0–15	0–4	28°	Very Loose
15–35	4–10	28–30°	Loose
35–65	10–30	30–36°	Medium
65–85	30–50	36–41°	Dense
7–85	50	41°	Very Dense

Note: φ values are to be increased by 5° for soils containing less than 5% silt.

Fine-grained soils

Soils containing clay-size particles and large proportion of silt have low permeability and their properties vary with the rate of load application. Under undrained condition, their strength is derived almost exclusively from cohesion. These soils often possess low shear strength and high compressibility, thus making them poor foundation material.

A cohesive soil is described as very soft, soft, medium, stiff, and so on according to its shear strength as determined from the unconfined compression test or from the undrained triaxial test. Attempts have also been made to correlate the standard penetration resistance (as above) with the shear strength of cohesive soils. Table 1.10 gives the classification of cohesive soils on the basis of their shear strength (Terzaghi and Peck 1967).

Table 1.10 Shear strength of cohesive soils
(Terzaghi and Peck 1967)

Consistency	Undrained shear strength, c_u (t/m²)	N (Blows per 30 cm)
Very Soft	0–1.25	0–2
Soft	1.25–2.50	2–4
Medium	2.50–5.00	4–8
Stiff	5.00–10.00	8–16
Very Stiff	10.00–20.00	16–32
Hard	> 20.00	32

Stiff clay often possesses cracks and fissures which affect the shear strength of the clay mass. These fissures are planes of weakness and are prone to softening by water. Laboratory tests on small specimens do not often give the properties of the soils in-situ (Bishop 1966, Burland et al. 1966, Marsland 1971).

Elastic parameters

Young's modulus and Poisson's ratio are important soil parameters that are required to study the deformation behaviour of a soil. When a saturated clay is loaded rapidly, no volume change of the clay occurs during loading and Poisson's ratio can be taken as 0.5. When there is volume change, typical values of Poisson's ratio may be taken as those in Table 1.11 (Barkan 1962).

Table 1.11 Poisson's ratio of different soils
(Barkan 1962)

Soil type	Poisson's ratio
Saturated clay (undrained)	0.50
Clay with sand and silt	0.30–0.42
Unsaturated clay	0.35–0.40
Loess	0.44
Sand	0.30–0.35

Young's modulus of a soil can be determined from the stress–strain relationship obtained from laboratory triaxial tests. However, these relationships are highly susceptible to sampling disturbances and the E value thus obtained is generally much lower than the in-situ modulus. In case of homogeneous deposits, determination of E by back calculation from field plate load test gives reliable data. Indirect estimate of E can also be made from empirical relation with the shear strength measured from the unconsolidated undrained triaxial test. Bjerrum (1964), and Bozuzuk and Leonards (1972) suggest the following approximate correlation:

$$E = (500 - 1000)c_u \qquad (1.29)$$

where c_u is the undrained shear strength of the clay.

In-situ compressibility

The pressure versus void ratio relationships of natural clays are very sensitive to sampling disturbances and the linear e versus $\log p$ relationship is not always obtained even for a normally consolidated clay. Also, there is a pseudo overconsolidation effect on the sample because of the removal of in-situ stresses by sampling. Consequently, e versus $\log p$ relationship obtained from laboratory consolidation tests on undisturbed samples generally takes the shapes as shown in Fig. 1.25. The curves move downwards as the sampling disturbances are increased. Schmertman (1953) observed that, irrespective of sampling disturbances, the straight line portions of all the curves meet at a void ratio equal to $0.42\ e_0$, where e_0 is the in-situ void ratio of the sample. Then joining this point with that corresponding to e_0 and p_0 (where $p_0 =$ in-situ effective overburden pressure) would give the virgin consolidation curve. For normally consolidated soil, the e_0–p_0 point lies to the right of the extension of the straight line in the laboratory e versus $\log p$ curve while for over-consolidated samples, the point would lie to the left. The field e versus $\log p$ relationship for the overconsolidated range is obtained by drawing a line parallel to the laboratory rebound curve and the point of intersection of this line with extended straight line of the laboratory curve would give the pre-consolidation pressure (Fig. 1.25).

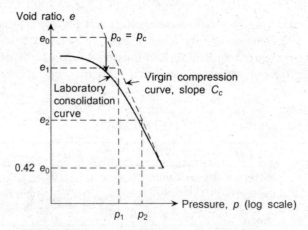

Fig. 1.25 Laboratory e versus $\log p$ relationship.

1.10 SOIL DEPOSITS OF INDIA

The soil deposits of India may be classified under most of the predominant geological formations (described earlier), namely

(a) Alluvial soils	(b) Marine deposits	(c) Black cotton soil
(d) Laterite soil	(e) Desert soil	(f) Boulder deposits

Figure 1.26 shows the distribution of predominant soil deposits in India (Ranjan and Rao 2000).

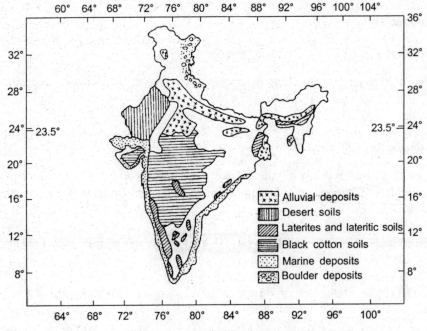

Fig. 1.26 Soil deposits of India.

Alluvial soils

Large parts of northern and eastern India lying in the Indo-gangetic plains and the Brahmaputra valley are covered by the sedimentary deposits of the rivers and their tributaries. They often have thickness greater than 100 m above the bed rock. The deposits mostly constitute layers of sand, silt, and clay depending on the position of the river away from the source.

Marine deposits

India has a long coast line extending along the Arabian sea, Indian ocean, and the Bay of Bengal. The deposits along the coast are mostly laid down by the sea. These marine clays of India are generally soft and often contain organic matter. They possess low shear strength and high compressibility.

Black cotton soil

The central part of India has extensive deposits of the expansive soil known as black cotton soil. This covers wide areas of Maharastra, Madhya Pradesh, Karnataka, Andhra Pradesh, Tamil Nadu, and Uttar Pradesh. The soil contains montmorillonite clay mineral which has high swelling potential.

Laterite soil

This soil covers wide areas of Kerala, Karnataka, Maharastra, Orissa, and parts of West Bengal. Laterites are residual soils formed by decomposition of rock which forms oxides of iron and aluminium.

Desert soil

Large areas of Rajasthan in the Thar desert are composed of wind blown deposits of desert soil, like loess. The sand dunes are often 15 m high and are formed under highly arid conditions.

Boulder deposits

Boulders are deposited is hilly terrains where the rivers flow with high velocity and carry large size boulders. These deposits are found in the sub-Himalayan regions of Uttar Pradesh and Himachal Pradesh.

===== **REFERENCES** =====

Barkan, D.D (1962), *Dynamics of Bases and Foundations*, McGraw Hill Book Co., New York.

Bishop, A.W. (1959), The Principle of Effective Stress, *Teknisk Ukebald,* Vol. 106, No. 39, pp. 859–863.

Bishop, A.W. (1966), Strength of Soils as Engineering Materials, *Geotechnique,* Vol. 16, pp. 89–130.

Bishop, A.W. and D.J. Henkel (1964), *The Measurement of Soil Properties in the Triaxial Test*, Edward Arnold, London.

Bjerrum, L. (1964), *Relasjon Melom Malte og Berengnede Setninger av Byggverk pa Leire og Sand*, Norwegian Geotechnical Institute, Oslo, Norway.

Bozuzuk and G.A. Leonards (1972), *The Gloucester Test Fill*, Proceedings ASCE Speciality Conference on performance of earth and earth retaining Structures, Lafayette, Indiana, USA.

Burland, L., F.G. Butler, and P. Dunican (1966), *The Behaviour and Design of Large Diameter Bored Piles in Stiff Clay*, Proceedings Symposium on Large Bored Piles, The Institution of Civil Engineers, London, pp. 51–71.

Chummar, A.V. (1976), *Foundation Problems in Cochin*, Proceedings Symposium in Foundations and Excavations in Weak Soil, Calcutta, Vol. 1, Paper No. C 4.

IS 1498 (1970), *Classification and Identification of Soils*, Bureau of Indian Standards, New Delhi.

Lambe, T.W. (1962), *Pore-pressures in a Foundation Clay 1962*, Proceedings ASCE, Soil Mechanics and Foundation Division, Vol. 88, pp. 19–47.

Marsland, A. (1971), *The Shear Strengths of Stiff Fissured Clays*, Proceedings Roscoe Memorial Symposium, Cambridge, U.K., pp. 59–68.

Menard, L. (1956), *An Apparatus for Measuring the Strength of Soils in Place, MS. Thesis*, University of Illinois, Urbana, Illinois, USA.

Ranjan, G. and A.S.R. Rao (2000), *Basic and Applied Soil Mechanics*, 2nd edition, New Age International (P) Ltd., New Delhi.

Schmertmann, J.H. (1953), *Undisturbed Consolidation Behaviour of Clays*, Transactions, ASCE, Vol. 120, p. 1201.

Scott, R.F. (1965), *Principles of Soil Mechanics*, Addison Wesley, Boston, USA.

Skempton, A.W. (1948), *The $\phi = 0$ Analysis of Stability and its Theoretical Basis*, Proceedings 2nd International Conference on SMFE, Vol. 2.

Skempton, A.W. and A.W. Bishop (1954), *Soils, their Elasticity and Inelasticity*, North Holland Publishing Co, Amsterdam.

Skempton, A.W. and R.D. Northey (1952), *The Sensitivity of Clays*, Geotechnique, Vol 3, pp. 30–54.

Taylor, D.W. (1948), *Fundamentals of Soil Mechanics*, John Wiley & Sons, New York.

Terzaghi, K. and R.B. Peck (1967), *Soil Mechanics in Engineering Practice*, 2nd edition, John Wiley and Sons, Inc, N.Y.

U.S. Department of Navy (1971), *Design Manual—Soil Mechanics, Foundations and Earth Structures*, NAVFAC, DM–7, U.S. Government Printing Office, Washington, D.C.

Site Investigation

2.1 INTRODUCTION

It is essential to carry out site investigation before preparing the design of civil engineering works. The investigation may range in scope from simple examination of the surface soils, with or without a few shallow trial pits, to a detailed study of the soil and ground water conditions for a considerable depth below the ground surface by means of boreholes and in-situ and/or laboratory tests on the soils encountered. The extent of the investigation depends on the importance of the structure, the complexity of the soil conditions, and the information already available on the behaviour of existing foundations on similar soils. Thus, it is not the normal practice to sink boreholes and carry out soil tests for single or two-storey dwelling houses since normally, there is adequate knowledge of the safe bearing pressure of the soil in any particular locality. Only in troublesome soils such as peat or loose fill would it be necessary to sink deep boreholes, possibly supplemented by soil tests. More extensive investigation for light structures is needed when structures are built on filled-up soil or in ground conditions where there is no information available on foundation behaviour of similar structures. A detailed site investigation involving deep boreholes and laboratory testing of soils is always a necessity for heavy structures, such as bridges, multi-storeyed buildings or industrial plants.

Thus, the major objectives of site investigation are:

(a) Knowing the general suitability of the site for proposed works.
(b) Assessing local conditions and problems likely to be encountered in foundation construction.
(c) Acquiring data for adequate and economic design of foundation.

2.2 INFORMATION EXTRACTED FROM SITE INVESTIGATION

A lot of information is extracted from site investigation to facilitate foundation design. This includes

1. General topography of the site which affects foundation design and construction, e.g., surface configuration, adjacent property, presence of water courses, and so on.

2. Location of buried services such as power lines, telephone cables, water mains, sewers pipes and so on.
3. General geology of the area with particular reference to the principal geological formations underlying the site.
4. Previous history and use of the site including information of any defects and failures of structures built on the site.
5. Any special features such as possibility of earthquake, flooding, seasonal swelling etc.
6. Availability and quality of local construction materials.
7. A detailed record of soil or rock strata, ground water conditions within the zone affected by foundation loading and of any deeper strata affecting the site conditions in any way.
8. Design data which comprises strength and compressibility characteristics of the different strata.
9. Results of chemical analysis on soil or ground water to determine possible deleterious effects on foundation structures.

2.3 STAGES OF SITE INVESTIGATION

Different stages of site investigation for a major civil engineering project may be summarised as shown in Table 2.1.

Table 2.1 Stages of site investigation

Reconnaissance Study	(a) Geological data
	(b) Pedological data
	(c) Areal photographs
	(d) Geophysical investigation
Detailed Investigation	(a) Boring
	(b) Sampling
	(c) Testing
	(i) Lab test
	(ii) Field test
	(d) Aerial photographs
	(e) Geophysical methods
Performance Study	(a) Further testing
	(b) Instrumentation
	(c) Performance evaluation

2.3.1 Reconnaissance Study

Reconnaissance study involves the preliminary feasibility study that is undertaken before any detailed planning is done—mainly for the purpose of selection of site. This is to be done at minimum cost and no large scale exploratory work is usually undertaken at this stage. The required data may be obtained from:

Geological survey reports and maps

Geological interpetation of land forms and underlying strata give the sequence of events leading to the formation of subsoil deposits. They help to define the properties of the material in a general way.

Pedological data

Many areas have been surveyed for agricultural purposes—usually to depths of 2 or 3 m. Materials are often classified according to colour, texture, chemical composition, and so on.

Aerial photographs/satellite images

Photographic representation of a portion of earth's surface taken from the air or space.

Geophysical methods

Application of the methods and principles of physics to determine the properties of subsurface materials. These methods are particularly useful for identifying bed rock. Seismic refraction method or electrical resistivity tests are usually done for this purpose.

2.4 BORING (DETAILED SOIL INVESTIGATION)

Detailed soil investigation is done through a series of boring, sampling, and testing to obtain the engineering properties of soil. The following subsections are devoted to boring which discuss different methods of boring in detail.

2.4.1 Trial Pits

Trial pits are the cheapest way of site exploration and do not require any specialized equipment. A pit is manually excavated to get an indication of the soil stratification and obtain undisturbed and disturbed samples. Trial pits allow visual inspection of any change of strata and facilitate in-situ testing. They are, however, suitable for exploration of shallow depth only. Figure 2.1 is a diagrammatic representation of trial pits.

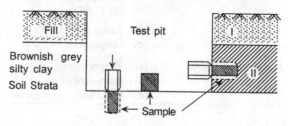

Fig. 2.1 Trial pit.

2.4.2 Wash Boring

A hole, usually 150–200 mm diameter, is advanced into the soil through a suitable cutter at the bottom of a drill rod. The soil is loosened and removed from the borehole by a stream of water or drilling mud, issuing from the lower end of the wash pipe which is worked up and down or rotated by hand in the borehole. Water or mud flow carries the soil up the annular space between the wash pipe and the casing, and it overflows at ground level where the soil in suspension is allowed to settle in a tank and the fluid is re-circulated or discharged to waste as required. Samples of the settled soil can be retained for identification purposes. Figure 2.2 shows the arrangement for wash boring.

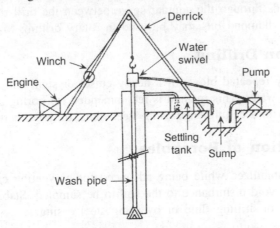

Fig. 2.2 Wash boring.

The method is simple and cheap. The structure of the soil below the boring apparatus is not disturbed and thus, both disturbed and undisturbed samples can be obtained.

2.4.3 Auger Boring

In this method, the borehole is advanced by turning an auger into the soil, withdrawing it and removing the soil for examination and test. The auger is re-inserted for further boring.

The auger may be manually or mechanically operated. Extensions are added to reach the desired depth. Disturbed samples may be obtained from the soil brought up by the auger while undisturbed samples are obtained by pushing sampling tubes at suitable intervals in the borehole. The apparatus for auger boring is shown in Fig. 2.3.

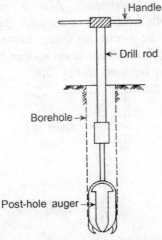

Fig. 2.3 Auger boring.

2.4.4 Rotary Drilling

Rotary drilling is done by rapidly rotating drilling bits attached to the bottom of the drill rod to cut and advance the borehole. Rotary drilling can be used in sand, clay or intact rocks with water or drilling mud being circulated through the drill rod to remove the cuttings as the mud returns upwards through the annular space between the drill rod and the side of the hole. Core barels with diamond bits may be used in rotary drilling to obtain rock cores.

2.4.5 Percussion Drilling

The soil is loosened by repeated blows of a heavy chisel or spud and the resulting slurry is removed by circulating water. This method is recommended for boring in rocks and hard soil. The hole is advanced with a cutting edge using steel shots, tungsten carbide, or diamond bits.

2.4.6 Stabilization of Boreholes

Boreholes need to be stabilized while being advanced for preventing caving in of sides and bottom of hole and to avoid disturbance to the soil to be sampled. Stabilization may be done by circulation of water or drilling fluid or by using steel casing.

Stabilization by water: Stabilization by water is not suitable in partly-saturated soils above G.W.T. because free water destroys the capillary forces and causes increase in water content. It is generally used in rock and stiff clays.

Stabilization by drilling fluid: Borehole is filled with drilling fluid or mud which, when circulated, removes the loose material from the bottom of the hole. Drilling mud is obtained by mixing locally available fat clays with water or by using commercially available bentonite.

Stabilizing effect of drilling mud is improved by higher specific gravity of the mud in comparison with water. Also, there is formation of a relatively impervious layer on the side of the borehole which gets liquefied again by resuming the agitation.

Stabilization by casing: Casing or lining a borehole with steel pipes provides the safest, though relatively expensive method of stabilization. After a certain depth or when difficult ground condition is reached, it is often difficult to advance the original casing. A smaller casing is then inserted through the one already in place. Lower end of casing is generally protected by a shoe or hardened steel with inside bevel so that the soil enters the casing and can be removed. This arrangement is depicted in Fig. 2.4.

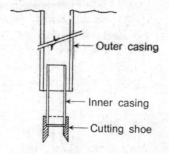

Fig. 2.4 Stabilization of borehole by casing.

Except when undisturbed samples are required in sensitive clays, the casing is generally driven by repeated blows of a drop hammer. Casing prevents side caving, but not always bottom caving. This can be achieved by filling the casing with water or drilling fluid. However, casing should not be filled with water if bottom of casing is above ground water table and undisturbed samples are required.

2.5 SAMPLING

The different types of sample obtained from boreholes are shown in Table 2.2.

Table 2.2 Types of sample

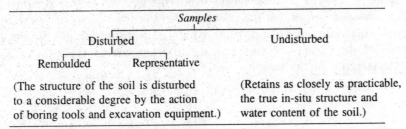

2.5.1 Sampling from Trial Pits

Block samples (refer Fig. 2.5) are hand cut from trial pits or open excavations. A block of clay is carefully trimmed with a sharp knife, taking care that no water comes into contact with the sample. Good quality samples can be obtained by this method, but if the soil does not possess any cohesion, it may be difficult, if not impossible, to obtain block samples.

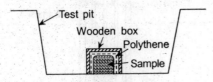

Fig. 2.5 Block sample of clay.

2.5.2 Sampling from Boreholes

Undisturbed samples may be obtained from boreholes by open drive samplers or piston samplers. An open drive sampler is shown in Fig. 2.6.

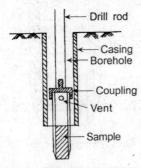

Fig. 2.6 Open drive sampler.

Open drive samplers consist of thin-walled tubes which are pushed or driven into the soil at the bottom of the hole and then rotated to detach the lower end of the sample from the soil as shown in Fig. 2.6. Most soft or moderately stiff cohesive soil can be sampled without extensive disturbance in thin-walled seamless steel tubes having diameter not less than 50 mm. The lower end of the tube is sharpened to from a cutting edge and the other end is machined for attachment to drill rods. The entire tube is pushed or driven into the soil at the bottom of the hole and is removed with the sample inside. The two ends of the tube are then sealed and the sample shipped to the laboratory.

Good quality undisturbed samples are obtained from piston samplers which use thin-walled sampling tubes with a piston inside. While the tube is being lowered to the bottom of the drill hole, the piston rods and the piston are held at the bottom of the sampler by means of a drill rod which rises to the top of the borehole. A piston sampler is shown in Fig. 2.7. The presence of the piston prevents excess soil from sequeezing into the tube and thus, maintains the integrity of the sample.

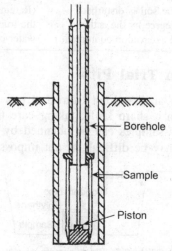

Borehole

Sample

Piston

Fig. 2.7 Piston sampler.

Table 2.3 lists the requirements of a good sampling tube (refer Fig. 2.8) as follows:

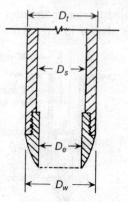

Fig. 2.8 Sampling tube.

Table 2.3 Requirements of sampling tube

(a) Area ratio, $C_a = \dfrac{D_w^2 - D_e^2}{D_e^2}$

This represents the amount of soil that is displaced when the sampler is forced into the ground. Thicker the tube, more is the disturbance. The area ratio of a good sampling tube should not exceed 15%.

(b) Inside clearance ratio, $C_i = \dfrac{D_s - D_e}{D_e}$

 For long samples, $0.75\% < C_i < 1.5\%$

 For short samples, $0 < C_i < 0.5\%$

The diameter of the sampling tube is kept slightly larger than the diameter of the cutting edge to minimise friction on the sample as it enters the sampling tube.

(c) Outside clearance ratio, $C_o = \dfrac{D_w - D_t}{D_t}$

 $\approx 2\text{--}3\%$

This clearance is provided to reduce the driving force required to penetrate the sampler into the soil. Diameter of samples should not be less than 38 mm. In general, 50–150 mm diameter undisturbed samples are obtained from boreholes.

2.5.3 Preservation of Samples

Undisturbed samples which are to be tested after some time should be maintained in such a way that the natural water content is retained and no evaporation is allowed.

Usually, two coats of 12 mm thick paraffin wax and petroleum jelly are applied in molten state on either end of the sample to keep the water content unchanged for considerable time when the sample is preserved in a humidity controlled room. In the absence of such facilities, the sampling tubes should be covered by hessian bags and sprinkled with water from time to time. Block samples may be coated with 6 mm thick paraffin wax and kept in air-tight box with saw dust filling the annular space between the box and the sample. Figure 2.9 shows some typical arrangements for preservation of samples.

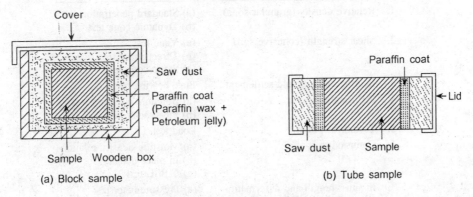

Fig. 2.9 Preservation of samples.

2.6 TESTING OF SOIL

Soil properties are determined from appropriate laboratory and field tests. The specifications regarding laboratory tests and field tests for routine soil investigation are given in Tables 2.4 and 2.5.

Table 2.4 Laboratory testing of soils

	Property of soil	*Type of test*	*Quality of sample*
Classification			
1.	Identification	Visual soil classification	R/D
2.	Grain size distribution	(a) Sieve analysis (b) Wet analysis	D
3.	Consistency limits of cohesive soils	(a) Liquid limit (b) Plastic limit (c) Shrinkage limit	R/D
4.	Moisture content	Moisture content	UD
5.	Unit weight	Specific gravity	D
Engineering properties			
1.	Shear strength	(a) Unconfined compression (b) Direct shear (c) Triaxial (UU/CU/CD)	UD
2.	Compressibility	(a) Oedometer test (b) Triaxial test	UD
3.	Permeability	(a) Constant head permeability test (b) Variable head permeability test	UD
4.	Compaction characteristics	(a) Proctor test (b) CBR test	R/D
5.	Chemical and mineralogical composition	(a) X-Ray diffraction (b) D.T.A. (c) Chemical test	R/D

R—Representative   D—Disturbed   UD—Undisturbed

Table 2.5 Field testing of soils

	Purpose of test	*Type of test*
1.	Relative density (granular soils)	(a) Standard penetration test (b) Dynamic cone test
2.	Shear strength (cohesive soil)	(a) Vane test (b) Direct shear test (c) Static cone test
3.	Bearing capacity and settlement	Plate bearing test
4.	Permeability	(a) Borehole/pumping test (b) Piezometer test
5.	Testing of piles	Load test
6.	Compaction control	(a) Moisture-density relation (b) In-place density (c) C.B.R. test
7.	In-situ strength and deformation characteristics of soil	(a) Pressuremeter test (b) Dilatometer test

2.7 FIELD TESTS

2.7.1 Standard Penetration Test (SPT)

It is extremely difficult to obtain undisturbed samples of granular soils, so in-situ SPT is performed at frequent intervals along the depth of a borehole. A standard split spoon sampler (Fig. 2.10) is driven 45 cm into the ground by means of a 65 kg hammer falling freely from a height of 75 cm. The total number of blows required to drive the second and third depth of 15 cm (i.e. total 30 cm) is called the standard penetration resistance (N blows per 30 cm). After the blow counts are recorded, the spoon is withdrawn and a representative sample is obtained for identification tests.

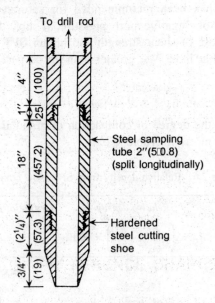

Fig. 2.10 SPT sampler.

For cohesive soils, a simple correlation between the standard penetration resistance (N) and the undrained shear strength, c_u has been proposed by Stroud (1974),

$$c_u = kN \tag{2.0}$$

where k is a constant having an average value of 4.5 kN/m². Similar relationship has been obtained by Sengupta (1984) who studied the correlation for some cohesive soils and obtained a value of 4.2 for the constant k. Terzaghi and Peck's relationship has been widely used to obtain the consistency of cohesive soils in terms of the undrained shear strength, as shown in Table 1.9 (Terzaghi and Peck, 1967).

In granular soils, the SPT blow count is affected by the effective overburden pressure, σ_v'. So, N value obtained from the field should be corrected to correspond to a standard value of σ_v'. Accordingly,

$$N_{cor} = C_n N_F \tag{2.1}$$

where

N_{cor} = corrected N value for a standard value of σ_v' (100 kN/m²)

C_n = correction factor

N_F = N value obtained from field

The correction factor C_n may be taken from the empirical relationship given by Skempton (1986),

$$C_n = \frac{2}{1 + 0.01\sigma_v'} \tag{2.2}$$

where σ_v' = vertical overburden pressure in kN/m²

A dilatancy correction has been recommended for saturated fine sands and silts to account for the development of negative pore pressure, if any, during driving of the SPT sampler and consequent increase of shear strength and higher SPT blow count (Terzaghi and Peck, 1967). For such soils that have N_{cor} greater than 15 as per Eq. (2.1), a correction for dilatancy may be made as,

$$N = 15 + 0.5(N_{cor} - 15) \tag{2.2a}$$

The relative density and the degree of compaction of granular soil can be obtained from Terzaghi's empirical correlation, as in Table 2.6.

Table 2.6 Relative density from SPT blow count

No. of blows (N/30 cm)	Relative density $R_D = \left(\dfrac{e_{max} - e}{e_{max} - e_{min}} \right) \times 100\%$	Degree of compaction
0–4	0–15%	Very loose
4–10	15–35%	Loose
10–30	25–65%	Medium
30–50	65–85%	Dense
> 50	> 85	Very Dense

Many attempts have been made to obtain empirical correlation between N_{cor} and the angle of shearing resistance of sand. The most recent attempt by Halanakar and Uchida (1996) appears to agree well with laboratory test data, which gives

$$\phi = \sqrt{20N_{cor}} + 17 \text{ degrees} \tag{2.3}$$

The modulus of elasticity is obtained by the relationship given by Mezenbach (1961) as

$$E = C_1 + C_2N \quad \text{kg/cm}^2 \tag{2.4}$$

where C_1 and C_2 are functions typical of the type of sand. Some C_1 and C_2 values corresponding to different soil types are given in Table 2.7.

Table 2.7 Modulus of elasticity of sand

	Soil type	C_1 (kg/cm^2)	C_2 (kg/cm^2/blow)
1.	Fine sand (above G.W.T)	52	3.3
2.	Fine sand (below G.W.T)	71	4.9
3.	Sand (Medium)	39	4.5
4.	Coarse sand	38	10.5
5.	Sand + gravel	43	11.8
6.	Silty sand	24	5.3
7.	Silt	12	5.8

A similar correlation between the compression modulus E and the SPT blow count N has been obtained by Papadopoulos (1992). This is given by

$$E = 75 + 8N \ (\text{kg/cm}^2) \qquad (2.4a)$$

Bowles (1988) also gives useful relations to evaluate the stress-strain modulus of sand from SPT blow count, as depicted in Table 2.8.

Table 2.8 Stress-strain modulus of sand (Bowles, 1988)

Type of sand	E (kg/cm^2)
Sand (normally consolidated)	$5(N + 15)$
Sand (saturated)	$2.5(N + 15)$
Sand (overconsolidated)	$7.5(N + 24)$
Sand with gravel	$12(N + 6)$ for $N > 15$
	$6(N + 6)$, $N \le 15$
Silty sand	$3 (N + 6)$

Although standard penetration test is basically a qualitative test, correct interpretation of data gives good evaluation of soil properties particularly in granular soil. The main sources of error include inadequate cleaning of borehole, eccentric hammer blow, and presence of large boulders and gravels which give erratic results.

2.7.2 Dynamic Cone Penetration Test (DCPT)

Dynamic cone penetration test is done by driving a standard 60° cone attached to a drill rod into the soil by blows of 65 kg hammer falling from a height of 750 mm. The blow count for every 30 cm penetration is made to get a continuous record of the variation of soil consistency with depth. The test does not need a borehole. It can be done quickly to cover a large area economically. The test helps to identify variability of subsoil profile and to locate soft pockets, such as filled up ponds. When DCPT is carried out close to a few boreholes, suitable correlations may be obtained for a particular site and the number of boreholes can be reduced.

The dynamic cone penetration test is done either by using a 50 cm diameter cone or a 65 mm diameter cone with circulation of bentonite slurry to eliminate friction on the drill rod. One of the proposed correlations between N_{cd} obtained from DCPT and N_{cor} obtained from SPT is

$$N_{cd} = (1.5 - 2\ )N_{cor} \tag{2.5}$$

These correlations can be used to obtain the SPT blow count, N from DCPT data.

2.7.3 Static Cone Penetration Test (SCPT)

The static cone penetration test (SCPT) is a direct sounding test which is done to obtain a continuous record of soil characteristics with depth and to estimate their engineering properties. The test does not need any borehole. A 60° cone having an apex angle of 60° and a base area of 10 cm² with a friction jacket above, is pushed into the ground at a steady rate of 20 mm/s. Modern static cone penetrometers have electrical measuring devices with wires from the transducers attached to the cone and the friction jacket giving continuous record of the cone and friction resistances as illustrated in Fig. 2.11.

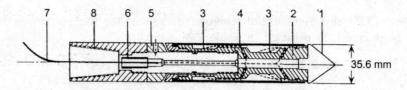

Fig. 2.11 Static cone penetrometer (electrical). 1. Conical point (10 cm²); 2. Load cell; 3. Strain gauges; 4. Friction sleeve (150 cm²); 5. Adjustment ring; 6. Waterproof bushing; 7. Cable; 8. connection with rods

The test measures the cone resistance, q_c, developed against the penetration of the cone and the frictional resistance, f_c developed between the sleeve and the surrounding soil as in Fig. 2.12.

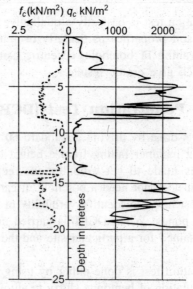

Fig. 2.12 Static cone penetrometer data.

Typical penetrometer test data give a continuous variation of the cone resistance and the frictional resistance with depth. In recent years, the static cone penetrometer has been modified to incorporate an electrical piezo-cone to give simultaneous measurement of tip resistance, side friction and the pore pressure as the cone is advanced in the soil. The development of pore pressure makes the interpretation of soil type more accurate in terms of permeability of the soil.

Lancellotta (1983) and Jamilkowski et al. (1985) proposed an empirical correlation between the relative density of normally consolidated sand, D_r and q_c,

$$D_r(\%) = -98 + 66 \log_{10} \left(\frac{q_c}{\sqrt{\sigma'}} \right) \tag{2.6}$$

where σ' = vertical effective stress at depth considered and both q_c and σ' are in units of tonnes per sq.m.

This relationship is based on the correlation obtained from several sands as depicted in Fig. 2.13.

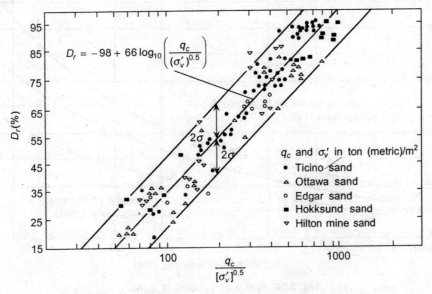

Fig. 2.13 Relationship between cone penetration resistance and relative density
(after Jamilkowski et al., 1985).

The peak friction angle ϕ' of normally consolidated sand may be obtained from the expression, (Kulhawy and Mayne 1990),

$$\phi' = \tan^{-1} \left(0.1 + 0.38 \log \frac{q_c}{\sigma'_v} \right) \tag{2.7}$$

For cohesive soil, Mayne and Kemper (1988) gave the following relations for the undrained shear strength c_u, preconsolidation pressure p_c, and the overconsolidation ratio, *OCR* as

$$c_u = \frac{q_c - \sigma_v}{20}$$

$$p_c = 0.243 \, (q_c)^{0.96} \tag{2.8}$$

$$OCR = 0.37 \left(\frac{q_c - \sigma_v}{\sigma_v'} \right)^{1.01}$$

where σ_v and σ_v' are the total and effective vertical stresses at the level of test respectively.

Some useful relationship between q_c and the SPT blow count have also been obtained by Robertson and Campanella (1983). The range of variation of q_c/N_F with mean grain size D_{50} is illustrated in Fig. 2.14.

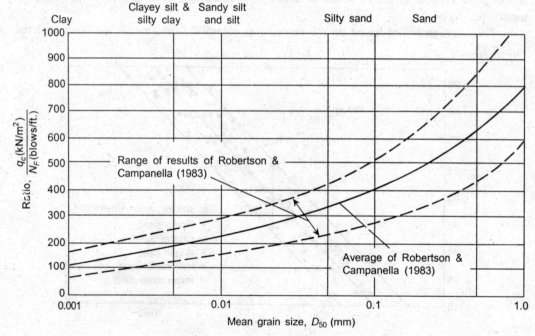

Fig. 2.14 Range of variation of q_c/N_F.

Meyerhof's (1965) simple correlation between q_c and N for fine to medium dense sand is also extensively used. This relation is expressed as

$$q_c = 4N \tag{2.9}$$

where q_c is in kg/cm^2.

2.7.4 Vane Shear Test

Vane shear test in cohesive soils obviates the difficulty of obtaining un-disturbed samples and are particularly suitable for sensitive clays. This test facilitates "averaging" the mass characteristics of soil in-situ.

A four bladed vane at the bottom of a drill rod is pushed into the soil and a torque is applied by turning a handle at the top to create a cylindrical shear surface, as depicted in Fig. 2.15.

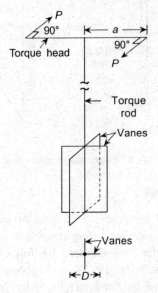

Fig. 2.15 Vane shear test arrangement.

At failure, the shear strength of the soil is related to the applied torque by the relationship

$$T = \frac{\pi D^2 H}{2}\left(1 + \frac{1}{3}\frac{D}{H}\right)\tau \qquad (2.10)$$

where,

D = diameter of sheared cylinder ≈ diameter of vane

H = Height of vane

τ = Shear strength acting along the surface as well as at the top and bottom of the sheared cylinder.

The assumption that the shear stress is uniformly distributed across the top and bottom is questionable but the variation due to any other assumed distribution is not great. Normally, 50 mm diameter × 100 mm long four bladed vane is used and the vane is rotated at the rate of 0.1 degree/s. Both undisturbed and remoulded strengths can be determined by first finding the undisturbed strength and then rotating the vane fully to obtain the remoulded strength.

2.7.5 Direct Shear Test (In-Situ)

In-situ direct shear test, depicted through Fig. 2.16, is particularly suitable where tests on small specimens are not representative of the performance of the in-situ soil, e.g. fissured clays. The test has been done extensively on London clay (Bishop 1966).

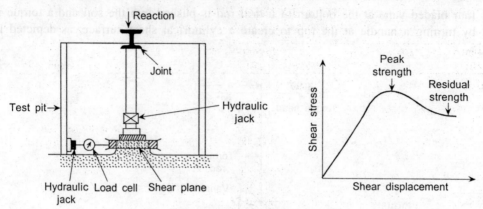

Fig. 2.16 In-situ direct shear test.

2.7.6 Plate Bearing Test

To study the bearing capacity and settlement behaviour of soils, a suitable method is to test a full scale foundation under its design load long enough to observe all settlement. However, this is rarely possible because of the time required for full consolidation and the heavy load required to produce a bearing capacity failure. As a substitute, small scale plate load tests are performed. The load can be applied by dead weight or by jacking against a reaction.

The test is carried out in a pit with either circular or square plates of width/diameter 300–750 mm. The size of the plate should be as large as possible and consistent with the capacity of the loading device. The load is increased in increments of about 1/10th of the estimated failure load or 1/5th of the proposed design load until complete bearing capacity failure or twice the design load is reached. A plot of settlement versus load intensity is then obtained as in Fig. 2.17.

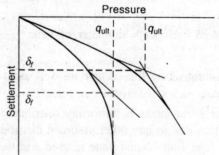

Fig. 2.17 Plate load test data.

The failure load is given by the intersection of initial and final tangent. If no well defined failure point is reached, the data are plotted on $\log \delta$ versus $\log p$ scale to obtain the point of intersection. Failure may also be taken to be the point corresponding to an arbitrarily chosen limit of settlement, depending on the requirement of the structure. Knowing the failure load and deformation characteristics from plate load test, the shear strength and modulus of elasticity of the soil may be obtained from correlation with bearing capacity and settlement equations of shallow foundations.

Limitations of load tests

The limitations of plate load test arise out of

(i) Extrapolation on the basis of theory of elasticity and/or empirical relation is only approximate due to non-homogeneity of the soil. Some agencies recommend the use of plates of different sizes and extrapolation for the actual foundation.

(ii) Load test data reflect the characteristics of a soil only within a depth approximately equal to twice the width of the plate.

(iii) Plate loading test is essentially a short term test (run in a few hours), so no indication of the long term consolidation behaviour is obtained.

(iv) The load test data alone do not give full indication of the properties of a subsoil. But, used judiciously in conjunction with other test data, it has valuable use in design, particularly in estimating the settlement of cohesionless soil.

2.7.7 Pressuremeter Test

Menard (1956) developed the pressuremeter test to measure the strength and deformation characteristics of soil in-situ. The test is done at different depths in a borehole with the help of a pressuremeter which consists of an expandable probe with a measuring cell at the centre and two guard cells at top and bottom. One such arrangement is shown in Fig. 2.18.

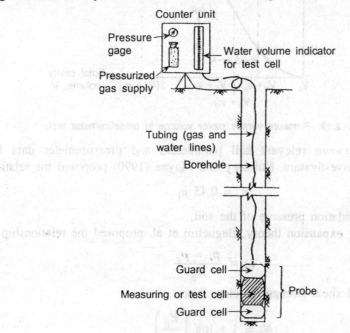

Fig. 2.18 Menard pressuremeter.

The probe is inserted in a pre-bored hole and is expanded in volume either by liquid or air pressure until the soil fails or the expanded volume of the measuring cell reaches twice the original volume of the cavity. The guard cells are used to minimise the end effect on the measuring cell. Table 2.9 gives the typical dimensions of the probe and borehole.

Table 2.9 Dimensions of pressuremeter probe and borehole

Hole designation	Diameter of probe (mm)	L_o (m)	L (m)	Borehole dia (mm) Nominal	Maximum
Ax	44	36	66	46	52
Bx	58	21	42	60	66
Nx	70	25	50	72	48

Figure 2.19 shows typical results of a pressuremeter test. The expanded volume of the measuring probe is plotted against the applied pressure. The curve is divided into three zones. Zone I represents reloading of the soil during which the soil is pushed back into the initial state of stress. The pressure p_o represents the in-situ overburden pressure. Zone II represents the pseudoelastic condition when the cell volume increases linearly with the pressure and p_f defines the yield stress. Zone III gives the plastic zone, p_l representing the limit pressure which is obtained by extrapolation.

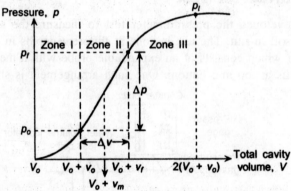

Fig. 2.19 Pressure versus cavity volume in pressuremeter test.

Correlations between relevant soil parameters and pressuremeter data have been developed by many investigators. Kulhawy and Mayne (1990) proposed the relationship,

$$p_c = 0.45 \, p_l \quad (2.11)$$

where p_c = preconsolidation pressure of the soil.

Based on cavity expansion theory, Baguélim et al. proposed the relationship

$$c_u = \frac{p_l - p_o}{N_p} \quad (2.12)$$

where c_u = undrained shear strength of clay. Also,

$$N_p = 1 + \log_e\left(\frac{E_p}{c_u}\right)$$

Here,

$$E_p = 2.66\left(V_o + \frac{V_o + v_f}{2}\right)\left(\frac{p_f - p_o}{v_f - V_o}\right)$$

E_p is pressuremeter modulus (**generally lies between 5 and 12**).

Further innovations in in-situ testing have been achieved through the flat-plate dilatometer test (Marchetti 1980, Schmertmann 1986). This is a further development of the pressuremeter test. But these tests are rather expensive and are yet to be adopted as a part of routine soil investigation.

The remaining field tests indicated in Table 2.5 are not directly relevant to foundation design. Field pumping tests are done to obtain the in-situ permeability of the soil which is required for working out a dewatering scheme. Compaction control tests are done to control the field compaction of soil in land reclamation and embankment construction. Load test on piles, which are required to check the safe load capacity of piles will be discussed in the chapter on pile foundations.

2.8 LABORATORY TESTS

A set of routine laboratory tests are required to be done to obtain the soil parameters for foundation design. These tests are indicated in Table 2.4. Care needs to be taken in choosing the appropriate tests for a particular soil type. Classification and identification tests are normally done on representative or disturbed samples while engineering properties are to be determined from tests on undisturbed samples. Sufficient number of tests should be done for each identified stratum to assess the relevant design parameters. The procedure for laboratory tests are given in I.S. Codes and other building codes.

2.9 GROUND WATER TABLE

The ground water table and seasonal fluctuations of the same are important parameters that are necessary for foundation design and for working out dewatering schemes for deep excavations. The position of ground water table is determined at the time of investigation by observations in open wells or boreholes allowing sufficient time for stabilization. Depending on the time of investigation, the measured ground water table may give the highest or lowest position of the same. To obtain the seasonal fluctuation of ground water table observations may be made in suitably placed piezometers at regular intervals of time.

2.10 PLANNING OF EXPLORATION PROGRAMME

2.10.1 Layout and Number of Boreholes

Whenever possible boreholes should be made as close as possible to the proposed foundations. This is particularly important where the subsoil is irregular in depth. First a preliminary layout is made, preferably on a suitable pattern of evenly spaced grid with supplementary boreholes as necessary. The number of boreholes depend on local conditions and the amount of fund allotted for site investigation. One may start with the minimum number, then go for supplementary boreholes if the subsoil conditions prove irregular. A minimum of two boreholes would be required for a foundation design. Some typical layout of boreholes are shown in Fig. 2.20.

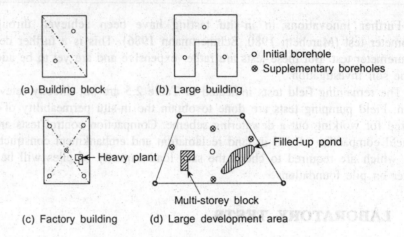

(a) Building block   (b) Large building

○ Initial borehole
⊗ Supplementary boreholes

(c) Factory building   (d) Large development area

Fig. 2.20 Layout of boreholes.

2.10.2 Depth of Boreholes

Depth of boreholes is governed by the depth of soil affected by the foundation loading. It should be at leats one and a half times the width of the loaded area. In case of narrow and widely spaced strip foundations, the borings may be comparatively shallow. But for large raft foundations or pile foundations, the borings have to be deep. Where foundations are extended to rock, it is necessary to prove that rock is, in fact, present at the assumed depth, so boreholes should be taken down to establish the depth to the rock surface. In general, unless hard soil bed rock is encountered at shallow depth the boring should be done to such depth that the net increase in soil pressure due to the foundation loading is less than 10% of the average foundation pressure or 10% of the vertical effective overburden pressure, as shown in Fig 2.21.

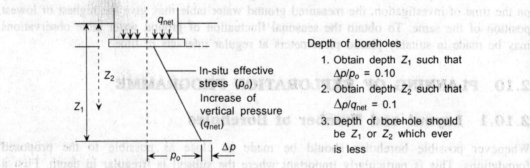

Depth of boreholes

1. Obtain depth Z_1 such that $\Delta p/p_0 = 0.10$
2. Obtain depth Z_2 such that $\Delta p/q_{net} = 0.1$
3. Depth of borehole should be Z_1 or Z_2 which ever is less

Fig. 2.21 Depth of boreholes.

As a rough indication, it is worthwhile to investigate the subsoil to a depth of at least twice the width of the anticipated largest size of foundation. If pile foundation is to be considered, the depth of boring should extend well into the bearing stratum so as to obtain the soil data necessary for evaluating the tip resistance of the piles.

REFERENCES

Bishop, A.W. (1966), Strength of Soils as Engineering Materials, *Geotechnique,* Vol. 16, pp. 89–130.

Hatanaka, M. and A. Uchida (1996), Empirical Correlation Between Penetration Resistance and Internal Friction Angle of Sandy Soils, *Soils and Foundations,* Vol. 36, No. 4, pp. 1–10.

Jamilkowski, M., C.C. Ladd, J.T. Germaine, and R. Lancellotta (1985), *New Developments in Field and Laboratory Testing of Soils,* Proceedings, 11th International conference on Soil Mechanics and Foundation Engineering, San Francisco, Vol. 1, pp. 57–153.

Kulhawy, F.H. and P.W. Mayne (1990), Manual on *Estimating Soil Properties for Foundation Design,* Electric Power Research Institute, Palo Alto, California.

Lancellotta, R. (1983), *Analisi Di Affidabilita in Ingegneria Geotecnia,* Atti Instituto Sciencza Costruzioni, No. 625, Politecnico di Torino.

Marchetti, S. (1980), *In-situ* Test by Flat Dilatometer, *Journal of Geotechnical Engineering Division,* ASCE, Vol. 106, GT3, pp. 299–321.

Mayne, P.W. and J.B. Kemper (1988), Profiling OCR in Stiff Clays by CPT and SPT, *Geotechnical Testing Journal,* ASTM, Vol. 11, No. 2, pp. 139–147.

Menard, L. (1956), *An Apparatus for Measuring the Strength of Soils in Place,* M.S. Thesis, University of Illinois, Urbana, Illinois, USA.

Meyerhof, G.G. (1965), Shallow Foundations, *Journal of Soil Mechanics & Foundation Division,* ASCE, Vol. 91, No. SM 2, pp. 21–31.

Robertson, P.K. and R.G. Campanella (1983), Interpretation of Cone Penetration Tests. Part I: Sand, *Canadian Geotechnical Journal,* Vol. 20, No. 4, pp. 718–733.

Schmertmann, J.H. (1986), Suggested Method for Performing the Flat Dilatometer Test, *Geotechnical Testing Journal,* ASTM, Vol. 9, No. 2, pp. 93–101.

Skempton, A.W. (1986), Standard Penetration Test Procedures and the Effect in Sands of Overburden Pressure, Relative Density, Particle-size, Aging and Over consolidation, *Geotechnique,* Vol. 36, No. 3, pp. 425–447.

Soil Data and Design Parameters

3.1 INTRODUCTION

The purpose of soil investigation is to provide the engineer with knowledge of the subsurface conditions at a given site for

- Safe and economic design of foundation and substructure.
- Overcoming construction problems that may be encountered at site.
- Investigation of failure/distress of engineering structures.

The extent and nature of investigation depends on the importance and type of the structure.

There should be a desired degree of interaction between the designer, the investigation agency, and the construction agency so that problems of design and construction may be identified in time and measures taken to tackle them before things get out of hand. Figure 3.1 is a schematic representation of interaction of various agencies involved in construction work.

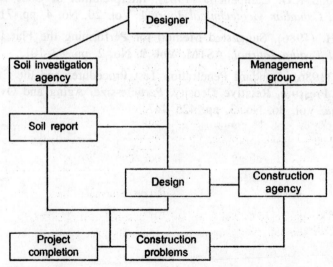

Fig. 3.1 Interaction of various agencies in construction work.

3.2 SOIL INVESTIGATION

3.2.1 Responsibility of Designer

The designer has major responsibility in ensuring proper execution of a soil investigation work. If adequate know-how is not available with the designer, he/she may engage the services of a consultant to advise him/her on all problems related to soil investigation and design.

The responsibilities of the designer in this respect may be summarized as follows:

(a) Draw up a comprehensive programme and specification of soil exploration work relevant to the project.
(b) Select a competent investigation agency.
(c) Ensure that field and laboratory tests are appropriate and done with care and thoroughness.
(d) Evaluate/interpret the soil report and select design parameters.
(e) Make the design.
(f) Interact with the contractor to overcome construction problems, if any.

3.2.2 Information Required from Soil Investigation

The objective of soil investigation is to obtain the following data pertaining to a given site:

(a) Engineering geology of the area
(b) General topography
(c) Past history and land use pattern, if any
(d) Soil stratification
(e) Depth to rock, if any
(f) Ground water and drainage
(g) Engineering properties of different strata
(h) Design recommendations, if the scope permits

3.2.3 Soil Test Report

A good soil test report should contain data regarding the following information:

(a) Project and site description
(b) Regional and site geology
(c) Dates of field and laboratory work
(d) Layout of structures and location of boreholes/field tests
(e) Method of investigation
   Field work
   Laboratory tests
(f) Details of field and laboratory work
(g) Ground water characteristics
(h) Field test data

    (i)   Laboratory test data

    (j)   Soil profile and stratification

    (k) Interpretation of data

    (l)   Design parameters

   (m) Design recommendations, if included in the scope of work

Some of these parameters are discussed in the remaining part of this section.

Project and site description

A soil report should give some background information pertaining to the project site for the designer to work out an economic design. Such information relates to

- General level of the site with respect to adjacent area
- Problems of water logging/drainage
- Surface configuration
- Pond, rock outcrop etc.
- Adjacent buildings
- Layout and type of structure, and
- Location of borehole and field tests

    A layout plan of the site showing location of proposed constructions, old structures to be demolished, if any, adjacent buildings/facilities and so on, should form an integral part of the soil report. This reveals, at a glance, the test locations in relation to the proposed structures and the problems to be encountered in making a deep excavation, close to an existing structure, for example. Such a plan also gives the information about possible weak spots in the site which may require special attention in design. A typical layout plan for a building project is shown in Fig. 3.2.

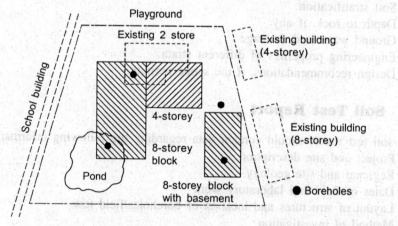

Fig. 3.2 Layout plan showing location of boreholes.

Method of investigation

Routine soil exploration is carried out through boreholes, in-situ standard penetration test within boreholes, and field tests such as static and dynamic cone penetration tests. Laboratory tests are carried out on disturbed/undisturbed samples collected from the boreholes. For a big project, a limited number of boreholes may be supplemented by dynamic cone penetration tests which are particularly useful for determining the depth of fill, if any, through appropriate correlation with borehole data. It is, however, necessary to do at least one dynamic cone test adjacent to each borehole. The location of a filled-up pond in the HUDCO project area in Ultadanga, Calcutta was demarcated in this manner with the help of dynamic cone test, as shown in Fig. 3.3.

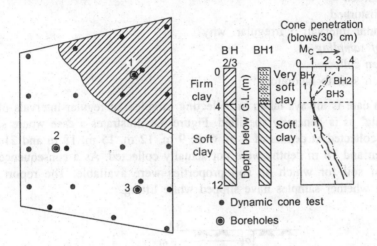

Fig. 3.3 Location of filled-up pond in HUDCO project site, Calcutta.

Date of investigation

The date of investigation is important in evaluating the fluctuation of ground water table (G.W.T.) at a given site. If there is seasonal fluctuation in water table, measurement at the time of investigation does not necessarily give the highest or lowest position of G.W.T. In case investigation is done in the dry season, local enquiry should be made about seasonal fluctuation for proper evaluation of G.W.T. The date of investigation becomes important in such an evaluation.

Boring and sampling

The method of boring adopted at a given site should be given due importance in evaluating the soil data. The report should clearly specify the technique—shell and auger, wash boring, or bentonite mud drilling—that has been used to make the boreholes. Also, the use of casing or chiselling done, should be clearly indicated in the report.

Sampling is an important aspect of soil investigation. The quality of samples determine the reliability or otherwise of the laboratory test data. The relevant data on sample and sample collection are:

(i) *Type of sampler used*
  • open drive sampler/piston sampler

    (ii) *Size of samplers*
        • length and diameter
        • area ratio
   (iii) *Method of advancing sampler*
        • pushing of driving
   (iv) *Schedule of sampling*
        • disturbed
        • undisturbed
        • regular interval; if irregular, why?
    (v) *Date of sampling*
        • when collected
        • when sent to laboratory

Not much care is always taken in collecting samples 'at regular intervals of depth or at changes of strata' as is generally specified. Figure 3.4 illustrates a case where samples were required to be collected at depths of 3 m, 6 m, 9 m, 12 m, 15 m, 18 m and 21 m but those from 6 m, 15 m and 18 m depths were not actually collected. As a consequence, there were some depths of soil for which no soil properties were available. The report should also clearly indicate whether samples have slipped while lifting.

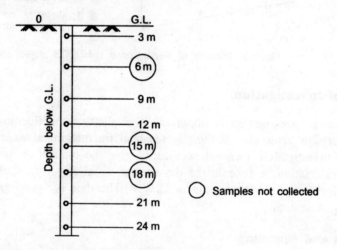

Fig. 3.4 Collection of undisturbed samples.

Testing

Laboratory tests on disturbed/undisturbed samples are done for the purpose of classification of strata and also for determination of their engineering properties. The schedule of tests should be drawn up with care, reflecting the soil type, and the nature of problem to be solved. This should better be done in consultation with the designer to avoid a random choice of the type of test from the schedule of tests given in the bill of quantities. Figure 3.5 gives a schedule of tests on samples from different strata of normal Calcutta deposit. The types of

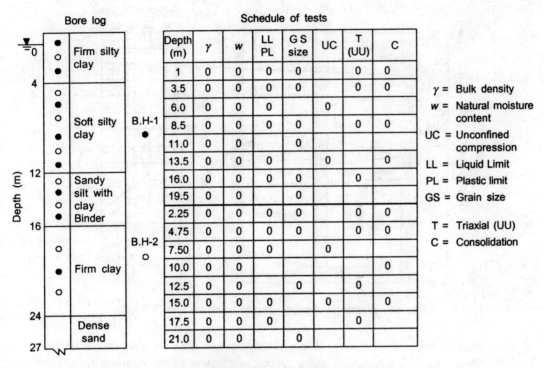

Fig. 3.5 Schedule of laboratory tests.

tests should be so chosen as to give the properties of all the strata relevant for design. One should avoid going for unconfined compression test in a predominantly silty soil.

Similarly, consolidation tests are required for cohesive strata only. For sandy strata, one has to rely more on field tests, such as SPT. Even undisturbed samples in sand do not provide much help.

Soil profile

Soil profiles should be drawn through a number of boreholes, if not through all of them, to give the subsoil stratification along a chosen alignment. Such soil profiles drawn for a number of carefully chosen alignments give a comprehensive picture of the variation of soil strata, throughout the site. Plotting separately for individual boreholes does not give the true picture at a glance. Therefore, the best way is to plot the soil profile on a desired alignment with respect to the variation of N value with depth. The relative consistency of different strata emerges clearly from such a diagram, as depicted through Fig. 3.6.

The position of ground water table at the time of investigation should be clearly indicated in the soil profile. Design of foundations should take appropriate note of any seasonal fluctuation in ground water table.

Laboratory test data

The laboratory test data are given in different forms in the soil test report. The interpretation

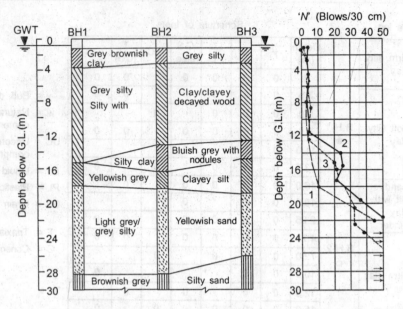

Fig. 3.6 Soil profile through selected boreholes and SPT data.

of data, thereby becomes difficult. While the basic test data should always be there, the engineer should be in a position to make proper assessment of the engineering properties of different strata to be used for design.

The basic test data may be summarized under the following categories for each major stratum of a subsoil deposit:

(a) *Classification data*
   Bulk density
   Natural moisture content
   Atterberg limits
   Grain-size distribution

(b) *Engineering properties*
   Shear strength parameters
   Permeability
   Consolidation test data: m_v, C_c, C_v
   Compaction characteristics

(c) *Chemical and mineralogical data*

(d) *Chemical test on water samples*

Data interpretation

The consistency data should be established before choosing the design parameters for a given problem. Further, the reliability of data and their consistency may be studied qualitatively from the results of individual tests. For example,

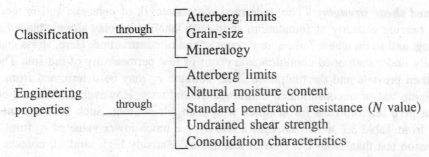

Classification   <u>through</u>   ⎡ Atterberg limits
                                     ⎢ Grain-size
                                     ⎣ Mineralogy

Engineering   ⎡ Atterberg limits
properties   <u>through</u>   ⎢ Natural moisture content
                                     ⎢ Standard penetration resistance (N value)
                                     ⎢ Undrained shear strength
                                     ⎣ Consolidation characteristics

Classification tests: A close examination of the grain-size distribution data and the Atterberg limits may reveal inconsistency in test results. In Table 3.1 which gives a set of test data, sample no. 4 with clay fraction of only 8% indicates a highly plastic clay with liquid limit 62% while sample no. 5 with a clay fraction as high as 48% has a liquid limit of 35% only. These data do not inspire confidence. It is necessary, therefore to check the Atterberg limits against the grain-size distribution of each sample to determine their reliability/consistency.

Table 3.1   Classification tests

Sample no.	Grain-size (%)			Atterberg limits	
	Sand	Silt	Clay	LL (%)	PL (%)
1	7	60	33	58	26.4
2	2	53	45	76	28.2
3	16	65	19	39	24.0
4	20	72	8	62	30.2
5	6	46	48	35	25.0

Consistency: The variation of Atterberg limits and natural moisture content with depth gives a clear indication of the relative consistency of different strata. A natural moisture content close to the plastic limit and a high bulk density confirms a firm to stiff clay whereas a natural moisture content approaching liquid limit should generally give a lower unit weight and thus, indicates a soft normally consolidated soil, Fig. 3.7.

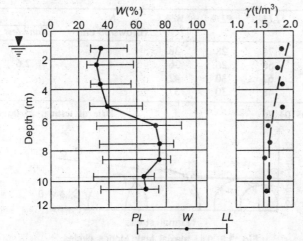

Fig. 3.7 Variation of natural moisture content and Atterberg limits with depth.

Undrained shear strength: The undrained shear strength of cohesive soil is required for to analyze bearing capacity of foundation, short-term stability of clay slopes, braced excavation, tunnelling, and so on where failure may occur at end of construction. Here, stress changes occur essentially under undrained condition as a result of low permeability of the soil. The condition $\phi_u = 0$ then prevails and the undrained shear strength c_u may be determined from unconfined compression test or unconsolidated undrained triaxial tests. However, unconfined compression does not give reliable data for samples with high silt content. Such inconsistent results are evident from Table 3.2 where samples 3 and 4 give much lower value of c_u from unconfined compression test than from UU triaxial test due to relatively high sand/silt content.

Table 3.2 Undrained shear strength

Sample no.	Grain-size (%)			c_u (t/m^2)	
	Sand	*Silt*	*Clay*	*UC test*	*UU test*
1	7	60	33	3.2	4.6
2	2	53	45	2.3	2.5
3	16	65	19	3.4	7.3
4	20	72	8	2.8	5.6

Time effect is an important parameter in evaluating the undrained shear strength of cohesive soils. Samples not preserved properly after sampling lose moisture content by evaporation and give higher strength from laboratory tests. Table 3.3 gives the typical data for samples tested after $2\frac{1}{2}$ months without proper preservation. The c_u values obtained from laboratory tests give much higher strength than in-situ vane shear test carried out during boring and sampling. Also, the Mohr envelopes obtained from UU triaxial tests are often shown to give a ϕ value for saturated cohesive soil. This is not theoretically permissible. If such results are obtained, the degree of saturation of the samples should be checked. Also, the friction in the loading piston if not eliminated properly may give misleading results. Figure 3.8 shows typical Mohr's circles from UU triaxial tests. In such situation it would be more appropriate to obtain an average c_u value for the sample (with $\phi = 0$) rather than trying to draw an envelope giving both C and ϕ.

Table 3.3 Time effect

Sample no.	Depth	LL%	PL%	W%	N (Blows/30 cm)	c_u(t/m^2)	
						Vane test	*Lab. test*
1	6	76	28	38			4.2
2	8.5	58	26	36	4	2.6	4.0
3	10.5	62	30	42			4.5
4	14.0	48	30	37			4.9

Date of sampling: 05 04 76   Date of testing: 25 06 76

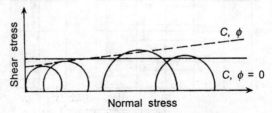

Fig. 3.8 UU triaxial test: Mohr's circles.

Effective stress parameters: The effective stress parameters are required for analysis of long term stability when all the excess pore pressures developed during construction have dissipated. However, in case of granular soils, effective stress parameters are to be used even for short-term stability because the high permeability of the soil generally ensures that all the excess pore-pressures get dissipated during construction itself. The effective shear parameters of a soil are determined from consolidated drained triaxial test or the consolidated undrained triaxial test with pore-pressure measurement. The latter is particularly useful in problems of stage construction of embankment where stability of the embankment is to be investigated for undrained loading after each stage of construction (Gangopadhyay and Som, 1974). The pore-pressure parameter A is also required for such an analysis.

For partially saturated soils and for soils with high silt content where partial drainage may occur even during load application, the total stress parameters of soil (c_u and ϕ_u) may have to be evaluated for stability analysis.

Consolidation: Consolidation tests are often done for pre-determined pressure ranges without any reference to the depth of the samples. In particular, if the sample is derived from deeper strata or it appears overconsolidated, the virgin compression curve has to be determined by loading the sample to sufficiently high pressures for establishing the field compression curve. Only then is it be possible to determine if a soil is normally consolidated or preconsolidated. In the latter case, the C_c value in the overconsolidation range and the preconsolidation pressure are important in selecting the design parameters. Moreover, calculating the m_v value for different stress ranges from the laboratory curve may not represent the field behaviour correctly because of sampling disturbances. It would be more appropriate to obtain the virgin curves from the test data—using Schmertmann's procedure, as shown in Fig. 3.9.

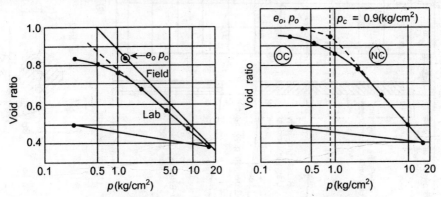

Fig. 3.9 Virgin consolidation curve (after Schmertmann 1953).

Design parameters

On the basis of field and laboratory test data, it should be possible to assign appropriate values of design parameters for each stratum. Considerable judgement is required to evaluate the data. Individual values may be erratic for various reasons described earlier. But an overall assessment for a particular stratum is necessary. Figure 3.10 gives the results of a controlled exploration programme for a failure investigation. The consistency of data becomes evident from such presentation and the results inspire confidence. The average engineering properties of each stratum can thus, be obtained without much difficulty.

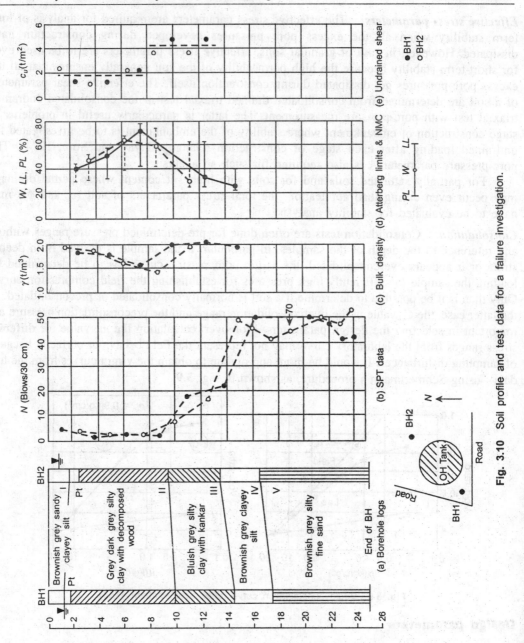

Fig. 3.10 Soil profile and test data for a failure investigation.

Gangopadhyay, C.R. and N.N. Som (1974), *An Approach to $\phi_u = 0$ Analysis for Stage Construction*, Proceedings ASCE, Vol. 100, GT6, pp. 699–703.

Foundations: Types and Design Criteria

4.1 INTRODUCTION

Foundation is that part of a structure which provides support to the structure and the loads coming from it. Thus, foundation means the soil or rock that ultimately supports the load and any part of the structure which serves to transmit the load into the soil. The design of foundation for a structure, therefore involves the following:

1. Evaluation of the capacity of the soil to support the loads and
2. Designing proper structural elements to transmit the super structure load into the soil.

Often the term *foundation* describes only the structural elements but this definition is incomplete, because the ability of the structural element to transmit the load is limited by the capability of the soil to support the load. Therefore, the problem should be considered as a whole and not in isolation. A foundation failure may destroy the superstructure as well while a failure in the superstructure might result only in localized damage and does not essentially mean failure of the foundation.

4.2 TYPES OF FOUNDATION

Foundations can be classified as shallow and deep foundations depending upon the depth of soil which is affected by the foundation loading and, consequently, affect the foundation behaviour. These can be further divided into different types of foundations which are normally adopted in practice. This classification is shown in Fig. 4.1.

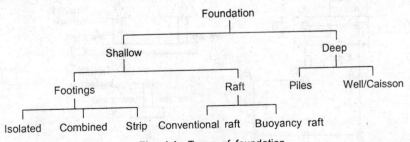

Fig. 4.1 Types of foundation.

4.2.1 Shallow Foundations

In shallow foundations, the load is transmitted to the soil lying immediately below the substructure, as shown in Fig. 4.2. Such foundations are used when the subsoil near the ground surface has adequate strength to support the load.

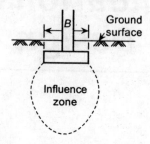

Fig. 4.2 Shallow foundation.

Footings

Isolated footing: Isolated footings are provided to support the columns of a building frame individually. Figure 4.3 depicts an isolated footing. Such footings behave independently of each other without being influenced by adjacent footings in any way.

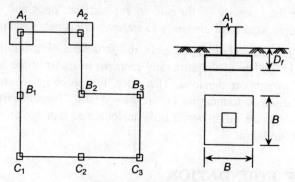

Fig. 4.3 Isolated footing.

Combined footing: Combined footings are designed to support two or more adjacent columns in a building frame where isolated footings either overlap or come very close to one another (refer Fig. 4.4).

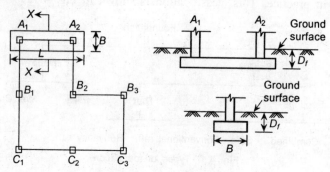

Fig. 4.4 Combined footing.

Strip footing: Strip footings support a load bearing wall or a number of closely spaced columns in a row. They form a long, narrow continuous foundation, with the width small compared to the length, as illustrated in Fig. 4.5.

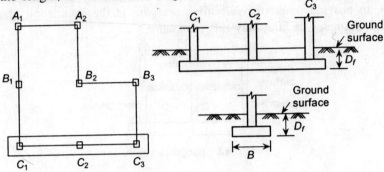

Fig. 4.5 Strip footing.

Raft or mat foundation

A large number of columns or often, the entire structure is founded on one single slab or raft. When individual column footings are, together, found to occupy more than 70% of the plan area of the building, raft foundations are provided. This is shown in Fig. 4.6. The basic difference between footings and rafts lies in their size—the latter being much larger and affecting a greater area of the soil in determining its behaviour.

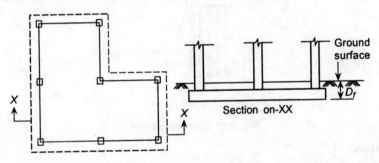

Fig. 4.6 Raft foundation.

Conventional raft: Conventional rafts are provided at shallow depth beneath the ground surface and backfilling is done on the raft to reach the original ground surface. Thereafter plinth filling is done to lay the ground floor of the building, as shown in Fig. 4.7.

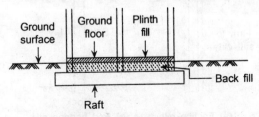

Fig. 4.7 Conventional raft.

Buoyancy raft: Buoyancy rafts are placed at some depth beneath the ground surface but no backfilling is done on the raft. A ground floor slab is provided at the desired height above the ground and the space between the ground floor slab and the foundation raft is kept void, refer Fig. 4.8, to provide relief of overburden pressure at the foundation level. Basement rafts are typical examples of buoyancy raft foundation.

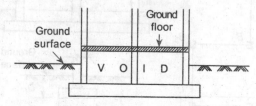

Fig. 4.8 Buoyancy raft.

4.2.2 Deep Foundations

In deep foundations, the load is transmitted well below the bottom of the substructure, as shown in Fig. 4.9. Deep foundations are provided when soil immediately below the structure does not have adequate bearing capacity but the soil at deeper strata have.

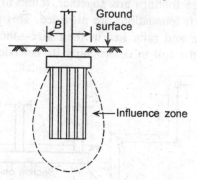

Fig. 4.9 Deep foundation.

Pile foundation

Piles transfer the load through soft upper strata either by end bearing on hard stratum or by friction between the soil and the pile shaft and are accordingly called end bearing piles or friction piles, Fig. 4.10.

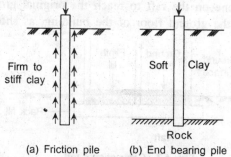

(a) Friction pile   (b) End bearing pile

Fig. 4.10 Pile foundations.

Usually a column load is supported on a group of piles through a pile cap, as shown in Fig. 4.11. Rafts and piles are sometimes combined to form a piled raft illustrated in Fig. 4.12.

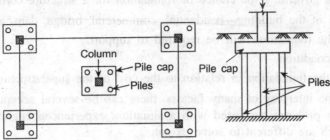

Fig. 4.11 Pile foundation.

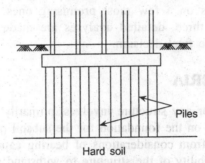

Fig. 4.12 Piled raft.

Piers and caissons

These are large diameter piles, in effect, which are used to support heavy structural load from bridges or very tall multistoreyed buildings. A pier or a well is a shaft drilled into the soil which is then filled with concrete, gravel, and so on. The bottom of the shaft may be undercut or belled out, either by hand or machine as depicted in Fig. 4.13 to afford a large bearing area. Wells are rigid structural elements and can take large lateral forces while piles are slender and liable to bend under flexural stress. Caissons are large diameter wells which are installed by special construction technique.

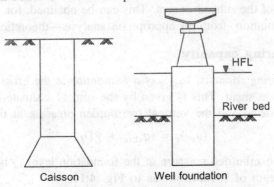

Fig. 4.13 Caisson and well foundation.

4.2.3 Choice of Foundation Type

The main criteria governing the choice of foundation for a structure comprise

(a) Function of the building—residential, commercial, bridge, dam, and so on.

(b) Loads, the foundation will be required to support.

(c) Subsoil condition.

(d) Cost of the foundation in relation to the cost of the superstructure.

On account of the interplay of many factors, there can be several acceptable solutions to a given foundation problem but faced with a situation, experienced engineers may arrive at conclusions which are different to some extent.

Often the choice of the type of foundation is arrived at by the process of elimination. An experienced engineer first discards, almost instinctively, the most unsuitable types of foundation and concentrates on a few most promising ones. When the choice has been narrowed down to two or three, detailed analyses are made and their relative economy studied before arriving at the final decision.

4.3 DESIGN CRITERIA

The design of foundation for any structure involves, primarily the determination of the net permissible bearing pressure on the foundation for the subsoil prevailing at the building site. This should be determined from considerations of bearing capacity, the magnitude and the rate of settlement, and the ability of the structure to withstand settlement. Foundations for a building should, therefore satisfy the following design criteria:

1. There must be adequate factor of safety against bearing capacity failure, and

2. The settlement of the foundation must be within permissible limits.

4.3.1 Bearing Capacity

Net ultimate bearing capacity

The *net ultimate bearing capacity,* $(q_{ult})_n$ of a foundation is the applied pressure at which complete shear failure of the subsoil occurs. This can be obtained, for a given foundation and for a given subsoil condition, from an appropriate analysis—theoretical or empirical.

Gross ultimate bearing capacity

The *gross ultimate bearing capacity,* $(q_{ult})_g$ *of a foundation* is the gross foundation pressure at which the subsoil fails in shear. This is given by the sum of ultimate bearing capacity of the soil at the depth considered and the vertical overburden pressure at that depth. Therefore,

$$(q_{ult})_g = (q_{ult})_n + \gamma D_f \tag{4.1}$$

where γD_f is the total overburden pressure at the foundation level (γ being the unit weight of the soil and D_f, the depth of foundation) as in Fig. 4.14.

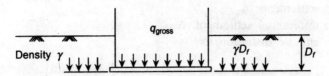

Fig. 4.14 Gross and net ultimate bearing capacity.

Allowable bearing capacity

The *allowable bearing capacity* of a foundation is the maximum allowable net pressure on the foundation determined from considerations of shear failure of the ground. This is obtained by dividing the net ultimate bearing capacity by a suitable factor of safety, that is,

$$q_{\text{all}} = \frac{(q_{\text{ult}})_n}{F}$$

where, F = factor of safety.

The determination of allowable bearing capacity of a foundation from shear failure consideration involves a bearing capacity analysis of the foundation with the relevant soil properties and the choice of an appropriate safety factor.

A factor of safety is applied on the ultimate bearing capacity of a foundation to safeguard against:

  (i)   natural variation in the shear strength of the soil.
 (ii)   uncertainties in the accuracy of test results to determine shear strength.
(iii)   uncertainties in the reliability of theoretical and empirical methods of determining bearing capacity.
(iv)   excessive yielding of the foundation when the soil approaches shear failure.

Of the above, natural variation in subsoil properties and uncertainty about the accuracy of test results are the primary reasons for requiring an adequate factor of safety in determining the allowable bearing capacity of a foundation. Subsoil properties at a site by their very nature, are heterogeneous and there is usually wide variation of test results. Therefore, a high degree of judgement is required in selecting the shear strength parameters for design. Any general guidance in this regard is neither possible nor always desirable, but a safety factor of 2.5–3.0 may be adopted to guard against the variations and uncertainties listed above. Lower factor of safety, say, 2.0 may be adopted for a temporary construction or on sites where subsoil condition is well known and uniform. Lowering the factor of safety even further may lead to local yielding and excessive shear deformation of the soil.

The first step in a foundation design is to determine the net allowable bearing capacity, as described above. It would, then be required to estimate the settlement of the foundation for a bearing pressure equal to the net allowable bearing capacity and then to see if the estimated settlement is within the permissible limits. If not, the foundation is to be redesigned.

4.3.2 The Settlement Criteria

Let us consider the columns in a building frame which were originally at the same level but have settled differentially after application of the building load, illustrated in Fig. 4.15. The different settlement criteria can then be stated as:

(a) Maximum settlement, ρ_{max}
(b) Maximum differential settlement, Δ
(c) Maximum angular distortion, δ/l

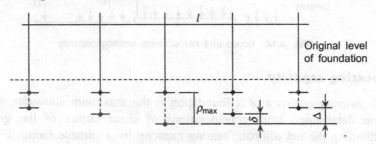

Fig. 4.15 Settlement of foundation and settlement criteria.

All these criteria can be evaluated from an adequate settlement analysis of the foundation. However, it is obvious that mere prediction of settlement is only of limited practical value unless some idea about how much settlement the building is going to tolerate without suffering damage, is obtained. If a building frame settles uniformly over its area, no matter by whatever amount, it has no adverse effect on the behaviour of its structural components. Maximum settlement is important in relation to access and services of the building and is generally of not much significance when it is within reasonable limits. Damage to structural components may, however, occur if there is excessive differential settlement.

Damages due to differential settlement may be classified under the following categories (Skempton and McDonald 1955),

(i) Structural damage involving frame members, namely, beams, columns, and their joints.
(ii) Architectural damage involving the walls, floors, and finishes.
(iii) Combined structural and architectural damage, and
(iv) Visual effects.

Structural damage: Building frames are generally designed to achieve uniform settlement. Since differential settlement occurs in most cases, secondary stresses are induced in the members of the framed structure, the evaluation of which is yet to become a standard practice, although with the advent of numerical analysis using computers, it is now possible to undertake theoretical analysis of complicated building frames for different conditions of total and differential settlement.

In steel frames, local failure may be prevented by yielding of the joints, provided mild steel is used because the relative rotation required to cause fracture is in most cases greater than that which can occur. However, with increasing use of welding in recent years, secondary stresses are of greater importance since yielding of welded connections would result in rupture and failure.

Architectural damage: This refers to cracks in the walls, floors, and finishes which is apparently a more immediate effect of differential settlement than the overstressing of structural members. Excessive cracking may cause damage to the functional aspects of the building. For example, major cracks are considered detrimental to hospital buildings, cold storage, and the like. Hence, in most cases, the cracks or the architectural damages are the guiding factor in determining the allowable settlement of buildings.

Combined architectural and structural damage: Usually, architectural damage occurs long before there is structural damage in beams and columns and consequently a structural damage is almost invariably accompanied by architectural damage.

Visual effect: Even before there is any architectural or structural damage, excessive differential settlement or tilt which can be recognised by naked eye cannot be accepted either psychologically or aesthetically.

Allowable settlement

The allowable settlement of a building can be determined from an integrated analysis of the building frame for given magnitudes of settlement. This is, however, laborious and time consuming and has to be done separately for each building. Therefore, attempts have been made to estimate the allowable settlement from statistical correlation between damage and settlement criteria.

The allowable settlement of buildings depends upon the type of construction, the type of foundation, and the nature of soil (sand or clay). The angular distortion appears to be the more useful criterion for establishing the allowable limits. Terzaghi (1938) studied the settlement pattern of a number of brick walls in Vienna. He found that the walls reached their ultimate strength when the angular distortion was 1/285 and concluded that an average settlement of 5–7.5 cm would be considered normal.

Skempton and McDonald (1955) derived a statistical correlation between damage and settlement of 98 buildings from different parts of the world and concluded that an angular distortion of 1/300 should be considered the allowable limit for conventional buildings. Jappeli (1965) observed the damages to a three-storeyed building on clay due to differential settlement and confirmed that an angular distortion of greater than 1/300 would lead to severe damages in walls of ordinary buildings. Whitman and Lambe (1964), on the other hand, studied the settlement pattern of buildings in MIT Campus and observed that an angular distortion of as little as 1/800 was sufficient to cause cracks in bricks and masonry elements. Mackinley (1964) made a study of more than fifteen structures damaged by settlement. He observed that there was no simple rule to define the tolerance of structures to settlement. A cement silo had collapsed in New York after only 5 cm of differential settlement whereas some public buildings in Mexico City had been in use even after differential settlement of more than 10 cm.

Rethati (1961) carried out an investigation of twelve buildings on fill. This time, the rigidity of the structure as represented by the number of storeys was also considered. While the critical angular distortion of 1/300 agreed well with those proposed by Skempton and McDonald, it was found that 91% of the buildings that suffered structural damage were two storeys or lower. Hence, the author concluded, the critical angular distortion should be related to the rigidity of the building. Some further studies on allowable settlement have been made by Feld (1964) and Grant et al. (1974).

While much work still remains to be done on the subject to recommend some readily acceptable values of allowable settlement, it seems there is a common agreement that an angular distortion of more than 1/300 may lead to damages in conventional load bearing wall and framed construction. There is, as yet, no agreed guideline as regards the allowable maximum settlement of a building. Skempton and McDonald (1955) have proposed some tentative damage limits which are shown in Table 4.1.

Table 4.1 Damage limits for load bearing walls in traditional type framed buildings

	Isolated foundation	*Raft foundation*
Sand	5 cm	5–7.5 cm
Clay	7.5 cm	10–12.5 cm

Indian Standard Code of Practice (IS 1904–1986) recommends the following (Table 4.2) permissible total settlement for RCC framed buildings on different types of foundation.

Table 4.2 Permissible settlement (IS 1904–1986)

		Isolated footings		*Raft foundation*	
		Sand/hard clay	*Plastic clay*	*Sand/hard clay*	*Plastic clay*
Steel structure	ρ_{max}	50	50	75	1/300
	$\delta/1$	1/300	1/300	1/300	1/300
RCC structure	ρ_{max}	50	75	75	100
	$\delta/1$	1/666	1/666	1/500	1/500
Multistorey buildings	ρ_{max}	60	75	75	125
RCC/steel framed Bldg.	$\delta/1$	1/500	1/500	1/400	1/300

An allowable limit of angular distortion of 1/300 has been proposed for framed buildings of both traditional and modern construction. This may not eliminate the chances of cracks in walls and floors altogether, but structural damage would, by and large, be eliminated. Bjerrum (1963) has given the damages limits for different performance criteria, which are shown in Table 4.3. To eliminate cracks in a building, the angular distortion should be less than 1/500.

Table 4.3 Damage limits of angular distortion for different settlement criteria

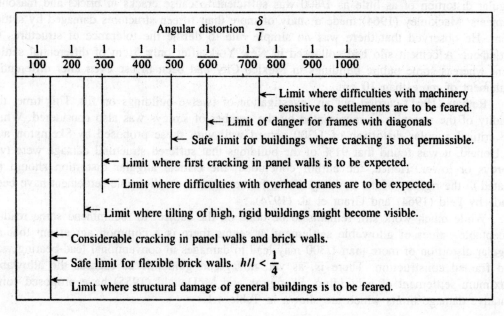

Net permissible bearing pressure

The net permissible bearing pressure on a foundation is to be determined from considerations of safe bearing capacity and permissible settlement, so as to satisfy both design criteria. Starting with the net allowable bearing capacity, if the estimated settlement appears to go beyond the permissible limits, the bearing pressure should be correspondingly reduced until the criterion of allowable settlement is also satisfied although this may mean a higher factor of safety against bearing capacity failure. For soft clays, in general, the settlement criterion governs the choice of net permissible bearing pressure. This is particularly so for raft foundations.

REFERENCES

Bjerrum, L. (1963), *Relation between Obstressed and Calculated Settlement of Structures in Clay, and Sand*, Norwegian Geotechnical Institute Bulletin, Oct 1969.

Grant, R., J.T. Christian, and E. Van Marcke (1974), *Differential Settlement of Buildings*, Proceedings ASCE, Vol. 100, GT9, pp. 973–991.

I.S. 1904 (1986), *Code of Practice for Design of Shallow Foundations*, Bureau of Indian Standards, New Delhi.

Jappeli (1965), *Settlement Studies of Some Structures in Europe*, Proceedings 6th International Conference on Soil Mechanics and Foundation Engineering, Vol. 2, pp. 88–92.

Rethati (1961), *Behaviour of Building Foundations on Embankments*, Proceedings 5th ICSMFT, Vol. 1, p. 781.

Skempton, A.W. and D.H. McDonald (1955), *A Survey of Comparison Between Calculated and Observed Settlement of Structures in Clay*, Proceedings Conference on Correlation of Calculated and Observed Stresses and Displacement of Structures, Institute of Civil Engineers, London, Vol. 1, p. 38.

Terzaghi, K. (1938), *Settlement of Structures in Europe and Methods of Observation*, Proceedings ASCE, Vol. 103, p. 1432.

Whitman, R.V. and T.W. Lambe (1964), *Soil Mechanics*, McGraw-Hill Publication, New York.

5 Stress Distribution in Soils

5.1 INTRODUCTION

An essential step in foundation design is to determine the magnitude and distribution of stresses that are developed in the soil due to the application of structural load. It is these stresses which not only cause settlement of the foundation but determine its stability against shear failure.

The stresses and strains in soil mass depend on the stress–deformation characteristics, anisotropy and non-homogeneity of the soil, and also on the boundary conditions. But the task of analyzing stresses taking all these factors into consideration is extremely complex and, therefore, the attempts that have been made to date are based on simplifying assumptions. The most widely used method of analysis is based on the consideration of soil as homogeneous, isotropic, elastic medium.

It is well understood that the assumption of linearity of the stress–strain relationship which forms the basis of elastic behaviour is a questionable simplification because soils in their behaviour are essentially non-linear. No other widely acceptable theory has yet been developed for practical use to describe the response of soils to stress changes. Also within the comparatively small range of stresses that are normally imposed by structural loads, the assumption of linearity, for most soils, may be considered to be reasonably valid. Also, limited field evidence reported by Plantema (1953) and Turnbull et al. (1961) show that measured stresses correspond fairly well to those predicted by elastic theory. Therefore, refinement of the methods of stress analysis based on the theory of elasticity—still assuming the validity of the linear stress–strain relationship, but taking into consideration the variations of properties within the soil mass—has often been attempted.

5.2 IN-SITU STRESS

The stresses in the subsoil due to the over burden are called in-situ or geostatic stresses. Figure 5.1 shows the in-situ stresses in a soil element at a depth z, below the ground surface.

Total vertical stress, $$\sigma_v = \gamma z \tag{5.1}$$

where, γ = unit weight of soil, and
z = depth below ground surface.

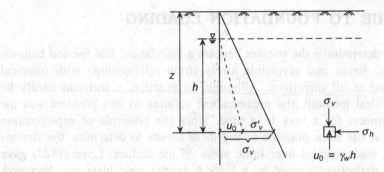

Fig. 5.1 In-situ stresses in soil.

Total horizontal stress, $\qquad\qquad\sigma_h = K\sigma_v = K\gamma z$ $\qquad\qquad$ (5.2)

where, K is the coefficient of lateral pressure at rest with respect to total stress.

Pore water pressure, $\qquad\qquad\qquad u_0 = \gamma_w h$ $\qquad\qquad$ (5.3)

where, γ_w = unit weight of water, and

$\qquad$ h = depth of the point below water table.

Effective vertical stress, $\qquad\qquad\sigma_v' = \sigma_v - u_0$

$\qquad\qquad\qquad\qquad\qquad\qquad = \gamma z - \gamma_w h$ $\qquad\qquad$ (5.4)

Effective horizontal stress, $\qquad\sigma_h' = \sigma_h - u_0$

$\qquad\qquad\qquad\qquad\qquad\qquad = K\gamma z - \gamma_w h$

$\qquad\qquad\qquad\qquad\qquad\qquad = K_o\sigma_v'$ $\qquad\qquad$ (5.5)

where K_o = coefficient of earth pressure at rest with respect to effective stress.

The value of K_o depends on the type of soil and its stress history. For normally consolidated soils, it varies from 0.4–0.7. For over consolidated soils, K_o depends on the overconsolidation ratio and generally becomes greater than 1 for overconsolidation ratio exceeding 4 (Som 1974).

Some empirical formulae for computing K_o values for soils are as follows:
For sand and normally consolidated clays, Jaky (1944) gave a relationship between K_o and the angle of shear resistance, ϕ',

$$K_o = 1 - \sin\phi' \qquad\qquad (5.6)$$

This was subsequently modified by Brooker and Ireland (1965) as

$$K_o = 0.95 - \sin\phi' \qquad\qquad (5.7)$$

For overconsolidated soils, Alpan (1967) gave the relationship,

$$(K_o)_{OC} = (K_o)_{NC} (OCR)^\lambda \qquad\qquad (5.8)$$

where λ is a factor depending on the plasticity index of the soil and is given by

$$I_p = -281 \log(1.85\lambda) \qquad\qquad (5.9)$$

Ladd (1977) suggested the value of 0.41 for λ.

5.3 STRESSES DUE TO FOUNDATION LOADING

It is generally assumed, in determining the stresses beneath a foundation, that the soil behaves as an elastic medium (i.e. linear and reversible stress–strain relationship) with identical properties at all points and in all directions. Although, in practice, a soil can hardly be approximated to such an ideal medium, the mathematical solution to this problem was the only one available to engineers for a very long time. Since the principle of superposition holds for such a medium, it has been possible to use these results to determine the stresses and deflections caused by loads applied over finite areas on the surface. Love (1923) gave equations for stresses and deflections caused by a loaded circular rigid plate and Newmark (1942) derived the expression for the stresses under the corner of a uniformly loaded rectangular area. The tables and charts prepared by Newmark and later by Fadum (1948) are extensively used to calculate the vertical stresses beneath a foundation. The case of a uniformly loaded strip was solved by Carothers (1920) and Jurgenson (1934). Bishop (1952) used stress functions and relaxation technique to calculate the stresses in and underneath a triangular dam. The most complete pattern of stresses, strains and deflections beneath a uniform circular load on a homogeneous half space can be obtained from tables prepared by Ahlvin and Ulery (1962). From all these results, it can be seen that the vertical stresses in a homogeneous, isotropic elastic body is a function only of the dimensions of the loaded area and independent of the elastic properties of the soil. However, this is not true in case of the lateral stresses and displacements.

5.3.1 Boussinesq Analysis: Point Load

Boussinesq (1985) (see Terzaghi 1943) was the first to obtain solution for the stresses and deformations in the interior of a soil mass due to a vertical point load applied at the ground surface, refer Fig. 5.2. He considered the soil mass as a half space bounded on the top by a horizontal plane (ground surface) and extending to infinity along depth and width.

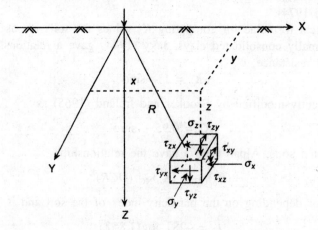

Fig. 5.2 Stress in the soil due to point load at surface (rectangular coordinates).

Considering soil as a homogeneous, isotropic, and elastic medium, Boussinesq obtained the expressions for stresses at a point (x, y, z) located at a distance, R from the origin of coordinates which is also the point of application of the vertical load, Q.
The stress components in Cartesian coordinates are given as:

$$\sigma_x = \frac{3Q}{2\pi}\left[\frac{x^2 z}{R^5} - \frac{1-2v}{3}\left\{-\frac{1}{R(R+z)} + \frac{(2R+z)x^2}{(R+z)^2 R^3} + \frac{z}{R^3}\right\}\right]$$

$$\sigma_y = \frac{3Q}{2\pi}\left[\frac{y^2 z}{R^5} - \frac{1-2v}{3}\left\{-\frac{1}{R(R+z)} + \frac{(2R+z)y^2}{(R+z)^2 R^3} + \frac{z}{R^3}\right\}\right]$$

$$\sigma_z = \frac{3Q}{2\pi}\frac{z^3}{R^5}$$

$$\tau_{yz} = \frac{3Q}{2\pi}\frac{yz^2}{R^5}$$

$$\tau_{xz} = \frac{3Q}{2\pi}\frac{xz^2}{R^5}$$

$$\tau_{xy} = \frac{3Q}{2\pi}\left[\frac{xyz}{R^5} - \frac{1-2v}{3}\left\{\frac{(2R+z)xy}{(R+z)^2 R^3}\right\}\right] \tag{5.10}$$

Here, $R = \sqrt{(x^2 + y^2 + z^2)}$ and

v = Poisson's ratio of the soil

It may be observed that the vertical stress is independent of both the stress–strain modulus and Poisson's ratio. The lateral stresses and shear stresses, however, depend on Poisson's ratio but even these are independent of stress–strain modulus. Values of $v = 0.5$ for saturated cohesive soils under undrained condition (no volume change) and 0.2–0.3 for cohesionless soils, are generally valid.

In cylindrical coordinates, the stress components (refer Fig. 5.3), are:

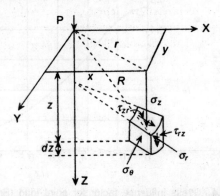

Fig. 5.3 Stresses in the soil due to point load at surface (cylindrical coordinates).

The stress components in cylindrical coordinates are written as:

$$\sigma_z = \frac{3Q}{2\pi}\frac{z^3}{R^5}$$

$$\sigma = \frac{Q}{2\pi}\left[\frac{3zr^2}{R^5} - \frac{1-2v}{R(R+z)}\right] \qquad (5.11)$$

$$\sigma_\theta = \frac{Q}{2\pi}(1-2v)\left[\frac{1}{R(R+z)} - \frac{2}{r^3}\right]$$

$$\tau_{rz} = \frac{3Q}{2\pi}\frac{z^2r}{R^5}$$

The above expressions for stresses are valid only at distances, away from the point of load application. At the point of load application, the stresses are theoretically infinite.

For foundation analysis, the vertical stresses on horizontal plane (σ_z) are mostly required.

Putting

$$R = \sqrt{(x^2 + y^2 + z^2)} = \sqrt{(r^2 + z^2)}$$

$$\sigma_z = \frac{3Q}{2\pi} I_B \qquad (5.12)$$

where $I_B = \dfrac{3}{2\pi\,[1 + (r/z)^2\,]^{5/2}}$

I_B is the influence coefficient for vertical stress at any point within the soil mass, for the Boussinesq problem. The values of I_B for different values of r/z are given in Fig. 5.4.

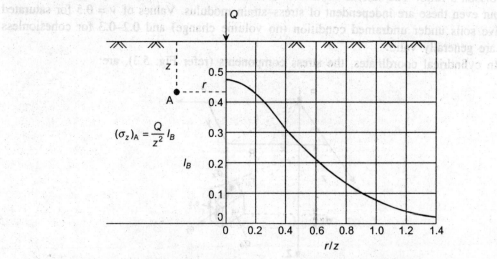

Fig. 5.4 Stress influence factor for point load (Boussinesq).

If a number of point loads Q_1, Q_2, and Q_3 are applied to the surface of soil, then the vertical stress on horizontal plane at any point M, is obtained by adding the stresses caused by the individual point loads, Fig. 5.5.

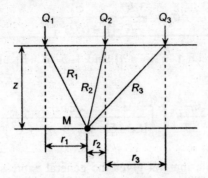

Fig. 5.5 Vertical stress in the soil due to a number of point loads.

In this case,

$$\sigma_z = \frac{Q_1}{z^2} I_{B1} + \frac{Q_2}{z^2} I_{B2} + \frac{Q_3}{z^2} I_{B3} \tag{5.13}$$

where the coefficients I_{B1}, I_{B2}, and I_{B3} are obtained from Fig. 5.4 for the corresponding ratios, r/z.

5.4 VERTICAL STRESSES BELOW UNIFORM RECTANGULAR LOAD

The vertical stress at the point M at depth z below the corner of a rectangular area of length $2a$ and width $2b$, due to a uniform vertical pressure q per unit area, can be obtained by integration of Boussinesq equation, which is given as Eq. (5.10). Figure 5.6 depicts this arrangement effectively.

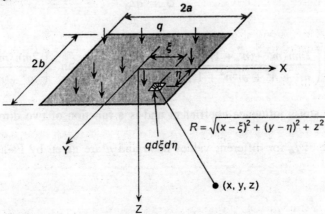

Fig. 5.6 Stresses below the corner of rectangular loaded area.

The equivalent point load on the infinitesimal area $dx\,dy$ is given by $dQ = q\,dx\,dy$
The vertical stress at M due to point load dQ (Eq. 5.12)

$$d\sigma_z = \frac{3\,dQ}{2\pi}\frac{z^3}{[1 + (r/z)^2]^{5/2}} \tag{5.14}$$

Integrating Eq. (5.14) between $[-a$ to $+a]$ and $[-b$ to $+b]$ along the x and y directions respectively we get,

$$\sigma_z = \frac{3qz^3}{2\pi}\int_{-a}^{a}\int_{-b}^{b}\frac{d\xi\,d\eta}{[(x-\xi)^2 + (y-\eta)^2 + z^2]^{5/2}} \tag{5.15}$$

Evaluation of this double integral gives the general expression for vertical stress at any point within the soil mass. Let us now consider the vertical stress at the origin ($x = y = 0$). Then,

$$\sigma_z(0,0,z) = \frac{2q}{\pi}\left(\frac{abz(a^2 + b^2 + 2z^2)}{(a^2 + z^2)(b^2 + z^2)\sqrt{a^2 + b^2 + z^2}} + \sin^{-1}\frac{ab}{\sqrt{a^2 + z^2}\sqrt{b^2 + z^2}}\right) \tag{5.16}$$

Now taking one quarter of this expression, the vertical stress below the corner of a flexible rectangular area ($a \times b$) (i.e. one quarter of the original rectangle $2a \times 2b$) is obtained as

$$\sigma_z = \frac{q}{4\pi}\left[\left(\frac{2mn(m^2 + n^2 + 1)^{1/2}}{m^2 + n^2 + m^2n^2 + 1}\right)\left(\frac{m^2 + n^2 + 2}{m^2 + n^2 + 1}\right) + \tan^{-1}\left(\frac{2mn(m^2 + n^2 + 1)^{1/2}}{m^2 + n^2 - m^2n^2 + 1}\right)\right] \tag{5.17}$$

where $m = \dfrac{a}{z}$ and $n = \dfrac{b}{z}$

or

$$\sigma_z = qI_\sigma \tag{5.18}$$

where

$$I_\sigma = \frac{1}{4\pi}\left[\left(\frac{2mn(m^2 + n^2 + 1)^{1/2}}{m^2 + n^2 + m^2n^2 + 1}\right)\left(\frac{m^2 + n^2 + 2}{m^2 + n^2 + 1}\right) + \tan^{-1}\left(\frac{2mn(m^2 + n^2 + 1)^{1/2}}{m^2 + n^2 - m^2n^2 + 1}\right)\right]$$

I_σ is called the stress influence coefficient and is a function of two dimensionless parameters m and n.

The value of I_σ for different values of m and n are given by Fadum (1948), are shown in Fig. 5.7.

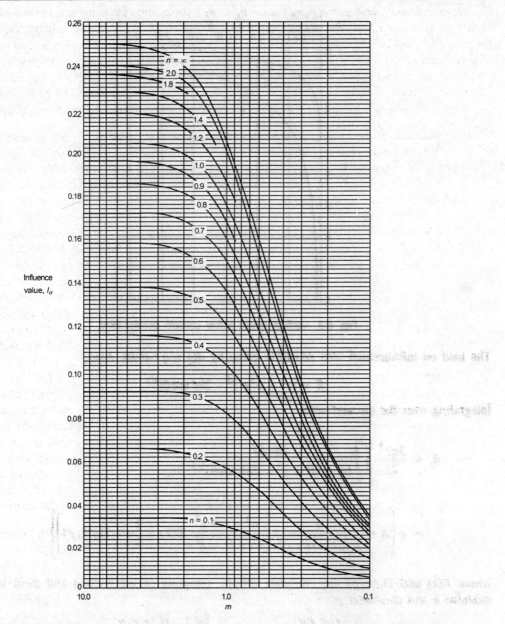

Fig. 5.7 Influence factor for vertical stress below corner of rectangular load, Fadum (1948).

5.5 VERTICAL STRESSES BELOW UNIFORM CIRCULAR LOAD

Following the same principle of superposition as for rectangular load, the vertical stresses below a uniformly distributed circular load may be obtained as in Fig. 5.8.

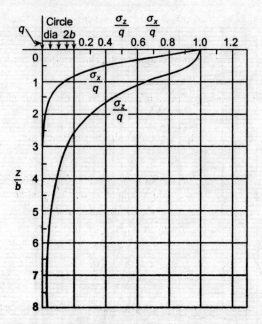

Fig. 5.3 Vertical stress below uniform circular load.

The load on infinitesimal area $r\,d\,r d\theta$ is given by $dQ = q\,r\,dr d\theta$. Also,

$$R = (r^2 + b^2 + z^2 - 2br\cos\theta)^{1/2}$$

Integrating over the circular area,

$$\sigma_z = \frac{3qz^3}{2z} \int_0^a \int_0^{2\pi} \frac{r dr d\theta}{[r^2 + b^2 + z^2 - 2br\cos\theta)^{5/2}} \tag{5.19}$$

$$= q\left\{A - \frac{n}{\pi\sqrt{n^2 + (1+t)^2}}\left[\frac{n^2 - 1 + t^2}{n^2 + (1 - t^2)}E(k) + \frac{1-t}{1+t}\Pi_0(k,p)\right]\right\} \tag{5.20}$$

where $E(k)$ and $\Pi_0(k,p)$ are complete elliptic integrals of the second and third kind of modulus k and parameter p.

$$t = r/a$$

$$n = z/a \qquad A \begin{cases} = 1 & \text{if } r < a \\ = \dfrac{1}{2} & \text{if } r = a \\ = 0 & \text{if } r > a \end{cases}$$

$$k^2 = \frac{4t}{n^2 + (t+1)^2}$$

For the special case of the points beneath the centre of the load, $r = 0$

$$(\sigma_z)_{r=0} = q\left\{1 - \frac{1}{[1 + (a/z)^2]^{3/2}}\right\} \tag{5.21}$$

The vertical stresses beneath the centre of a uniform circular load are shown in Fig. 5.8. The stress influence coefficient for a circular loaded area for different values of r/a and z/a is given in Appendix A.

5.6 OTHER COMMON LOADING TYPES

5.6.1 Uniform Line Load

The vertical stress in the soil due to a line load p per unit length, applied at the surface is given by

$$\sigma_z = \frac{2p}{\pi}\frac{z^3}{(x^2 + z^2)^2} \tag{5.22}$$

$$= \frac{p}{z}\left(\frac{2}{\pi(1 + x^2/z^2)^2}\right)$$

The stress distribution directly beneath the load $(x = 0)$ is shown in Fig. 5.9(a). The variation of $\sigma_z/(p/z)$ with x/z is shown in Appendix A.

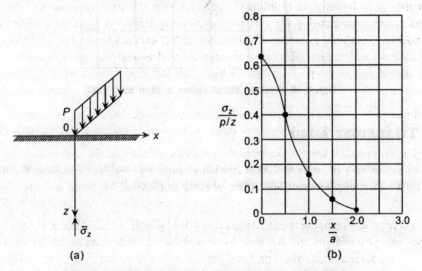

Fig. 5.9   Vertical stresses in the soil due to a line load.

5.6.2 Uniform Strip Load

The vertical stresses in the soil due to a uniformly distributed strip load is given by the expression,

$$\sigma_z = \frac{q}{\pi}\left[\tan^{-1}\left(\frac{z}{x-a}\right) - \tan^{-1}\left(\frac{z}{x+a}\right) - \frac{2az(x^2 - z^2 - a^2)}{(x^2 + z^2 - a^2)^2 + 4a^2x^2}\right] \tag{5.23}$$

The corresponding stress distribution below the centre line and the edge are shown in Fig. 5.10. The variation of σ_z/q with x/a and z/a is tabulated in Appendix A.

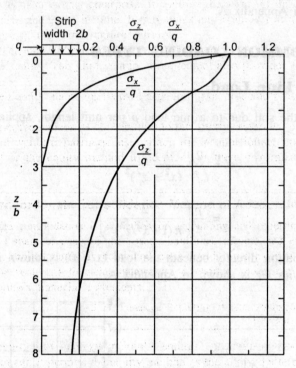

Fig. 5.10 Vertical stress below uniform strip load.

5.6.3 Triangular Load

The vertical stresses in the soil due to a triangular load increasing from zero at the origin to q per unit area at a distance a [refer Fig. 5.11(a)] is given by,

$$(\sigma_z)_A = \frac{qx}{\pi a}\left[\tan^{-1}\left(\frac{z}{x-a}\right) - \tan^{-1}\left(\frac{z}{x}\right)\right] - \left[\frac{qz}{\pi}\frac{x-a}{(x-a)^2 + z^2}\right] \tag{5.24}$$

or

$$\frac{\sigma_z}{q} = \frac{1}{\pi}\left[\frac{x}{a}\left\{\tan^{-1}\left(\frac{\frac{z}{a}}{\frac{x}{a}-1}\right) - \tan^{-1}\frac{z}{a}\frac{a}{x}\right\} - \frac{z}{a}\frac{\frac{x}{a}-1}{\left(\frac{x}{a}-1\right)^2 + \left(\frac{z}{a}\right)^2}\right]$$

For $x = a$, that is, for points below B,

$$\sigma_z = qI'$$

where $I' = \dfrac{1}{\pi}\left(\dfrac{\pi}{2} - \tan^{-1}\dfrac{z}{a}\right).$

The variation of I' with z/a is shown in Fig. 5.11(b). The tabulated values of σ_z/q for different values of x/a and z/a are given in Appendix A.

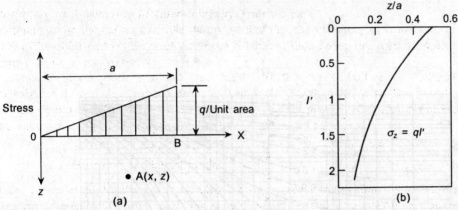

Fig. 5.11 (a) Vertical stress due to a triangular load. (b) Variation of I' with z/a (stress below point B).

5.6.4 Embankment Type Loading

For an embankment of height H, Fig. 5.12(a), the vertical stress at any point below B is given as, Das (1997)

$$\Delta p = \frac{q_o}{\pi}\left[\frac{B_1 + B_2}{B_2}(\alpha_1 + \alpha_2) - \frac{B_1}{B_2}\alpha_2\right] \tag{5.25}$$

where $q_o = \gamma H$

γ = unit weight of embankment soil

H = Height of embankment

$$\alpha_1 = \tan^{-1}\frac{B_1 + B_2}{z} - \tan^{-1}\frac{B_1}{z} \text{ rad}$$

$$\alpha_2 = \tan^{-1}\frac{B_1}{z}$$

$$\Delta p = q_o I' \tag{5.26}$$

where $I' = f\left(\dfrac{B_1}{z}, \dfrac{B_2}{z}\right).$

The variation of I' with $\dfrac{B_1}{z}$ and $\dfrac{B_2}{z}$ is shown in Fig. 5.12.

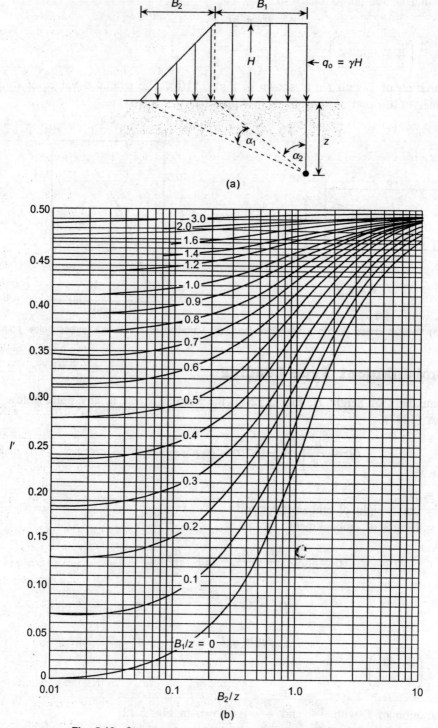

Fig. 5.12 Stresses due to embankment type loading, Das (1997).

5.7 STRESS AT ANY POINT BELOW RECTANGULAR LOAD

The principle of superposition allows determination of vertical stress at any point below a rectangular loaded area. For point A, for example, Fig. 5.13, the loaded area may be divided into four smaller rectangles keeping the point A at the corner of each rectangle. Then,

$$(\sigma_z)_A = q[I_{\sigma_I} + I_{\sigma_{II}} + I_{\sigma_{III}} + I_{\sigma_{IV}}] \tag{5.27}$$

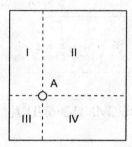

Fig. 5.13 Stresses below point within a loaded rectangular area.

where q is the applied load intensity and I_{σ_I}, $I_{\sigma_{II}}$, $I_{\sigma_{III}}$ and $I_{\sigma_{IV}}$ are the stress influence factors for points below the corner of respective rectangular areas.

Similarly, for points outside the loaded rectangle, Fig. 5.14, the vertical stress below point B may be obtained as,

$$(\sigma_z)_B = q[I_{\sigma_{I+II}} + I_{\sigma_{I+IV}} - I_{\sigma_{III}} - I_{\sigma_{IV}}] \tag{5.28}$$

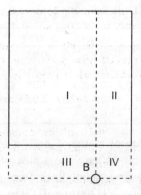

Fig. 5.14 Stresses below point outside a loaded rectangular area.

5.8 NEWMARK'S CHART

Newmark (1942) derived a simple graphical calculation for determining the vertical stress at any point within a soil mass for any shape of loaded area.

Let us consider the stress beneath the centre of a loaded circular area, Fig. 5.15.

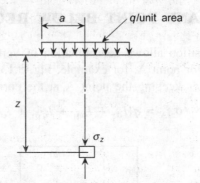

Fig. 5.15 Stress beneath centre of a loaded circular area: construction of Newmark's chart.

$$\sigma_z = q\left\{1 - \frac{1}{[1 + (a/z)^2]^{3/2}}\right\} \tag{5.29}$$

On rearranging the above equation,

$$\frac{a}{z} = \left(1 - \frac{\sigma_z}{q}\right)^{-2/3} - 1 \tag{5.30}$$

The interpretation of this equation is that a/z ratio is the relative size of a circular loaded area in terms of the depth z such that when loaded it gives a unique pressure ratio, σ_z/q on the soil at that depth. By substituting different values of σ_z/q in Eq. (5.30), corresponding values of a/z can be obtained, as in Table 5.1.

Table 5.1 Relationship between a/z and σ_z/q

a/z	σ_z/q	a/z	σ_z/q
0.27	0.1	0.92	0.6
0.40	0.2	1.11	0.7
0.52	0.3	1.39	0.8
0.64	0.4	1.91	0.9
0.77	0.5	∞	1.0

Then, taking an arbitrary value of z (say, 1 cm), a series of concentric circles of radius 0.27 cm, 0.4 cm, and so on can be drawn. The series of rings is further subdivided into a number of units (say 200) by drawing radial lines from the centre, as shown in Fig. 5.16. The value of each unit then becomes $(1/200)q = 0.005q$.

To obtain the stress at a point A located at depth z below a footing, the loaded area is drawn on a tracing paper to a scale z equal to the scale for which the Newmark's chart to be used has been drawn. The plan on the tracing paper is placed on the Newmark's chart such that the point A is placed at the centre of the chart. Then, the number of units of the Newmark's Chart (N) enclosed within, are counted. The stress at A is given by,

$$(\sigma_z)_A = q \times N \times \text{value of each unit.}$$

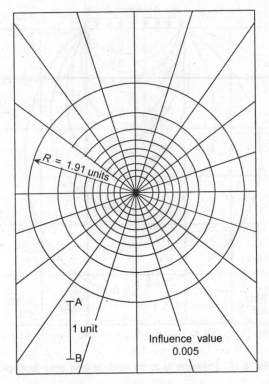

Fig. 5.16 Newmark's chart.

By moving the plan of the building and bringing different points at the centre of the Newmark's chart, the stresses at different points may be obtained. For different depths, the plan of the building is to be drawn fresh to the appropriate scale.

5.9 PRESSURE BULB

When a soil is subjected to a foundation load, it is important to know the zone of soil beneath the foundation which is significantly stressed by the applied load. This is generally expressed graphically by isobars or pressure bulb.

An *isobar* for a given surface load is a curve which connects all points below the ground surface having equal stress. The procedure for obtaining isobars is as follows:

1. Divide the half space in the vicinity of the load area into sufficient number of grid points.
2. Compute vertical stress at each grid point using an appropriate formula/table/chart.
3. Draw contours of equal vertical stresses, say $0.8q$, $0.5q$, $0.2q$, $0.1q$ and so on.

The bulb formed by the set of isobars is called a pressure bulb. Figure 5.17 shows pressure bulbs for vertical stress due to uniform circular and strip loads. It can be seen that for strip loading, the depth of pressure bulb upto which significant vertical stress exists ($\sigma_z/q > 0.1$) is about three times the width of the load area. For circular loading, this depth is about twice the width of load area.

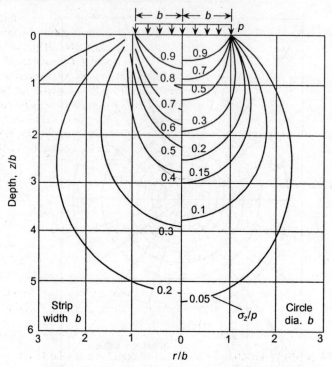

Fig. 5.17 Pressure bulb for circular and strip footings.

The pressure bulb gives the zone of soil which influenced by the foundation loading and is useful in planning soil exploration programmes, and in the study of settlement and interference of footing.

5.10 RIGIDITY OF FOOTINGS: CONTACT PRESSURE

Most footings possess a definite rigidity. It is important to estimate the effect of rigidity on the distribution of pressure on the footing base (contact pressure), and on the stress distribution within the soil mass.

According to Boussinesq analysis, the vertical deformation of a point on the surface of an elastic half space under the action of a concentrated load P is given by

$$W = \frac{P}{\pi R}(1 - v^2)E \tag{5.31}$$

For an arbitrary loading area (Fig. 5.13), the vertical displacement of the point A is given by

$$w_z = \frac{1 - v^2}{\pi E} \iint_F \frac{p(\xi - \eta)\, d\xi\, d\eta}{\sqrt{(x - \xi)^2 + (y - \eta)^2}} \tag{5.32}$$

where F is the loading area over which integration has to be done.

With an absolutely rigid foundation, all points on the surface of contact will have the same vertical displacement. Thus, the condition of absolute rigidity of a foundation is,

$$w_z = \frac{1-v^2}{\pi E} \int\int_F \frac{p(\xi-\eta)d\xi d\eta}{\sqrt{(x-\xi)^2 + (y-\eta)^2}} = \text{constt} \tag{5.33}$$

The solution of the integral equation for an absolutely rigid circular footing with a central load gives the contact pressure at the point M as:

$$p(x, y) = \frac{p_m}{2\sqrt{[1-(\zeta/a)^2]}} \tag{5.34}$$

where,

    a = radius of foundation base

    ζ = distance from the centre of base to a given point ($\zeta \le a$)

    p_m = mean pressure per unit area of the base

It can be seen from Eq. (5.34) that for $\zeta = a$ (i.e. at footing edge), $p(x, y) = \infty$ and for $\zeta = O$ (i.e. at footing centre), $p(x, y) = p_m/2$.

For a strip footing, the contact pressure at the point M is given by,

$$P(x, y) = \frac{2p_m}{\pi\sqrt{[1-(y/b)^2]}} \tag{5.35}$$

where,

    y = horizontal distance of the point M from the centre of footing

    b = half width of footing

The distribution of contact pressure for an absolutely rigid footing on an elastic half space will have a saddle like shape with infinite pressure at the ends, as shown in Fig. 5.18. However, in actual practice, there is redistribution of stresses over the base (the stresses at the edge of footing cannot exceed the bearing capacity of the soil) and the contact pressure at the base of a rigid footing becomes much more uniform as shown by curves in Fig. 5.18.

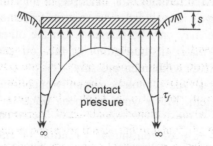

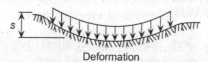

Fig. 5.18 Rigid footing: contact pressure.

For foundation of finite rigidity, the contact pressure can be obtained by solving the integral equation, Eq. (5.32) together with the differential equation for bending of plates. Using such an approach, Borowicka (1936, 1938) obtained the contact pressure distribution for a uniformly loaded circular/strip footing on a semi-infinite elastic mass. The shearing stress along the base of the foundation was assumed to be zero. It was found that the distribution of contact pressure is strongly dependent on a dimensionless factor, termed flexibility factor, of the form,

$$K = \frac{1}{6}\left(\frac{1 - v_S^2}{1 - v_F^2}\right)\left(\frac{E_F}{E_S}\right)\left(\frac{T}{b}\right)^3 \tag{5.36}$$

where,

v_S = Poisson's ratio of soil

v_F = Poisson's ratio of footing material

E_F, E_S = Young's Modulus of footing material and soil, respectively

b = radius for circular footing

T = Thickness of footing

It should be noted that for an elastic footing, the distribution of contact pressure depends on the elastic properties of the supporting medium, on the flexural rigidity of the footing, and on distribution of loads on the footing.

The nonuniform distribution of contact pressure influences stress distribution in the soil only upto a small depth from the base, and the pressure bulb is slightly affected. As a result, the influence of rigidity of footing on settlement is relatively small.

5.11 NON-HOMOGENEOUS SOILS

The engineering properties of a soil are not generally uniform throughout its mass. This non-uniformity may manifest itself in both spatial (non-homogeneous) and directional (anisotropic) variation of the modulus of deformation. The variation of soil properties with depth may be due to many factors. Often the subsoil consists of different geological formations with different characteristics, for example, a clay deposit underlain by sand or rock. If the underlying stratum is well below the surface of the clay relative to the size of the loaded area, its influence may only be marginal. On the other hand, even in a deep layer of apparently homogeneous material, the rigidity of the soil generally increases with depth due to the increase in effective overburden pressure.

In dealing with the first type of non-homogenity mentioned above, a subsoil is often considered as a layered system. Much work has been done on this subject, particularly in connection with the design of pavements and runways. (Biot 1938, Pickett 1938, Burmister 1943, Poulos 1967). In the case of continuous variation of elastic parameters with depth, Gibson's analysis (Gibson 1967, 1968) may be referred to.

5.11.1 Two-layer System

A simple two-layer system, Fig. 5.19, may consist of either two elastic layers with different engineering properties or single elastic layer on a rigid base.

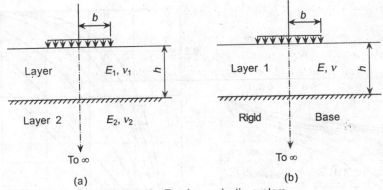

Fig. 5.19 Two-layer elastic system.

Two layers with different elastic parameters

This situation is often encountered in the case of pavements where stiffer layers are placed on a soft subgrade. However, in the case of foundations, the situation is often reversed and one may encounter a layer of soft soil overlying a stronger deposit.

Biot (1935) and Pickett (1938) were among the first to solve the problem of stress distribution in the two-layer rigid base system. However, their results could only be used to determine the stresses at the surface of the base layer. In a series of papers in 1943 and 1945, Burmister (1943, 1945a, b, c) presented the general theory of stresses and displacements in layered soils from which exact solutions could be obtained for axi-symmetric loading. Using Burmister's analysis, Fox (1948) published tabulated values of stresses due to a uniform circular loading with or without friction at the interface for the case of Poisson's ratio, $v = 1/2$. The case of the line or strip loading was analyzed by Lemcoe (1961) who developed equations of stresses for a general two-layer system and tabulated numerical values for the particular case of $E_1/E_2 = 50$ and $v_1 = v_2 = 1/4$.

In the general two-layer system for a circular load, the stresses depend on the values of v_1 and on the two parameters (refer Fig. 5.19).

$$a = \frac{b}{h} \quad \text{and} \quad K = \frac{E_1}{E_2} \tag{5.37}$$

where,

    b = radius of the loaded area

    h = thickness of the top layer and

    E_1, E_2 are the elastic modulii of, respectively the top and bottom layers.

In Fig. 5.20 are plotted the distribution of vertical stresses beneath the centre of a circle for the special case of $a = 1$ and where the upper layer is stiffer than the lower. It can be seen that the presence of the stiff upper layer has a considerable influence on the stresses, particularly in the vicinity of the interface. For example, a rigid upper layer which is five times stiffer than the subgrade (i.e. $E_1/E_2 = 5$) reduces the stress at the interface to 60% of the Boussinesq value. This load spreading capacity of the stiff upper layer has been successfully employed in the design of pavements on soft subgrades.

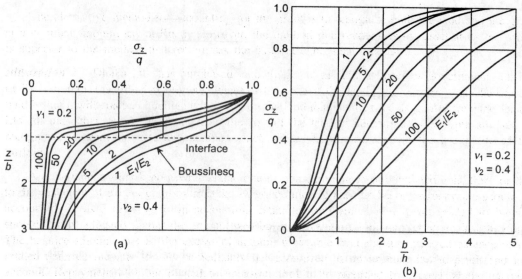

Fig. 5.20 Two-layer elastic system: Effect of rigidity of upper layer on the vertical stress beneath centre of a uniform circular load (after Burmister 1963).

The effect of relative size of loaded areas and thickness of the upper layer on the vertical stresses at the interface is shown in Fig. 5.19b. The upper layer is most effective in spreading the load when its thickness lies between b and $3b$ while for very thin and very thick layers, they approach the Boussinesq values.

The case of a foundation where a soft layer is underlain by a stiffer deposit ($E_1/E_2 < 1$) has not been evaluated but from extrapolation, it can be concluded that the stresses in the upper layer will, if anything, be greater than those for a homogeneous medium.

Single elastic layer on a rigid base

This is a special case of the above problem with the elastic modulus of the bottom layer $E_2 = \infty$. The problem was first solved by Burmister (1956) who extended his earlier work to analyze the stresses and strains in the upper layer of a two-layer rigid base system. From his influence charts, it is possible to obtain the complete pattern of stresses and displacements under the corner of a uniformly loaded rectangle for Poisson's ratio, $v = 0.2$ and 0.4.

The same problem was considered in detail by Poulos (1967), who used Burmister's theory to compute a set of influence factors for stresses and surface displacements due to a point load, for values of Poisson's ratio, $v = 0$, 0.2, 0.4, and 0.5. By integration of these point load factors, he then calculated the corresponding influence factors for different loading types.

The vertical and radial stresses beneath the centre of a loaded circle for values of $h/b = 1, 2, 4$, and 8 and $v = 1/2$ are shown in Fig. 5.21. The stresses for the homogeneous half-space (Boussinesq) are also plotted for comparison. It can be seen that the presence of a rigid layer at a shallow depth relative to the size of the loaded area significantly alters the stress pattern. For small values of h/b just underneath the load, vertical stresses may even be greater than the applied pressure. However, with increasing depth, the effect of the rigid base gradually diminishes and for $h/b > 8$, the stresses are almost indistinguishable from the Boussinesq values.

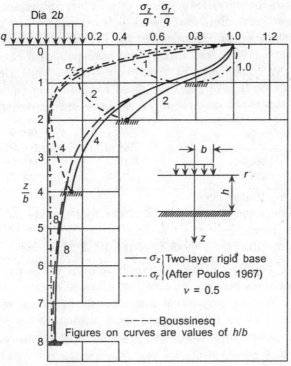

Fig. 5.21 Distribution of stresses in two-layer rigid base system (after Poulos, 1967).

5.11.2 Three-layer Systems

The analyses of three-layer soil systems (Fig. 5.22) are much more complex than for two-layers and solutions have only been obtained for stresses and deflections beneath a uniform circular load. Burmister (1945) was the first to develop the general theory for such a system with both rough and smooth interfaces.

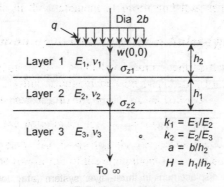

Fig. 5.22 Three-layer elastic system.

Burmister's equations were used by Acum and Fox (1951) to calculate the stresses at the interfaces (for $v_1 = v_2 = 0.5$ and for full continuity between the layers). Schiffman (1957) presented methods for numerical solutions of influence values and tabulated results for a particular case. In later years, Burmister's work was extended to compute the stresses and deflections for any combination of thicknesses of the individual layers and size of the loaded area. Jones (1962), Peattie (1962), and Kirk (1966) have published charts and tables giving the stress factors for any combination of three-layer systems while De Barros (1966) and Uyeshita and Meyerhof (1967) have published those for deflection factors.

The stress and strains in a three-layer medium are governed by the following dimensionless parameters:

$$a = \frac{b}{h_2}; \quad H = \frac{h_1}{h_2}; \quad k_1 = \frac{E_1}{E_2}; \quad \text{and} \quad k_2 = \frac{E_2}{E_3} \quad (5.38)$$

where,

  b is radius of the loaded circle

  h_1 and h_2 are thicknesses respectively of the first and second layers (The bottom layer is semi-infinite).

  E_1, E_2, and E_3 are the elastic modulii of the 1st, 2nd, and the 3rd layers respectively.

Figure 5.23 shows the effect of the relative thicknesses of the two stiffer upper layers on the stresses and deflections beneath the centre of a loaded circular area, for the particular case of $h_1 + h_2 = 2b$. For the purpose of comparison, the Boussinesq stresses and the deflections calculated on the basis of Boussinesq stress distribution are also plotted. It can be seen that the actual values are lower than those given by Boussinesq although the maximum discrepancy in deflection is not more than 25%.

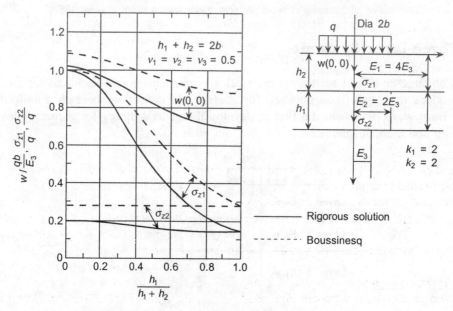

Fig. 5.23 Stress and displacement in three-layer system (after Jones 1962, Uyeshita and Meyerhof 1967).

For the situation where the layers become successivley stiffer with depth ($k_1 = 0.2$, $k_2 = 0.2$), the assumption of homogeneity will underestimate the stresses by up to 30% for $h_1/h_1 + h_2 \to 1$ while for smaller relative thicknesses of the top layer, the error is considerably less, as illustrated in Fig. 5.24.

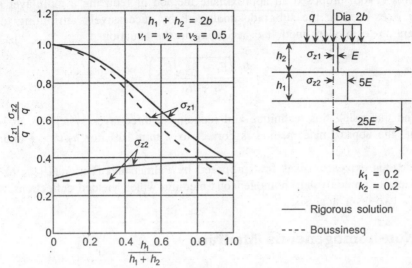

Fig. 5.24 Stress in a three-layer system (after Jones 1962).

5.11.3 Multilayer Systems

The problem of multilayer system involves immense complexity and to date no analytical solution is available for anything consisting of more than three layers. Vesic (1963) has suggested an approximate method of calculating the elastic settlement of a foundation on a multilayered medium assuming Boussinesq stress distribution but using the proper elastic modulus for the respective layers. His charts and method of calculation are shown in Fig. 5.25.

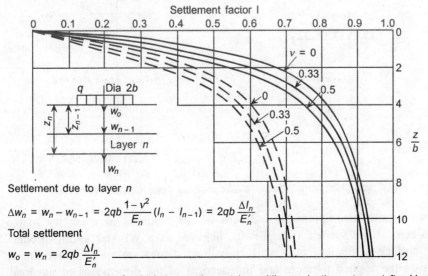

Settlement due to layer n

$$\Delta w_n = w_n - w_{n-1} = 2qb\frac{1-\nu^2}{E_n}(I_n - I_{n-1}) = 2qb\frac{\Delta I_n}{E'_n}$$

Total settlement

$$w_0 = w_n = 2qb\frac{\Delta I_n}{E'_n}$$

Fig. 5.25 Approximate method of calculating settlement in multilayer elastic systems (after Vesic 1963).

Vesic observed that in three-layered systems, the shape of the deflected surface computed by this approximate technique agrees better with measured deflections of pavements than the more rigorous analyses.

De Barros (1966) proposed an approximate method of reducing a multilayer system to a three-layer one, keeping the subgrade unaltered, by successively attributing to the two adjacent layers an 'equivalent modulus' according to the equation

$$E_{1,2} = \left(\frac{h_1 \sqrt[3]{E_1} + h_2 \sqrt[3]{E_2}}{h_1 + h_2} \right)^3 \tag{5.39}$$

He found that using this technique and reducing a three-layer system to an equivalent two-layer one, the approximate method is correct to within 10% for $h_2/b > 1$ and 14% for $h_2/b > 2$.

An analogous expression was first proposed by Palmer and Barber (1940) to reduce a two-layer system to an equivalent homogeneous medium which yielded deflections very close to Burmister's two-layer analysis.

5.11.4 Non-homogeneous Medium

The problem of the non-homogeneous soil medium whose modulus of elasticity varies as a continuous function of depth has received only limited attention so far. Korenev (1957), Sherman (1959), Golecki (1959), Hruban (1959), and Lekhnitskii (1962) have studied particular problems of non-homogeneity, but no comprehensive theory had been presented until Gibson developed the theory (Gibson 1967, 1968) of stresses and displacements in a non-homogeneous, isotropic elastic half-space subjected to strip or axially symmetric loading normal to its place boundary, Fig. 5.26.

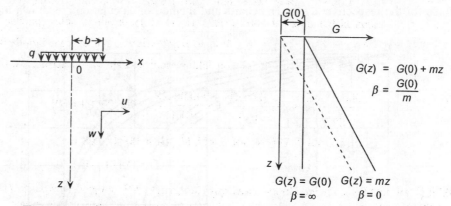

Fig. 5.26 Non-homogeneous elastic medium (after Gibson 1967, 1968).

A semi-infinite incompressible medium whose modulus of elasticity increases with depth from zero at the surface (i.e. $\beta = 0$), behaves as a Winkler spring model. In other words, the surface settlement of a uniformly loaded area on such a medium is directly proportional to the applied pressure and independent of the dimensions of the load.

The distribution of stresses in a semi-infinite medium is not significantly affected by this type of non-homogeneity. Figure 5.27 shows the stress distribution for $\beta/b = 0.1$ and 10. Indeed, the two limiting cases $\beta/b = 0$ and $\beta/b = \infty$ give exactly the same stresses while in the intermediate range $0 < \beta/b < \infty$, both the vertical and horizontal stresses tend to be a little higher than the corresponding stresses for the homogeneous medium, though the difference is never greater than 10%. However, over most of the range ($0 < \beta/b < 0.5$ and $5 < \beta/b < \infty$), the discrepancy is less than 5%. This observation is true of vertical and horizontal stresses due to both axi-symmetric and strip loadings, as depicted in Fig. 5.28.

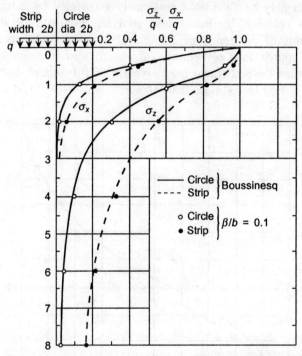

Fig. 5.27 Vertical stress beneath centre of foundations: non-homogeneous medium (Som 1968).

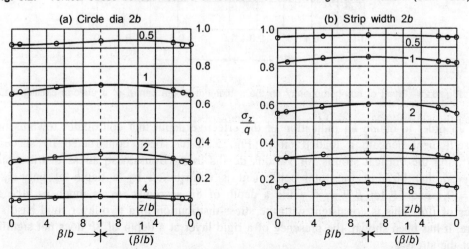

Fig. 5.28 Effect of non-homogeneity on the vertical stresses beneath centre of foundations (Som 1968).

Extensive field measurements of stresses reported by Plantema (1953), Turnbull et al. (1961), and the Waterways Experimental Station (1953, 1954) have shown very close agreement with the predictions based on Boussinesq analysis even though the soil media varied from non-homogeneous deposits (Plantema) to fairly homogeneous test sections of clayey silt and sand (Turnbull et al.). This is certainly in agreement with the theoretical results presented here.

If the foregoing conclusions are correct, the settlement of a foundation on a non-homogeneous medium may be calculated reasonably accurately by assuming that the stress distribution could be obtained by Boussinesq analysis. Figure 5.29 shows a comparison between the settlements of the centre of a uniformly loaded circle obtained from rigorous computation and the approximate settlements calculated from Boussinesq stress distribution. The latter underestimates the settlement for all values of β/b though the maximum error (in the range $0.5 < \beta/b < 1.5$) is no more than 10%.

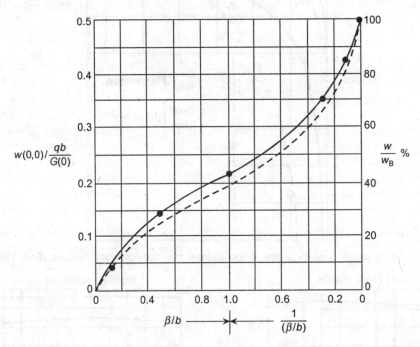

Fig. 5.29 Effect of non-homogenity on the settlement of the centre of a uniform circular load.

In order to obtain an indication of the effective depth that contributes towards most of the settlement beneath a loaded circle, Fig. 5.30 has been constructed by successively integrating the vertical strains for various depths using Boussinesq stress distribution. It is observed that 80% of the total settlement is contributed by a depth of only $1.5b$ for $\beta/b = 0.1$ and $5b$ for $\beta/b = \infty$ while a depth of $8b$ accounts for as much as 90% for all values of β/b. This is consistent with the stress-distribution in a two-layer rigid base system, where, it has been shown, the presence of a rigid layer at a depth of $8b$ does not significantly affect the stresses.

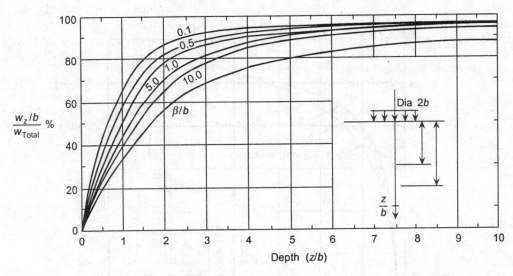

Fig. 5.30 Effective depth of soil beneath a circular foundation: non-homogeneous medium.

5.12 NONLINEAR SOIL

The problem of stress analysis in a soil medium with a nonlinear stress–strain relationship is immensely complex and no general solution is available yet. Huang (1968) presented a method of analyzing stresses and displacements beneath a circular load in a nonlinear soil medium whose modulus of elasticity is a function of the stresses.

$$E = E_0 [1 + c_\beta (\sigma_z + \sigma_r + \sigma_\theta + c_\gamma \gamma_z/b)] \tag{5.40}$$

where c_β (the nonlinear coefficient) and c_γ (the body force coefficient) are material parameters. He divided the semi-infinite medium into a multilayer system assuming a rigid base at a depth of $100b$ (see Fig. 5.28) and assigned to each layer, a modulus corresponding to the stresses at the midpoint. Employing Burmister's boundary and continuity conditions (Burmister 1943, 1945) and using the method of successive approximations, Huang calculated the stresses and displacements until two consecutive iterations gave the same modulus. His results are shown in Fig. 5.31. Again a close agreement with the Boussinesq stress distribution in noted. Comparison between the actual settlements calculated rigorously by Huang and the approximate settlements calculated on the basis of Boussinesq stress distribution but using the proper variation of E with depth show that the two methods do not differ by more than 10%, as shown in Fig. 5.32.

However, the assumption made by Huang, that each layer has uniform modulus means that the problem is, in effect, reduced to a multilayer Burmister problem with the elastic modulii determined by the stresses in the centre of the individual layers as shown in Fig. 5.32.

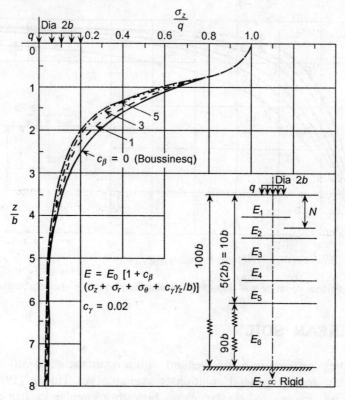

Fig. 5.31 Stresses in nonlinear soil medium (after Huang 1968).

$$E = E_0 [1 + c_\beta (\sigma_z + \sigma_r + \sigma_\theta + c_\gamma \gamma_z/b)]$$

$$c_\gamma = 0.02$$

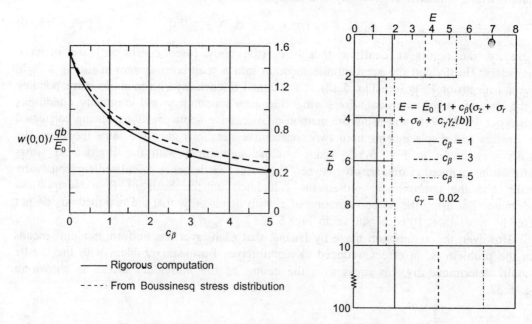

Rigorous computation

From Boussinesq stress distribution

$$E = E_0 [1 + c_\beta(\sigma_z + \sigma_r + \sigma_\theta + c_\gamma \gamma_z/b)]$$

——— $c_\beta = 1$

-------- $c_\beta = 3$

—·—·— $c_\beta = 5$

$c_\gamma = 0.02$

Fig. 5.32 Settlement at centre of circular load: nonlinear medium.

At present, the problem of nonlinear soil medium is best solved by numerical methods using computers. Finite element modelling is most widely used (Zienkiwicz 1971). However, a detailed treatment of the finite element analysis is beyond the scope of this book. The reader may refer to the relevant literature for further details (Desai and Abel 1972, Desai and Christian 1977, Smith 1988).

The distribution of vertical and horizontal stresses with depth, along the centre line of a model circular footing resting on a saturated normally consolidated clay, with a load intensity, $q/\sigma'_{vo} = 0.35$, 0.77, and 1.05 (q = applied pressure, σ'_{vo} = effective vertical consolidation pressure), as obtained from a nonlinear finite element analysis ((Das 1975) and (Das and Gangopadhyay 1978)), is shown in Fig. 5.33. On the same figure are plotted the stresses, obtained from elastic analysis and those obtained by measurement at $q/\sigma'_{vo} \cong 1$, using pressure cells. It can be seen that vertical stresses do not change much with load increments and elastic theory can predict the stresses quite well but horizontal stresses are significantly dependent on load increment. When the loading is small ($q/\sigma'_{vo} = 0.35$ in this case), the elastic theory may be capable of predicting horizontal stresses but at higher loads, non-linear analysis is necessary to predict horizontal stresses satisfactorily.

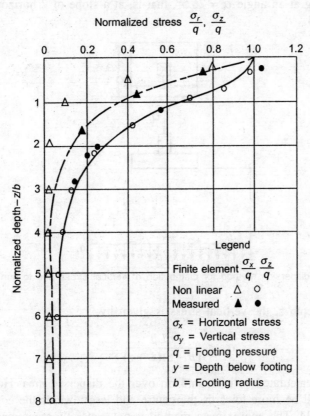

Fig. 5.33 Comparison of stresses obtained from linear and non-linear analysis (Das 1975).

Therefore, it appears that any deviation from the classical problem of homogeneous elasticity, either in terms of nonlinearity of stress–strain relationship or in terms of non-homogeneity of the medium will only have a marginal effect on the stress distribution so long as the medium is subjected to certain boundary stresses. The displacements will, of course, be significantly affected but from the foregoing it can be deduced that the settlement can be obtained with reasonable accuracy, by assuming the Boussinesq stress-distribution but taking account of the proper stress–strain relationship and/or non-homogeneity in calculating the strains. Also, as long as the pressure bulb due to a footing load is restricted within the upper layer of a soil deposit, the Boussinesq analysis should be reasonably valid, notwithstanding any non-homogeneity below that layer.

5.13 APPROXIMATE METHOD OF DETERMINING VERTICAL STRESS

The pressure on footing founded at or near the ground surface gets dispersed to a wider area at a depth. In the approximate method, the stress dispersion is assumed along lines through the edge of the footing at an angle $\alpha = 26.5°$, that is, at a slope of 2 horizontal to 1 vertical, as shown in Fig. 5.34.

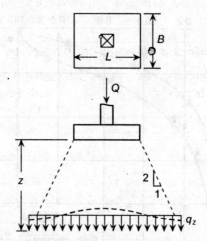

Fig. 5.34 2:1 dispersion method for distribution of vertical stress due to surface load.

Accordingly at any depth z, the vertical stress is given by,

$$\sigma_z = \frac{qLB}{(B + z)(L + z)} \tag{5.41}$$

The stress thus calculated will be uniform over the dispersed area. However, in actual practice, the stress will be more towards the centre and less towards the edges as shown by broken line in Fig. 5.34. The approximate method is not generally recommended for detailed design, although the method comes handy for rough calculations in the absence of necessary charts and tables.

Example 5.1

Figure 5.35 shows four vertical loads of 1000 kN each placed at the corners of a square of side 5 m. Determine the increase of vertical stress 5 m below

(a) each load
(b) midpoint between adjacent loads
(c) the centre of the rectangle

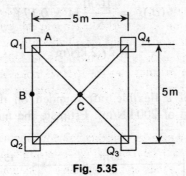

Fig. 5.35

Solution

From Fig 5.4, vertical stresses due to a point load

$$\sigma_z = \frac{Q}{z^2} I_B, \text{ where } I_B = f(r/z)$$

Point A

$Q_1 = 1000 \text{ kN}; \ r/z = 0; \ I_B = 0.478$

$Q_2 = Q_4 = 1000 \text{ kN}, \ r/z = \dfrac{5}{5} = 1.0; \ I_B = 0.084$

$Q_3 = \sqrt{(5^2 + 5^2)} = 7.07 \text{ m}; \ \dfrac{r}{z} = \dfrac{7.07}{5} = 1.41; \ I_B = 0.031$

$$\therefore \quad (\sigma_z)_A = \frac{100}{5^2}[0.478 + 2(0.084) + 0.031] = 27.1 \text{ kN/m}^2$$

Point B

$Q_1 = Q_2 = 1000 \text{ kN}; \ \dfrac{r}{z} = \dfrac{2.5}{5} = 0.5; \ I_B = 0.27$

$Q_3 = Q_4 = 1000 \text{ kN}; \ \dfrac{r}{z} = \dfrac{\sqrt{5^2 + 2.5^2}}{5} = 1.12; \ I_B = 0.062$

$$\therefore \quad (\sigma_z)_B = \frac{100}{5^2} \ [2(0.27) + 2(0.062)]$$

$$= 26.6 \text{ kN/m}^2$$

Point C

$$Q_1 = Q_2 = Q_3 = Q_4 = 1000 \text{ kN}$$

$$\frac{r}{z} = \frac{\sqrt{2.5^2 + 2.5^2}}{5} = 0.71; \ I_B = 0.17$$

$$\therefore \quad (\sigma_z)_C = \frac{1000}{5^2}[4 \times 0.17]$$

$$= 27.2 \text{ kN/m}^2$$

Example 5.2

Figure 5.36 shows the plan of a flexible raft founded on the ground surface. The area supports a uniform vertical load of 200 kN/m². Estimate the increase in vertical stress 15 m below point A.

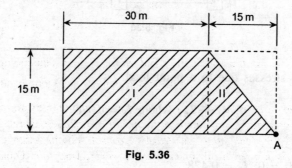

Fig. 5.36

Solution

Consider two rectangles (I + II) and II with the point A at the corner of each rectangle.

$$(\sigma_z)_A = q\left(I_{\sigma I+II} - \frac{1}{2} I_{\sigma II}\right)$$

Rectangle (I + II): 15 m × 45 m

$$m = \frac{45}{15} = 3.0; \quad n = \frac{15}{15} = 1.0, \ I_{\sigma I+II} = 0.201$$

Rectangle II: 15 m × 15 m

$$m = n = \frac{15}{15} = 1.0; \ I_{\sigma II} = 0.174$$

$$\therefore \quad (\sigma_z)_A = 200\left[0.201 - \frac{1}{2} \times 0.174\right]$$

$$= 22.8 \text{ kN/m}^2$$

Example 5.3

Figure 5.37 shows a raft foundation (10 m × 20 m) built 5 m from a tower. Determine the increase of stress 5, 10, 15, and 25 m below the tower and draw the stress distribution. (Take $q_n = 100$ kN/m^2)

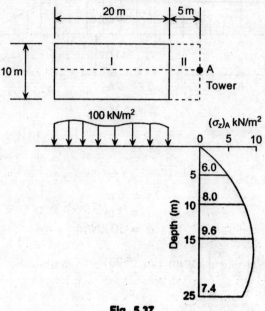

Fig. 5.37

Solution

Consider rectangles (I + II) and II with point A below each rectangle.

Rectangle (I + II): 25 m × 5 m

Rectangle II: 5 m × 5 m

Depth (m)		Influence factor			$I_\sigma = 2(I_{\sigma_{I+II}} - I_{\sigma_{II}})$	$(\sigma_z)_A = 100 I_\sigma$ kN/m^2
		$I_{\sigma_{(I+II)}}$		$I_{\sigma_{(II)}}$		
5 *m*	5	0.204	1.0	0.174	0.06	6.0
n	1		1.0			
10 *m*	2.5	0.136	0.5	0.096	0.08	8.0
n	0.5		0.5			
15 *m*	1.67	0.094	0.33	0.046	0.096	9.6
n	0.33		0.33			
25 *m*	1.0	0.054	0.2	0.017	0.074	7.4
n	0.2		0.2			

Example 5.4

Figure 5.38 shows the section of an earth dam. Determine the increase in vertical stress 3 m below points A and B.

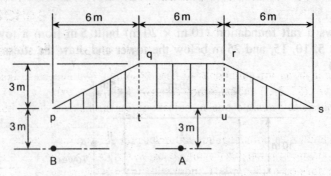

Fig. 5.38

Solution

Point A

Stress due to strip load qrut:

$$2a = 6 \text{ m}$$
$$z = 3 \text{ m}$$
$$q = 50 \text{ kN/m}^2$$

$$\sigma_z = q(I_\sigma) \text{ where } I_\sigma = f\left(\frac{z}{a}, \frac{x}{a}\right), \text{ from Eq. (5.23)}$$

Here, $\dfrac{x}{a} = 0$; $\dfrac{z}{a} = \dfrac{3}{3} = 1.0$; $I_\sigma = 0.96$

Stress due to triangular load, pqt

$$a = 6 \text{ m}$$
$$z = 3 \text{ m}$$
$$q = 50 \text{ kN/m}^2$$

$$\sigma_z = q(I_\sigma) \text{ where } I_\sigma = f\left(\frac{z}{a}, \frac{x}{a}\right), \text{ from Eq. (5.24)}$$

Here, $\dfrac{x}{a} = \dfrac{9}{6} = 1.5$; $\dfrac{z}{a} = \dfrac{3}{6} = 0.5$; $I_\sigma = 0.06$

$$\therefore \quad (\sigma_z)_A = 50[0.96 + 2(0.06)]$$
$$= 54 \text{ kN/m}^2$$

Point B

Stress due to strip load: qrut

$$\frac{x}{a} = \frac{9}{3} = 3.0; \quad \frac{z}{a} = 1.0; \quad I_\sigma = 0.003$$

Stress due to triangular load: pqt

$$\frac{x}{a} = 0; \quad \frac{z}{a} = \frac{3}{6} = 0.5; \quad I_\sigma = 0.13$$

Stress due to triangular load: rus

$$\frac{x}{a} = \frac{18}{6} = 3; \quad \frac{z}{a} = 0.5; \quad I_\sigma = 0.002$$

$$\therefore \quad (\sigma_z)_B = 50(0.003 + 0.13 + 0.002)$$

$$= 6.75 \text{ kN/m}^2$$

Example 5.5

Figure 5.39 shows 20 m dia × 15 m high steel storage tank is founded on RCC raft (500 mm thick) 5 m below GL. Determine the increase in vertical stress along the centre of the tank 10 m, 20 m and 30 m below the foundation when the tank is filled of water. Draw the distribution of vertical stress increase with depth and superimpose the same on the in-situ stresses to obtain the variation of stress increment ratio $\Delta p/p_0$ with depth. Take unit weight of soil = 18 kN/m^3.

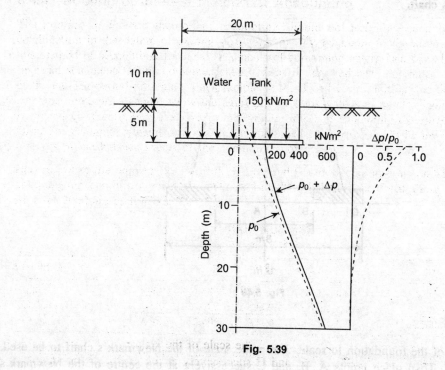

Fig. 5.39

Solution

Water load = 15 × 10 = 150 kN/m^2

Self weight of foundation = 0.5 × 24 = 12 kN/m^2

q_{gross} = 150 + 12 = 162 kN/m^2

Pressure reduced by excavation = 18 × 5 = 90 kN/m^2

q_{net} = 162 – 90 = 72 kN/m^2

Depth (m)	$\dfrac{r}{a}$	$\dfrac{z}{a}$	I_σ	$\Delta p = q_{net}\, I_\sigma$ (kN/m²)	p_o (kN/m²)	$\dfrac{\Delta p}{p_o}$
0	0	0	1.0	72.0	90	0.8
10	0	1.0	0.65	46.8	180	0.26
20	0	2.0	0.28	20.2	360	0.06
30	0	3.0	0.13	9.4	540	0.17
40	0	4.0	0.08	5.8	720	0.008

Example 5.6

A reinforced concrete tower is provided on a ring foundation of inner diameter 6 m and outer diameter 12 m, as shown in Fig. 5.40. If the foundation carries a distributed load of 150 kN/m², determine the vertical stress distribution at a depth of 6 m below the foundation. Use Newmark's chart.

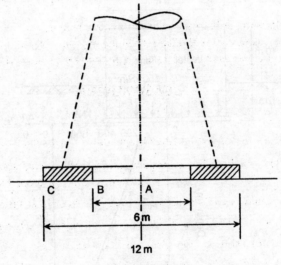

Fig. 5.40

Solution

Draw the plan of the foundation to scale, 3 m = the scale of the Newmark's chart to be used, as in Fig. 5.41. Then place points A, B, and C successively at the centre of the Newmark's chart and count the number of units enclosed in each case. Then,

$$\sigma_z = n(0.005) \times q_n$$

Vertical stress below
$A = 70 \times 0.005 \times 150 = 52.5$ kN/m²
$B = 60 \times 0.005 \times 150 = 45.0$ kN/m²
$C = 46 \times 0.005 \times 150 = 44.5$ kN/m²

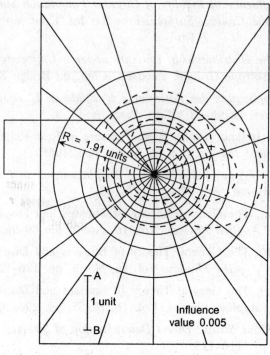

Fig. 5.41

Example 5.7

Solve Example 5.2 by Newmark's chart.

Solution

Draw the plan of the raft on tracing paper to a scale, 15 m = scale of the Newmark's chart, Fig. 5.40. Place point A at the centre of the Newmark's chart and count the number of units enclosed by the diagram. Then,

$$(\sigma_z)_A = 23.5 \times 0.005 \times 200$$
$$= 23.5 \text{ kN/m}^2$$

REFERENCES

Ahlvin, R.G. and H.H. Ulery (1962), *Tabulated Values for Determining the Complete Pattern of Stresses, Strains, and Deflections Beneath a Uniform Circular Load on a Homogeneous Half Space*, Highway Research Board Bulletin No. 342, p. 1.

Biot, M.A. (1935), *The Effect of Certain Discontinuities on the Pressure Distribution in a Loaded Soil*, Physics, Vol. 6, No. 12, p. 367.

Bishop, A.W. (1952), *The Stability of Earth Dams*, Ph.D. Thesis, University of London.

Borowicka, H. (1936), *Influence of Rigidity of Circular Foundation Slab on the Distribution of Pressures over the Contact Surface*, Proc 1st Int. Conf. on S M & F E, Vol 2 pp. 144–149.

Borowicka, H. (1938), *The Distribution of Pressure under a Uniformly Loaded Elastic Strip Resting on Elastic-isotrope Ground*, Zued Int. Conf. on Bridge & Str. Engs, Berlin.

Boussinesq, J. (1985), *Application des Potential a L' Etude de L' equilibre et tu Movement des Solides Elastiques*, Gauthiers Villars, Paris.

Brooker, E.W. and H.O. Ireland (1965), Earth Pressure at Rest Related to Stress History, *Canadian Geotechnical Journal*, Vol. 2, No. 1, Feb. 1965.

Burmister, D.M. (1943), *Theory of Stresses and Displacements in Layered Systems*, Proc Highway Research Board, Vol. 22, pp. 127–148.

Burmister, D.M. (1945 a), The General Theory of Stresses and Displacements in Layered Systems, *Journal of Applied Physics*, Vol. 16, No. 2, pp. 89–96.

Burmister, D.M. (1945 b), The General Theory of Stresses and Displacements in Layered Systems, *Journal of Applied Physics*, Vol. 16, No. 3, pp. 126–127.

Burmister, D.M. (1945 c), The General Theory of Stresses and Displacements in Layered Systems, *Journal of Applied Physics*, Vol. 16, No. 5, pp. 296–302.

Carothers, S.D. (1920), *Plane Strain: Direct Determination of Stresses*, Proc . Royal Society Series A: Vol. 97, pp. 110–123.

Das, B. (1997), *Advanced Soil Mechanics*, 2nd ed., Taylor and Francis, Washington, D.C.

De Barros, S.T. (1966), *Deflection Factor Charts for Two and Three Layer Systems*, Highway Research Record No. 145, p. 83.

Desai, C.S. and J.F. Abel (1972), *Introduction to Finite Element Methods*, Van Nostrand Reinhold, New York.

Desai, C.S. and J.T. Christian (1977), *Numerical Methods in Geotechnical Engineering*, McGraw-Hill, New York.

Das, S.C. (1975), *Predicted and Measured Values of Stress and Displacements Downloading and Construction under Circular Footings Resting on Saturated Clay Medium*, Ph.D Thesis, Jadavpur University, Calcutta.

Das, S.C. and C.R. Gangopadhyay (1978), *Undrained Stresses and Deformations under Footings as Clay*, Proc. ASCE, Vol. 104, GT 1, pp. 11–25.

Fadum, R.E. (1948), *Influence Values for Estimating Stresses in Elastic Foundations*, Proc 2nd ICSMFE, Vol. 3, p. 77.

Fox (1948), *The Mean Elastic Settlement of a Uniformly Loaded Area at a Depth Below Ground Surface*, Proc. 2nd ICSMFE, Rotterdam, Vol. 1, p. 129.

Gibson, R.E. (1967), Some Results Concerning Displacements and Stresses in Non homogeneous Elastic Half Space, *Geotechnique*, Vol. 17, No. 1.

Gibson, R.E. (1968), *Letter to Geotechnique*, Vol. 18, No. 2.

Golecki, J. (1959), *On the Foundation of the Theory of Elasticity of Plane Incompressible Non-homogeneous Bodies*, Proc. IUTAM Symposium, Pergamon Press, London.

Hruban, K. (1959), *The Basic Problem of a Nonlinear and a Nonhomogeneous Half Space, Nonhomogeneity in Elasticity and Plasticity,* Pergamon Press, London.

Huang, Y.H. (1968), Stresses and Displacements in Nonlinear Soil Medium, *Journal ASCE Soil Mechanics and Foundation Division,* Vol. 94, SM 1.

Jaky, J. (1944), *The Coefficient of Earth Pressure at Rest* (In Hungarian), Proc. 2nd Int. Conf. on Soil Mechanics and Foundation Engineering, Rotterdam, Vol. 2, pp. 16–20.

Jones, A. (1962), *Table of Stresses in Three-layer Elastic Systems,* Highway Research Board Bulletin No. 342.

Jurgenson, L. (1934), Application of Theories of Elasticity and Plasticity to Foundation Problems, *Journal Boston Soc. of Civil Engineering,* Vol. 21, No. 3.

Kirk, J.M. (1966), *Tables of Radial Stresses in Top Layer of Three-layer Elastic Systems at Distance from Load Axis,* Highway Research Record No. 145.

Korenev, B.G. (1957), *A Die Resting on an Elastic Half Space, the Modulus of Elasticity of which is an Exponential Function of Depth,* Dokl. Nank S.S.R., Vol. 112.

Ladd (1977), *Stress Defamation and Strength Characteristics*, State-of-the art report, Proc. 7th ICSMFE, Vol. 2, pp. 421–494.

Lekhnitskii, S.G. (1962), Radial Distribution of Stresses in a Wedge and in a Half Plane with Variable Modulus of Elasticity, *PMM,* Vol. 26, No. 1.

Lemcoe, M.M. (1961), *Stresses in Layered Elastic Solids,* Trans., ASCE, Vol. 126, p. 194.

Love, A.E.H. (1928), *The Stress Produced in Semi-infinite Solid by Pressure on Part of the Boundary,* Phil. Trans. of Royal Society, Series A, Vol. 228, p. 377.

Newmark, N.M. (1942), *Influence Charts for Computation of Stresses in Elastic Foundations,* Circular No. 24, Eng. Exp. Stn. Univ. of Illinois, USA.

Plantema (1953), *Soil Pressure Measurements during Loading Tests on a Runway,* Proc. 3rd ICSMFE, Zurich, Vol. 1, p. 289.

Pickett, G. (1938), *Stress Distribution in a Layered Soil with Some Rigid Boundaries,* Proc. Highway Research Board, Vol. 18, Part 2, 1938.

Poulos, H.G. (1967), *Control of Leakage in the Triaxial Test,* Harvards Soil Mechanics Series No. 71, Cambridge, Mass.

Peattie, K.R. (1962), *Stress and Strain Factors for Three-layer Systems,* Highway Research Board Bulletin No. 342.

Schiffman, R.L. (1957), *Consolidation of Soil under Time-dependent Loading and Variable Permeability,* Proc. Highway Research Board, Vol. 37, pp. 584–617.

Sherman, D.I. (1959), *On the Problem of Plane Strain in Nonhomogeneous Media,* Nonhomogeneity in Elasticity and Plasticity, Pergamon Press.

Som, N. (1968), *The Effect of Stress Path on the Deformation and Consolidation of London Clays,* Ph.D. Thesis, University of London.

Som, N. (1974), *Lateral Stresses during One-dimensional Consolidation of an Overconsolidated Clay,* Proc. 2nd S. E. Asian Conf. on Soil Mechanics and Foundation Engineering, Singapore, pp. 295–307.

Terzaghi, K. (1943), *Theoretical Soil Mechanics*, John Wiley and Sons Inc.

Turnbull, W.J., A. Maxwell, and R.G. Ahlvin (1961), *Stresses and Deflections in Homogeneous Soil Mass,* Proc. 5th ICSMFE, Paris, Vol. 2, p. 337.

Uyeshita, K. and G.G. Meyerhof (1967), Deflections of Multilayered Soil System, *Journal ASCE SMFE,* Vol. 93, SM 5.

Vesic, A.B. and R.D. Barksdale, (1963), *On Shear Strength of Sand at Very High Pressures,* ASTM STP No. 361, p. 301.

Waterways Expt. Station, Vicksburg (1953, 1954), *Investigations of Pressures and Deflections for Flexible Pavements,* Report 3 (Sept. 1953), Report 4 (Dec. 1954).

Zienkiwicz, O.C. (1977), *The Finite Element Method, in Engineering Science,* McGraw-Hill Book Co., London.

6 Bearing Capacity of Shallow Foundations

6.1 INTRODUCTION

The bearing capacity of foundation is the maximum load per unit area which the soil can support without failure. It depends on the shear strength of soil as well as on the type, size, depth, and shape of the foundation.

Figure 6.1 shows a typical load versus settlement relationship of a footing. As the footing load is increased, the settlement also increases. Initially the settlement increases almost linearly with load indicating elastic behaviour of the soil. Thereafter, the settlement increases more rapidly and then continues to increase even without any appreciable increase of load. The foundation is then said to have failed, that is, the soil has reached its capacity to bear superimposed load.

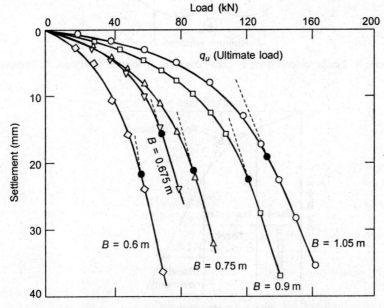

Fig. 6.1 Load versus settlement relationship of footing. (Test data from Das 1999)

In order to make the bearing capacity analysis of a foundation it is necessary to know the actual mechanism of failure. There are different methods to analyze bearing capacity, each based on different assumptions on the mechanism of failure and mobilisation of shear strength of soil.

6.2 FAILURE MECHANISM

From model studies, it has been observed that there are possibly four stages of deformation in the soil leading ultimately to a bearing capacity failure. These include:

1. Elastic deformation or distortion
2. Local cracking around the perimeter of the loaded area
3. Formation of a wedge below the footing which moves downward, pushing the soil side ways.
4. Formation of a rupture surface which may extend upto the level of foundation. The footing then sinks rapidly into the soil followed by bulging or heaving at the top.

These stages are illustrated in Fig. 6.2(a) and the corresponding load–settlement curve is shown in Fig. 6.2(b) where a method of determining failure load is also indicated.

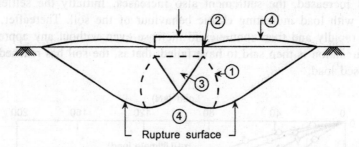

(a) Stages of failure: 1. Elastic deformation; 2. Local cracking; 3. Wedge formation; 4. Rupture surface.

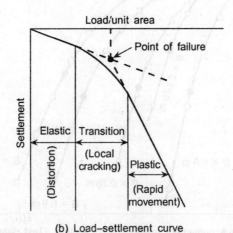

(b) Load–settlement curve

Fig. 6.2 Failure mechanism: General shear.

The load–settlement curve has three distinct parts:

(a) *Elastic*, where there is distortion of soil.

(b) *Transition*, where there is local cracking.

(c) *Plastic*, indicated by rapid movement.

This type of failure is called *General Shear Failure* where large settlement of footing is not required for the development of rupture surface. This is generally observed when the ratio of depth of footing to the width of footing is relatively small and the soil consists of medium to stiff clay or medium to dense sand.

In loose sand, much larger settlement would be required for the shear surface to develop upto the level of footing. The footing may be considered to have failed by excessive settlement before that stage is reached. This type of failure is known as *Local Shear Failure* and is illustrated in Fig. 6.3(a). The corresponding load–settlement curve is shown in Fig. 6.3(b). The point on the load–settlement curve where the slope becomes steep and almost constant is considered as the failure point. If the settlement corresponding to this point is more than a chosen critical value, say 10% of the width of the footing, then the load corresponding to that critical settlement is considered as the failure load.

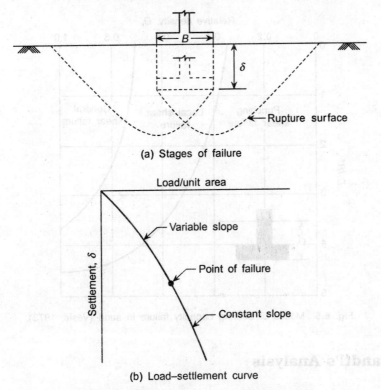

(a) Stages of failure

(b) Load–settlement curve

Fig. 6.3 Failure mechanism: Local shear.

When the soil is predominantly soft and cohesive, the settlement increases at a very rapid rate and reaches the critical value even before the rupture surface is fully developed. This type of failure is known as *punching shear failure*, refer Fig. 6.4.

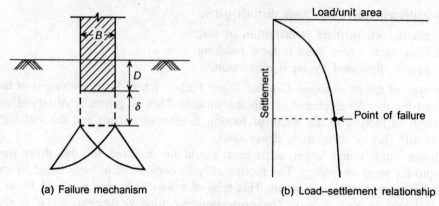

(a) Failure mechanism

(b) Load–settlement relationship

Fig. 6.4 Punching shear failure.

Vesic (1973) gives the range of relative density of granular soil for punching shear, local shear, and general shear failure, depicted in Fig. 6.5.

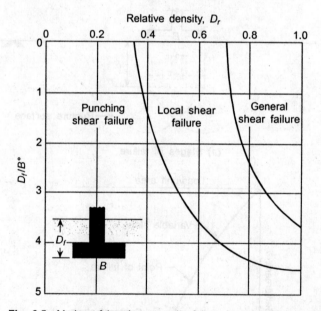

Fig. 6.5 Modes of bearing capacity failure in sand (Vesic, 1973).

6.2.1 Prandtl's Analysis

A method for analysis of bearing capacity, considering an apparently realistic mechanism of failure was first suggested by Prandtl in his plastic equilibrium theory (Terzaghi, 1943). According to Prandtl, the failure mass consists of three zones, with the failure surface given by a logarithmic spiral. Figure 6.6 shows Prandtl's bearing capacity analysis.

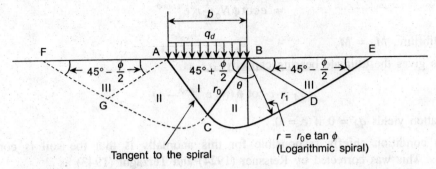

Fig. 6.6 Prandtl's bearing capacity analysis: failure surface and zones of failure.

When the applied footing pressure reaches the ultimate bearing pressure, the soil fails along a surface ACDE or BCGF. The failure mass consists of three zones. Zone I, immediately below the footing is under active pressure where failure surfaces AC and BC develop, inclined at $(45° + \phi/2)$ to horizontal. The soil is assumed to be weightless, and the stresses in this zone are assumed to be hydrostatic. As the wedge ABC tends to move downward along with the footing, it pushes the surrounding soil side ways. Passive state develops in zone III where failure planes FG, AG or BD, and DE, all inclined at $(45° - \phi/2)$ to the horizontal, develop. Zone II is the zone of radial failure planes. The surface CG or CD is assumed to be a logarithmic spiral, with B or A as the pole. The equation of the spiral is $r = r_0 e^{\theta \tan\phi}$ where r_0 = BC or AC, and r is any radius BX at a spiral angle CBX = θ.

From the Mohr's circle for $c - \phi$ soil, the normal stress corresponding to the cohesion intercept, is $\sigma_i = c \cot\phi$. This is termed as *initial stress*, which acts normally to BC and AC (assumed hydrostatic pressure in zone I). Also the applied pressure, q_d is assumed to be transferred normally on to BC or AC.

Thus, the force on BC or AC, $\quad P_I = (\sigma_i + q_d)r_0.$

The disturbing moment of this force about B is

$$M_d = r_0 (\sigma_i + q_d)r_0/2$$

$$= (c \cot\phi + q_d)r_0^2/2 \tag{6.1}$$

The passive resistance, P_p on the face BD is given by

$$P_p = \sigma_i N_\phi \overline{BD}$$

where, $N_\phi = \tan^2(45° + \phi/2)$

This is because σ_i, due to cohesion alone, is transmitted by the wedge BDE. The resisting moment M_γ is given by its moment about B as,

$$M_\gamma = \frac{P_p BD}{2} = \frac{\sigma_i N_\phi (BD)^2}{2}$$

$$= c \cot \phi \, N_\phi \frac{1}{2} r_0^2 e^{\pi \tan \phi} \tag{6.2}$$

For equilibrium, $M_d = M_\gamma$.

This gives the ultimate bearing capacity, q_d, as

$$q_d = c \cot \phi \, (N_\phi e^{\pi \tan \phi} - 1) \tag{6.3}$$

This equation yields $q_d = 0$ if $c = 0$.

The condition chiefly responsible for this anomaly is that the soil is considered weightless. This was corrected by Reissner (1924) and Terzaghi (1943) as

$$q_d = c \cot \phi + \left(\frac{1}{2} \gamma B \sqrt{N_\phi} \right) (N_\phi e^{\pi \tan \phi} - 1) \tag{6.4}$$

For soil with $\phi = 0$, the logarithmic spiral becomes a circle, and Prandtl's Eq. (6.3) or (6.4) gives on application of L'Hospital's rule,

$$q_d = (\pi + 2)c = 5.14c \tag{6.5}$$

It may be noted that Prandtl's expression is independent of the width of the footing. Also the assumed shape of the failure surface does not resemble the actual failure surface because of the compressibility of soil, roughness of contact surface and other factors.

6.2.2 Terzaghi's Analysis

Terzaghi (1943) considered the roughness of the footing and also the weight of the soil above the horizontal plane through the base of the footing, and modified the expression derived by Prandtl. The corresponding failure surface and failure mass are shown in Fig. 6.7.

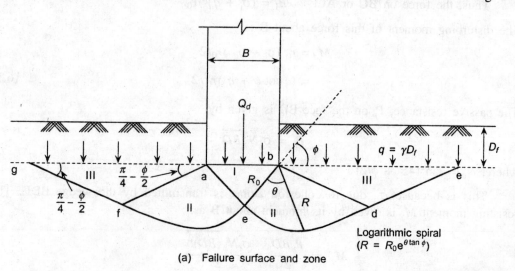

(a) Failure surface and zone

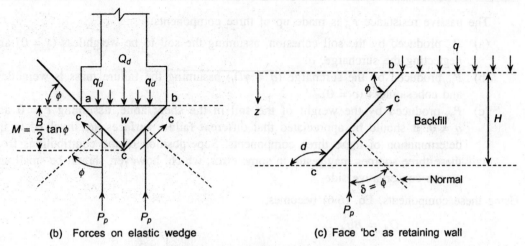

(b) Forces on elastic wedge   (c) Face 'bc' as retaining wall

Fig. 6.7 Terzaghi's bearing capacity analysis.

The soil mass above the failure surface consists of three zones:

Zone I: Due to the friction and adhesion between the soil and the base of the footing, this zone cannot spread laterally. It moves downward as an elastic wedge and the soil in this zone behaves as if it is a part of the footing. The two sides of the wedge ac and bc make angle ϕ with the horizontal.

Zone II: The zones aef and bed are zones of radial shear. The soil in this zone is pushed into zone III.

Zone III: These are two passive **Rankine** zones. The boundaries of these zones make angles $(45° - \phi/2)$ with the horizontal.

As the footing sinks into the ground, the faces ac and bc of the wedge abc push the soil to the sides. When plastic equilibrium is reached, the forces acting on the wedge abc are the ones shown in Fig. 6.6. The forces are

(a) The ultimate footing load $Q_d = B q_d$

(b) The weight of the wedge, $W = (1/4)\gamma B^2 \tan\phi$

(c) The passive resistance P_p acting on the faces ac and bc. Since the soil shears along these planes, and the shearing is between soil and soil, P_p is inclined at an angle ϕ to the normal. Since ac and bc are inclined at ϕ to the horizontal, P_p acts vertically.

(d) The cohesive force on the faces ac and bc: $C_a = cB/(2\cos\phi)$ where c = unit cohesion.

For equilibrium,

$$Q_d + \frac{1}{4}\gamma B^2 \tan\phi - 2P_p - Bc\tan\phi = 0$$

or,

$$Q_d = 2P_p = Bc\tan\phi - \frac{1}{4}\gamma B^2 \tan\phi \qquad (6.6)$$

An expression for P_p may be obtained by considering the face bc as a retaining wall.

The passive resistance P_p is made up of three components:

(a) P_{pc} produced by the soil cohesion, assuming the soil to be weightless ($\gamma = 0$) and neglecting the surcharge, q.

(b) P_{pq} produced by the surcharge ($q = \gamma D_f$) assuming the failure mass is weightless and cohesionless ($c = 0$).

(c) $P_{p\gamma}$ produced by the weight of the soil in the shear zone, assuming $c = 0$ and $q = 0$. It should be appreciated that different failure surfaces are involved in the determination of these three components. Superposition of the contributions from these three sources may result in some error, which, however, should be small and on the conservative side.

Using these components, Eq. (6.6) becomes,

$$Q_d = Bq_d$$

$$= 2(P_{pc} + P_{pq} + P_{p\gamma}) + Bc\tan\phi - \frac{1}{4}\gamma B^2 \tan\phi \qquad (6.7)$$

Let $2P_{pc} + Bc\tan\phi = BcN_c$

$$2P_{pq} = BqN_q$$

$$2P_{p\gamma} - \frac{1}{4}\gamma B^2 \tan\phi = B\frac{1}{2}\gamma BN_\gamma$$

Eq. (6.7) can now be written as

$$q_d = cN_c + qN_q + 0.5\gamma BN_\gamma \qquad (6.8)$$

Eq. (6.8) is the Terzaghi's bearing capacity equation for a strip footing corresponding to general shear failure. N_c, N_q, and N_γ are dimensionless bearing capacity factors which depend on ϕ only.

The values of Terzaghi's bearing capacity factors for general shear failure given in Table 6.1.

Table 6.1 Terzaghi's bearing capacity factors for general shear failure

Q	N_c	N_q	N_γ
0	5.7	1.0	0.10
5	7.0	1.6	0.14
10	9.5	2.7	0.7
15	13.0	4.5	2.0
20	17.0	7.5	4.8
25	24.0	13.0	9.8
30	37.0	23.0	20.0
35	58.0	42.0	43.0
40	98.0	77.0	98.0

For footings on saturated cohesive soil, critical condition for stability generally occurs at end of construction. Here, undrained condition exists, for which $\phi_u = 0$.
Corresponding values of bearing capacity factors (refer Eq. 6.8) are $N_c = 5.7$, $N_q = 1$ and $N_\gamma = 0$

Thus,

$$q_d = 5.7c + q = 5.7c + \gamma D_f \qquad (6.9)$$

6.2.3 Skempton Method

Although Terzaghi's method has been used widely in practice, other methods of bearing capacity analysis have been proposed. Some of these are presented below:

Skempton (1951) suggested a very practical method of obtaining bearing capacity of footings on saturated clay.

$$q_d = cN_c + \gamma D_f \qquad (6.10)$$

On the basis of theory, laboratory tests and field observations, Skempton obtained expressions for N_c for various shape and depth of foundation. These are:

For strip footings:

$$N_c = 5\left(\frac{1 + 0.2D_f}{B}\right), \quad \text{for} \quad N_c \leq 7.5 \qquad (6.11)$$

For rectangular, square or circular footings:

$$N_c = 6\left(1 + 0.2\frac{B}{L}\right)\left(1 + 0.2\frac{D_f}{B}\right), \quad \text{for} \quad N_c \leq 9 \qquad (6.12)$$

where, D_f = depth of footing
B = width (or diameter) of footing
L = Length of footing

6.2.4 Meyerhof's Method

Meyerhof (1951) suggested a method of analysis in which the failure surface would extend upto the ground surface, unlike that of Terzaghi's analysis, in which failure surface was considered to extend upto the base level of the footing. The failure surface and the zones of shear considered by Meyerhof are shown in Fig. 6.8.

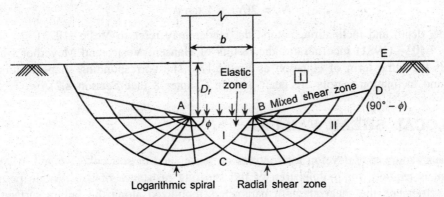

Fig. 6.8 Meyerhof's bearing capacity analysis.

Meyerhof's bearing capacity equation for strip footing is similar in form to that of Terzaghi, but the values of N_c, N_q, and N_γ are different. This method generally over estimates bearing capacity for sandy soil, because footings may fail from settlement consideration much before the failure surface reaches the ground surface. However, for clay soil the method gives quite good results.

6.2.5 Hansen's Method

Brinch Hansen (1957, 1970) proposed a general expression for bearing capacity considering effects of shape and depth of footing and inclination of applied load:

$$q_d = cN_c s_c d_c i_c + qN_q s_q d_q i_q + 0.5\gamma\, BN_\gamma s_\gamma d_\gamma i_\gamma \qquad (6.13)$$

where s_c, s_q, and s_γ are empirical shape factors.
d_c, d_q, and d_γ are empirical depth factors.
i_c, i_q, and i_γ are empirical inclination factors.

The recommendations of Hansen for N_c and N_q are identical to those of Meyerhof, and are the result of those of Prandtl (1921) and Reissner (1924).
These are :

$$N_c = (N_q - 1)\cot\phi$$
$$N_q = (e^{\pi\tan\phi})\tan^2(45° + \phi/2) \qquad (6.14)$$
$$N_\gamma = 1.5(N_q - 1)\tan\phi$$

6.2.6 Vesic's method

The failure surface considered by Vesic is similar to that of Terzaghi's with the exception that the zone I below the footing is in active Rankine state, with inclined faces of the wedge at $(45° + \phi/2)$ to the horizontal. The bearing capacity equation is the same as Eq. (6.13). The factors N_c and N_q are identical to those of Meyerhof and Hansen. N_γ, given by Vesic, is a simplified form of that given by Caqṳt and Kerisal (1948).

$$N_\gamma = 2(N_q + 1)\tan\phi \qquad (6.15)$$

For shape, depth and inclination factors, the reader may refer to Vesic (1973).
I.S. 6403—(1981) incorporates the results of Hansen, Vesic, and Meyerhof's analyses and gives the same form of equation as Eq. (6.13). The corresponding shape factors, depth factors, and inclination factors are dealt with in Chapter 8 (see Section 8.2).

6.3 LOCAL SHEAR FAILURE

Local shear failure may develop for footings on loose sand or soft cohesive soil, where large settlement is required for mobilization of full shearing resistance of soil. Within permissible limit of settlement, the shear strength parameters mobilized along the failure surface are c_m and ϕ_m. Terzaghi proposed that for local shear failure, c_m and ϕ_m should be used in the bearing capacity equation and factors N_c', N_q', and N_γ' should be determined on the basis of ϕ_m instead of ϕ. Terzaghi empirically suggested that

$$\phi_m = (2/3)\phi \quad \text{and} \quad c_m = (2/3)c$$

Thus, the bearing capacity equation for local shear failure becomes,

$$q_d = (2/3)cN_c' + qN_q' + 0.5\gamma BN_\gamma' \tag{6.16}$$

where N_c', N_q', and N_γ' are bearing capacity factors for local shear failure depicted in Fig. 6.9.

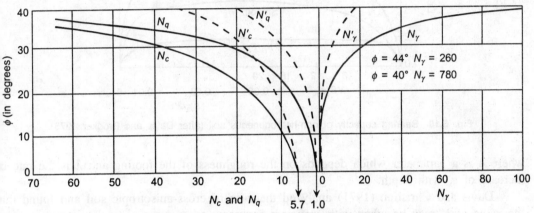

Fig. 6.9 Terzaghi's bearing capacity failure for general and local shear.

6.4 SQUARE AND CIRCULAR FOOTINGS

Equation (6.8) is valid for plane strain failure condition, as may occur in the case of strip footings. For square or circular footings, plastic zones would be three dimensional. So plane strain analysis is not strictly applicable. On the basis of experimental and field evidences, Terzaghi suggested the following modifications for circular and square footings.

Circular footing

$$q_d = 1.3cN_c + \gamma D_f N_q + 0.3\gamma DN_\gamma \tag{6.17}$$

where D is the diameter of the footing.

Square footing

$$q_d = 1.3cN_c + \gamma D_f N_q + 0.4\gamma BN_\gamma \tag{6.18}$$

where B is the width of footing.

6.5 BEARING CAPACITY OF NON-HOMOGENEOUS SOIL

Soft normally consolidated clays often show increase of undrained shear strength with depth because of increasing overburden pressure. Davis and Brooker (1973) gave solutions for a strip footing for undrained shear strength of the soil increasing linearly with depth, as illustrated in Fig. 6.10. The net ultimate bearing capacity is given by,

$$q_{ult} = A\left[(2 + \pi)c_{uo} + \lambda\frac{B}{4}\right] \tag{6.19}$$

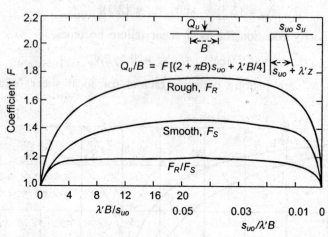

Fig. 6.10 Bearing capacity of non-homogeneous soil (after Davis and Brooker 1973).

where A is a parameter which depends on the roughness of the footing and λ is the rate o
increase of c_u with depth.

Davis and Christian (1971) analyzed the case of cross-anisotropic soil and found tha
the value of c_u may be taken with sufficient accuracy, as

$$c_u = 0.9\left(\frac{c_{uv} + c_{uh}}{2}\right) \tag{6.20}$$

where c_{uv} and c_{uh} are the undrained strength of the soil in the vertical and horizonta
direction respectively. Good prediction of bearing capacity of model footings in Boston blu
clay was obtained by using Eq. (6.20).

Vesic (1975) made detailed theoretical analysis of two-layer soil system, show
in Fig. 6.11, with the bearing stratum either softer or stiffer than the underlying stratum
In the first case, failure is partly by lateral plastic flow whereas in the second situation
failure is caused by punching shear. The net ultimate bearing capacity of the footing is give
by,

$$q_{ult} = c_1N_m \tag{6.21}$$

where N_m is a modified bearing capacity factor which depends on the ratio of shear strengt
of the two strata and the thickness of the bearing stratum and is given as

$$N_m = \frac{kN_c^*(N_c^* + \beta - 1)[(k + 1)N_c^{*2} + (1 + k\beta)N_c^* + \beta - 1]}{[k(k + 1)N_c^* + k + \beta - 1][(N_c^* + \beta)N_c^* + \beta - 1] - (kN_c^* + \beta - 1)(N_c^* + 1)} \tag{6.22}$$

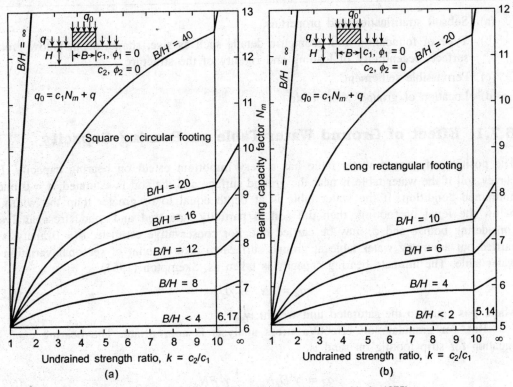

Fig. 6.11 Bearing capacity of stratified soil (after Vesic 1975).

With the development of non-linear finite element techniques, it is now possible to carry out rigorous analysis to determine the bearing capacity of footing for non-ideal field situations. Simple cases of bilinear stress–strain model, elastoplastic model or piecewise linear representation of the stress–strain behaviour have been adopted to obtain good prediction of the load-deformation behaviour of footings (D'Appolonia and Lambe 1970).

6.6 LIMITATIONS OF THEORETICAL ANALYSIS

Accurate prediction of bearing capacity by theoretical analysis often becomes difficult due to various reasons. This departure from accuracy may be because

(a) Correct estimation of in-situ soil properties are not always possible.
(b) Bearing capacity factors are sensitive to ϕ, which may change even during the process of failure.
(c) The unit weight of soil in the failure zones also changes during failure.
(d) The true shape of rupture surface is difficult to determine.

6.7 FACTORS AFFECTING BEARING CAPACITY

Bearing capacity depends on a number of factors. Some of the important factors are listed here.

(a) Subsoil stratification and properties.

(b) Type of foundation and geometric details such as size, shape, depth below ground surface, eccentricity of loading and rigidity of the structure.

(c) Permissible settlement.

(d) Location of ground water table.

6.7.1 Effect of Ground Water Table on Bearing Capacity

The position of ground water table has a very important effect on bearing capacity. For clayey soil if the water table is near the ground surface and the soil is saturated, $\phi = 0$ under undrained condition. If the water table is at depth equal to or greater than the width, B below the level of footing, then the soil is partially saturated and total stress analysis, considering both c and ϕ, may be carried out. For conservative estimate, $\phi = 0$ analysis is carried out assuming water table at ground surface to take account of seasonal variation in water table. The ultimate bearing capacity is taken as, Skempton (1951),

$$q_d = cN_c + \gamma D_f \tag{6.23}$$

where γ is equal to the saturated unit weight, γ_{sat}.

But, for granular soil effective stress analysis is appropriate. The bearing capacity equation for strip footing on sand is,

$$q_d = \gamma' D_f N_q + \frac{1}{2}\gamma' B N_\gamma \tag{6.24}$$

where γ' is the effective unit weight and N_q and N_γ depend on ϕ'. Here, γ' depends on the position of water table, as depicted in Fig. 6.12.

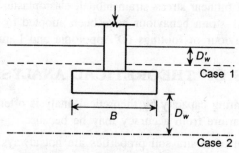

Fig. 6.12 Effect of ground water table.

For the strip footing shown in Fig. 6.12, the depth of failure surface does not extend beyond a depth equal to the width of the footing B below the footing base. If the water table is at or beyond this depth, it will have no effect on bearing capacity. If the water table is at the base of the footing or above, then in the second term of Eq. (6.24) $\gamma' = \gamma'_{sub}$, which is about half the value of γ_{sat}. The second term of Eq. (6.24) can now be written as $(1/2)R_w\gamma B N_\gamma$ where R_w is a water table correction which may vary from 0.5 to 1 and can be expressed as

$$R_w = 0.5\left(1 + \frac{D_w}{B}\right) \tag{6.25}$$

where D_w is the depth of water table below the base of footing.

Likewise, first term of Eq. (6.24) can be written as $(R'_w \gamma D_f N_q)$ where R'_w is a water table correction factor, which may lie in the range (0.5–1) and may be expressed as

$$R'_w = 0.5\left(1 + \frac{D'_w}{D_f}\right) \tag{6.26}$$

where D'_w is the depth of water table below ground surface.

Thus, to take effect of water table, the bearing capacity equation for a strip footing on granular soil may be expressed as:

$$q_d = R'_w \gamma D_f N_q + 0.5 R_w \gamma B N_\gamma \tag{6.27}$$

where γ is the bulk unit weight, and R_w, and R'_w are the water table correction factors which may vary from 0.5–1.

A practical method of considering the effect of ground water table is given in Chapter 8 (Section 8.2.3)

6.8 GROSS AND NET SOIL PRESSURE: SAFE BEARING CAPACITY

The total pressure transmitted to the subsoil by a foundation is the *gross soil pressure*, and the maximum gross pressure at which the soil fails is known as the *ultimate gross bearing capacity*. At the level of foundation, which is at a depth D_f below the ground surface, the overburden pressure is γD_f. The soil at this level was under this pressure prior to the application of footing load. The pressure transmitted by the foundation in excess of the overburden pressure is the *net bearing pressure*. The maximum net pressure at which the foundation fails is known as *ultimate net bearing capacity*. *Safe bearing capacity* is the maximum intensity of pressure which the soil can support without the risk of shear failure. This is obtained by dividing the ultimate bearing capacity by a factor of safety. Since the soil is originally subjected to the overburden pressure γD_f, there is no need to apply a factor of safety to the component of gross bearing capacity which is due to γD_f.

Therefore,

$$q_g\,(\text{all}) = \frac{q_{\text{ult}(n)}}{Fs} + \gamma D_f$$

It may be noted that there is a difference between safe bearing capacity and allowable bearing pressure. The allowable bearing pressure is the maximum net pressure that can be applied on the soil without the risk of shear failure or settlement beyond permissible limits.

6.9 BEARING CAPACITY FROM FIELD TESTS

For granular soils, estimation of field value of ϕ from laboratory test is extremely difficult. It is more convenient to estimate ϕ from results of penetration tests, for example, N value from SPT, N_c value from dynamic cone penetration test, or q_c from static cone penetration test. Methods are also available to compute bearing capacity directly from penetration test results. Some of these methods are as follows:

1. *Using charts given by Peck, Hansen, and Thornburn (1976):*

   Figure 6.13 is a plot of bearing capacity factors, N_q and N_γ against ϕ as well as N value (corrected). Considerations of both general and local shear failures are incorporated in the chart. This chart can be used directly for N_q and N_γ (according to Terzaghi's method) for use in bearing capacity equation, Eq. (6.8).

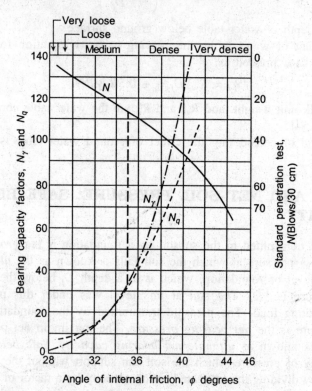

Fig. 6.13 Bearing capacity of footings based on 'N' value.

2. *Teng's Formulae:*

   Using the concept of Peck, Hansen, and Thornburn, Teng (1962) developed the following expression for calculating net ultimate bearing capacity on granular soil. For strip footing,

   $$q_{nd} = \frac{1}{60}[3N^2 BW_\gamma' + 5(100 + N^2)D_f W_q'] \qquad (6.28)$$

   For square or circular footings,

   $$q_{nd} = \frac{1}{30}[N^2 BW_\gamma' + 3(100 + N^2)D_f W_q'] \qquad (6.29)$$

where q_{nd} = net ultimate bearing capacity in t/m² or kN/m².
   N = average N value corrected for overburden pressure.
   D_f = depth of footing in metres; If $D_f > B$, take $D_f = B$.
   W_q' and W_γ' = correction factors for water table.

3. *Static cone test:*

IS:6403—1981 gives a method of determining net ultimate bearing capacity of strip footings on cohesionless soil using static cone resistance, q_c. The chart is presented in Fig. 6.14. q_c values at different depth are obtained for selected locations at the site. The field values are corrected for the dead weight of the sounding rod. Then the average q_c value is obtained for each of the locations by averaging the values between the base of the footing and depth below the base 1.5–2 times the width of the footing. The smallest of the average values is used in Fig. 6.11 to find bearing capacity factors N_q and N_γ, which may be used for computing the bearing capacity.

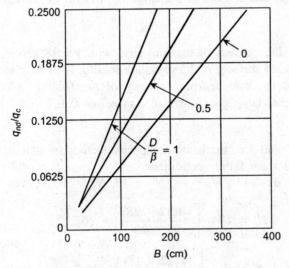

Fig. 6.14 Bearing capacity from static cone test.

Example 6.1

A square footing of width 2 m rests at a depth of 1 m at a site where the subsoil consists of soft to medium clay down a depth of 8 m below ground level, which is underlain by a dense coarse sand deposit. The water table is at 1.5 m below G.L. The clay has c_u = 30 kN/m², $\phi = 0$ and γ = 19 kN/m³.

   Determine the net load the footing can safely carry with a factor of safety of 3 against shear failure. What will be the safe load if the water table rises to the ground surface?

Solution

Since the thickness of clay layer beneath the level of footing base is more than the width of the footing, the bearing capacity will be governed by upper layer of clay.

According to Skempton's formula

$$(q_{ult})_{net} = c_u N_c \qquad \text{Here,} \quad N_c = 6(1 + 0.2 D_f/B)$$

$$= 30 \times 6.6 \text{ kN/m}^2 \qquad = 6\left[1 + 0.2\left(\frac{1}{2}\right)\right]$$

$$= 6.6$$

$$\therefore \qquad (q_{safe})_{net} = \frac{30 \times 6.6}{2.5} \quad (Fs = 2.5)$$

$$= 80 \text{ kN/m}^2$$

∴ Net safe carrying capacity of the footing $= 80 \times 2 \times 2$

$$= 320 \text{ kN}$$

Since this is a total stress analysis, there will be no change in the safe load if the water table rises to the ground surface, unless there is a change in strength of the clay.

Example 6.2

The subsoil at a building site consists of medium sand with $\gamma = 18$ kN/m³, $c' = 0$, $\phi' = 32°$ and water table at the ground surface. A 2.5 m square footing is to be placed at 1.5 m below ground surface. Compute the safe bearing capacity of the footing. What would be safe bearing pressure if the water table goes down to 3 m below G.L.?

Solution

Since ϕ' lies between 28° and 35°, the bearing capacity factors are obtained by interpolation between local and general shear failure conditions.

Referring to Fig. 6.7, for $\phi' = 32°$, $N_q = 28$, $N'_q = 10$, $N_\gamma = 30$, $N'_\gamma = 6$

we get,

$$N_q = 10 + \left[\frac{18(32 - 28)}{35 - 28}\right] = 20.3$$

and

$$N_\gamma = 6 + \left[\frac{24(32 - 28)}{35 - 28}\right] = 19.7$$

Case I (water table at G.L.)

Using Eq. (6.27) modified for square footing, ultimate gross bearing capacity

$$(q_{ult})_{gross} = \gamma' D_f N_q + 0.4\gamma' B N_\gamma$$

$$= 8.0 \times 1.5 \times 20.3 + 0.4 \times 8.0 \times 2.5 \times 19.7$$

$$= 243.6 + 157.6 = 401.2 \text{ kN/m}^2.$$

and ultimate net bearing capacity,

$$(q_{ult})_{net} = (q_{ult})_{gross} - \gamma D_f$$

$$= 401.2 - 18 \times 1.5 = 384.2 \text{ kN/m}^2$$

Safe net bearing capacity,

$$(q_{safe})_{net} = 384.2/2.5 = 152 \text{ kN/m}^2 \quad (Fs = 2.5)$$

and gross bearing capacity,

$$(q_{safe})_{gross} = 152 + 18 \times 1.5 = 179 \text{ kN/m}^2$$

Case II (Water table at 3 m below G.L., that is, at a depth greater than width footing)

$$(q_{ult})_{gross} = 18 \times 1.5 \times 20.3 + 0.4(14 \times 2.5)19.7$$

$$= 548.1 + 275.8 = 823.9 \text{ kN/m}^2$$

$$(q_{ult})_{net} = 823.9 - 18 \times 1.5 = 796.9 \text{ kN/m}^2$$

∴   safe net bearing capacity,

$$(q_{safe})_{net} = 796.9/2.5 = 319 \cong 320 \text{ kN/m}^2$$

and safe gross bearing capacity,

$$(q_{safe})_{gross} = 320 + 18 \times 1.5 = 347 \cong 350 \text{ kN/m}^2$$

(This problem may also be solved using water table correction factors W'_q and W'_γ with marginal change in the result. Also Fig. 6.10 may be used for obtaining relevant N_q and N_γ values.)

Example 6.3

A rectangular footing, with a plan area of 1.4 m × 2 m is to be placed at a depth of 2 m below the ground surface. The footing would be subjected to a load inclined at 10° to the vertical. The subsoil is clayey, sandy silt with saturated unit weight of 18 kN/m³, and $c' = 10 \text{ kN/m}^2$ and $\phi' = 30°$. Assuming the rate of loading is such that drained condition prevails, compute the magnitude of load the footing can carry if the water table is at the base of the footing. Use IS: 6403—1981 recommendations and take $Fs = 3$.

Equation (6.13) gives,

$$q_{nd} = cN_c s_c d_c i_c + q(N_q - 1)s_q d_q i_q + 0.5\gamma BN_\gamma s_\gamma d_\gamma i_\gamma W'$$

Here $c = c' = 10 \text{ kN/m}^2$, $\phi = \phi' = 30°$

$$N_q = (e^{\pi \tan \phi}) \tan^2(45° + \phi/2) = 18.38$$

$$N_c = (N_q - 1) \cot \phi = 17.38 \cot 30° = 30.10$$

$$N_\gamma = 2(N_q + 1) \tan \phi = 22.37$$

$$s_c = 1 + 0.2 B/L = 1.14$$

$$s_q = 1 + 0.2 B/L = 1.14$$

$$s_\gamma = 1 - 0.4 B/L = 0.72$$

$$d_c = 1 + 0.2 D_f/B \tan (45° + \phi/2) = 1 + 0.2 \times 2/1.4 \times 1.732 = 1.5$$

$$d_q = d_\gamma = 1 + 0.1D_f \tan(45° + \phi/2) = 1 + 0.35/1.4 = 1.25$$

$$i_c = i_q = (1 - \alpha/90)^2 = 0.79$$

$$i_\gamma = 0.44$$

$$W' = 0.5$$

Hence,

$$q_{nd} = (10 \times 30.1 \times 1.14 \times 1.5 \times 0.79) + (18 \times 2 \times (18.38 - 1) \times 1.14 \times 1.25 \times$$
$$0.79) + (0.5 \times 18 \times 1.4 \times 22.37 \times 0.72 \times 1.25 \times 0.44 \times 0.5)$$

$$= 406.6 + 704.4 + 55.8 = 1166.8 \text{ kN.}$$

$$\therefore \quad (q_{net})_{safe} = 1166.8/3 = 388.9 \cong 380 \text{ kN/m}^2$$

Hence, safe load $= 380 \times 1.4 \times 2 = 1064 \cong 1060$ kN

Example 6.4

What will be the safe load in Example 6.3 if undrained condition prevails? Take $c_u = 30$ kN/m^2, $\phi_u = 0$, $N_c = 5.14$, $N_q = 1$, and $N_\gamma = 0$.

Solution

$$q_{nd} = c_u N_c s_c d_c i_c$$

$$= 30 \times 5.14 \times 1.14 \times 1.5 \times 0.79$$

$$= 208.3 \text{ kN/m}^2$$

$$\therefore \quad (q_{net})_{safe} = 208.3/3 = 69.4 \cong 70 \text{ kN/m}^2$$

$$\therefore \quad \text{Safe load} = 70 \times 1.4 \times 2 = 196 \text{ kN} \cong 200 \text{ kN}$$

REFERENCES

Brinch Hansen, J. (1961), *A General Formula For Bearing Capacity*, Danish Geotechnical Institute, Bulletin No. 11, Copenhagen.

Brinch Hansen, J. (1970), *A Revised and Extended Formula for Bearing Capacity*, Danish Geotechnical Institute, Bulletin No. 28, Copenhagen.

Davis and Brooker (1973), The Effect of Increasing Strength with Depth on the Bearing Capacity of Clays, *Geotechnique,* Vol. 23, pp. 551–563.

D'Appolonia, D.J. and T.W. Lambe (1970), Method of Predicting Initial Settlement, *Journal Soil Mechanics and Foundation Division*, ASCE, Vol. 96, pp. 523–544.

IS 6403 (1981), *Code of Practice for Determination of Bearing Capacity of Shallow Foundations*, Bureau of Indian Standards, New Delhi.

Meyerhof, G.G. (1951), The Ultimate Bearing Capacity of Foundations, *Geotechnique,* Vol. 2, pp. 301–332.

Peck, R.B., W.E. Hansen, and W.H. Thornburn (1962), *Foundation Engineering*, 2nd Edition, John Wiley & Sons, New York.

Prandtl, L. (1921), Uber die Eindvingungs Festingkite von Schneiden, *Zeitschrift fur Angebandte Mathematik and Mechanik,* Vol. 1, No. 1, pp. 15–22.

Reissner (1924), *Zuns Erddruck-problem*, Proc. First Int. Congress on Applied Mechanics, Dept. pp. 295–311.

Skempton, A.W. (1951), *The Bearing Capacity of Clays*, Building Research Congress, England.

Terzaghi, K. (1951), *Theoretical Soil Mechanics*, John Wiley, New York.

Vesic, A.S. (1973), Analysis of Ultimate Loads of Shallow Foundations, *Journal of Soil Mechanics and Foundation Division* ASCE, Vol. 99, No. S M 1, pp. 45–73.

Vesic, A.S. (1975), Bearing Capacity of Shallow Foundations, Chapter 3 of *Foundation Engineering Handbook,* Van Nostrand Reinhold Co., New York.

7 Settlement Analysis

7.1 INTRODUCTION

Foundations on soft clay are liable to undergo excessive settlement and prediction of this settlement is an important aspect of foundation design. A part of this settlement, known as the immediate settlement, is a result of undrained shear deformation of the soil and takes place during loading. A major part of the settlement, however, occurs due to volume change of the clay caused by the dissipation of excess pore pressure developed from imposed loads. This is a gradual process and is known as *primary consolidation*. In addition, there may be settlement due to secondary consolidation caused by volume change of the clay apparently under constant effective stress. Except in clays which exhibit high secondary effect, settlement analysis usually involves an evaluation of the immediate and primary consolidation settlement only.

Soft clays, in general, are normally consolidated and their behaviour differs somewhat from overconsolidated clays and clay shales. A survey of case records of settlement of shallow foundations on clay (Simons and Som 1970) reveals that structures founded on N.C. clays, in general, experience more settlement than those founded on overconsolidated clay. This is only to be expected because the main effect of overconsolidation is to reduce the compressibility of the clay. However, it is more important, that for normally consolidated clay, only 15–20% of the total settlement occurs during construction, whereas for overconsolidated clays, as much as 50–60% of the total settlement may occur during construction. This shows the importance of consolidation on the settlement of foundations on clay and evaluation of this settlement becomes a major part of the settlement analysis.

7.1.1 Definitions

Figure 7.1 shows the development of net pressure and the associated settlement of a foundation on clay during different stages of construction. Before application of the building load, there is usually an excavation down to the foundation level during which there is a heave of the subsoil. The magnitude of this heave is generally small, unless the excavation is deep and is kept open for a long time. After excavation, the foundation load is gradually applied and there comes a time when the applied pressure on the soil equals the pressure released by excavation and the net foundation pressure becomes zero. The corresponding settlement due to the restoration of the excavation load may be assumed to be equal to the prior heave. Any further increase of load is a net increase in foundation pressure and the associated settlement is the net settlement of the foundation.

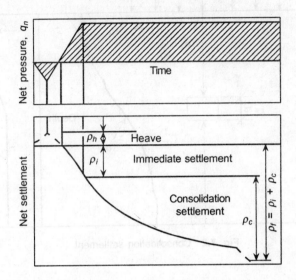

Fig. 7.1 Time–settlement relationship of foundations on clay.

When the construction is complete and the total building load has been applied, the immediate settlement, ρ_i has already occurred. If the construction is rapid this settlement is a result of shear deformation of the clay. For a saturated clay, this occurs essentially under the condition of zero volume strain. Any consolidation settlement at this stage is small.

After the construction is over, the clay undergoes consolidation and the settlement gradually increases with time until, theoretically, after infinite time, the consolidation is complete. The total consolidation settlement, ρ_c is then added to the immediate settlement, ρ_i, to give the final settlement, ρ_f of the foundation. Therefore,

$$\rho_f = \rho_i + \rho_c \tag{7.1}$$

7.1.2 Methods of Settlement Analysis

The principle of effective stress and Terzaghi's theory of one-dimensional consolidation have been the essential basis for settlement analysis of foundations founded on clay. The earliest method of settlement analysis for a foundation placed on a layer of clay is shown in Fig. 7.2. The settlement is given by the equation,

$$\rho_c = \int_0^z m_v \, \Delta\sigma_z \, dz \tag{7.2}$$

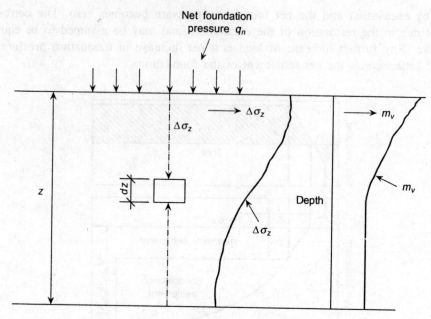

Fig. 7.2 Consolidation settlement.

where,

m_v is the compressibility of the clay as determined from one-dimensional consolidation test,

$\Delta\sigma_z$ is the increase of vertical stress at the centre of the layer, and

dz is the thickness of the clay layer.

The method was proposed by Terzaghi (1929) for obtaining the consolidation settlement of clayey strata subjected to lateral confinement, that is, where all settlement is due to one-dimensional compression of the clay. Although this method, is valid only in cases where the condition of no lateral strain is at least approximately valid it has been extended to cases in the field where the foundation rests on a deep bed of clay (Taylor 1948). In such cases, there is lateral deformation during load application which gives rise to 'immediate' settlement. Therefore, the conventional method seems physically inadequate to describe completely the behaviour of a clay that undergoes important lateral deformations.

Immediate settlement

It is now the common practice to obtain the 'immediate settlement', that is, the settlement that takes place under undrained condition due to the shear deformation of the clay, from the equations of elasticity. This has the general expression (given by Terzaghi (1943)),

$$\rho_i = \left(\frac{q_n B(1 - v^2)}{E}\right) I_\rho \tag{7.3}$$

where,

q_n is the net foundation pressure,

B is a width of the foundation,

E is the Young's modulus of the clay,

v is the Poisson's ratio, and

I_ρ is the influence coefficient whose magnitude depends on the geometry of the foundation.

The values of I_ρ for different values of L/B for rectangular foundations are given in Table 7.1. For rigid foundations, the influence coefficient may be taken as:

$$(I_\rho)_{\text{RIGID}} = 0.8(I_\rho)_{\text{CENTRE}} \qquad (7.4)$$

Table 7.1 Values of influence coefficient I_ρ for rectangle foundation, $L \times B$

$m = L/B$	Values of I_ρ	
	Centre	Corner
1.0	1.12	
1.5	1.36	
2.0	1.52	
3.0	1.78	1/2 centre
4.0	1.96	values
5.0	2.10	
6.0	2.24	
7.0	2.32	
8.0	2.42	
9.0	2.50	
10.0	2.53	
Circle: Dia B	1.00	0.64

Because of sample disturbance, values of E obtained from laboratory triaxial compression tests are often unreliable. It is preferable to obtain E values from plate load tests made in the field or from established empirical relations of E and the undrained shear strength, c_u, of the clay. Butler (1974) gives E/c_u ratio of 400 for overconsolidated London clay while Bjerrum (1973) gives the c_u/p ratios in the range of 500 to 1500 for normally consolidated clays, based on measurement of undrained shear strength of the clay by the vane shear test.

For foundations placed at some depth beneath the ground surface, a 'depth correction' may be applied to the settlement calculated by Eq. (7.3) (Fox 1948). The depth correction depends on the depth to width ratio and length to width ratio of the foundation. Values of the depth correction, given in Fig. 7.3, are defined as

$$\text{Depth correction factor} = \frac{\text{settlement of foundation at depth } D}{\text{settlement of corresponding surface foundation}}$$

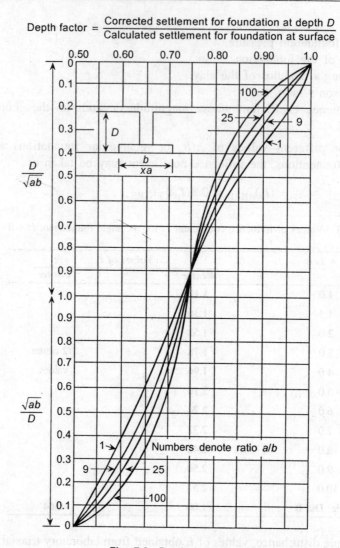

Fig. 7.3 Depth correction factor.

Consolidation settlement

While Eq. (7.3) still remains the most widely used expression for calculating the immediate settlement, the method of obtaining the consolidation settlement has undergone some changes. Skempton and Bjerrum (1957) recognized that there was lateral deformation in the clay during load application and that the consolidation settlement should be a function of the excess pore pressure set up by the applied load. Therefore, taking account of the shear stresses, a modified expression for the total settlement was given as,

$$\rho_f = \rho_i + \mu\rho_{oed} \tag{7.5}$$

where

μ is the pore pressure correction factor which depends on the pore pressure parameter A and the geometry of the foundation and

ρ_{oed} is the consolidation settlement obtained by the direct application of the oedometer test results (Eq. (7.2)).

The values of pore pressure correction factor as given by Skempton and Bjerrum (1957) are given in Fig. 7.4.

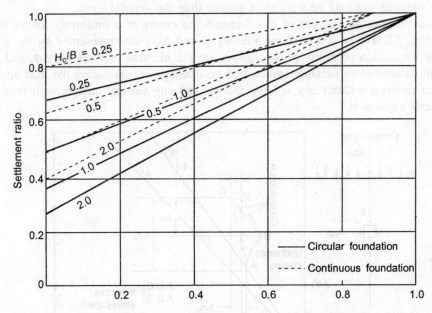

Fig. 7.4 Pore pressure correction factor for foundations on clay (after Skempton and Bjerrum 1957).

For routine design, the following values of the pore pressure correction factor may be used:

Normally consolidation clay	: 1.0–0.7
Overconsolidated clay	: 0.7–0.5
Heavily overconsolidated clay	: 0.5–0.3

7.2 STRESS-PATH

It is well-recognized that the deformation of an element of soil is a function not only of magnitude of the applied stresses but also of the manner of their application. In other words, a knowledge of the magnitude of stress increase is not, in itself, sufficient to indicate precisely how a soil element is going to deform. To obtain a more complete picture, we have also to know how the applied stresses change, at what rate and in what relation to one another (Som 1968).

A stress-path is essentially a curve drawn through points on a plot of stress changes. It shows the relationship between the components of stresses at various stages in moving from one stress point to another. In the present study, however, consideration is given only to cases where by virtue of symmetry, the intermediate and minor principal stresses are equal and where the vertical and horizontal stresses are the principal stresses, for example, in the laboratory triaxial test or along the centre line beneath a loaded circular area. In its natural condition, before any load is applied, an element of soil is in a K_o state of stresses (K_o = coefficient of earth pressure at rest). Depending on whether the soil is normally consolidated or overconsolidated the horizontal stress in-situ may be smaller or greater than the vertical stress.

Let us consider an element of clay beneath the centre of a uniformly loaded circle, as shown in Fig. 7.5. The in-situ effective stresses (p and $K_o p$) are represented by the point 'A'. Due to the foundation pressure, q, the stresses on the element increase by $\Delta\sigma_v$ and $\Delta\sigma_{h1}$. If the pressure is applied sufficiently fast so that no drainage occurs during the load application, the element deforms without any volume change and any vertical compression is associated with a lateral expansion.

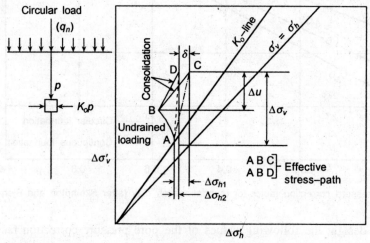

Fig. 7.5 Stress–path of an element of soil beneath foundation in the field.

Now, the increase of stresses $\Delta\sigma_v$ and $\Delta\sigma_{h1}$, which are the increase in principal stresses in this instance, will set up an excess pore–water pressure in the element according to the equation,

$$\Delta u = B \left[\Delta\sigma_{h1} + A \left(\Delta\sigma_v - \Delta\sigma_{h1} \right) \right] \tag{7.6}$$

where A and B are Skempton's pore-pressure parameters (Skempton, 1954).

If the clay is saturated, as all clays below the water table are, $B = 1$. Then,

$$\Delta u = \Delta\sigma_{h1} + A \left(\Delta\sigma_v - \Delta\sigma_{h1} \right) \tag{7.7}$$

Therefore, immediately after the load application, the effective stresses become:

$$\left. \begin{array}{l} \sigma'_v o = p + \Delta\sigma_v - \Delta u \\ \sigma'_h o = K_o p + \Delta\sigma_{h1} - \Delta u \end{array} \right\} \tag{7.8}$$

Since for most clays, the value of A is positive in the range of stresses normally encountered in practice, the excess pore-pressure Δu is greater than $\Delta\sigma_{h1}$. So while the effective vertical stress increases on load application, the effective horizontal stress decreases and the stress point moves from 'A' to 'B'. The vertical strain during loading, that is, the immediate settlement, is, therefore, a function of the stress-path AB.

The element now begins to consolidate. At the early stages of consolidation, the increase in effective horizontal stress is only a recompression until the original value $k_o p$ is restored. Beyond this point, any further increase of horizontal stress is a net increase while there is a net increase in vertical stress during the entire process of consolidation.

Now during load application, in undrained condition, a saturated clay behaves as an incompressible medium with Poisson's ratio, $v = 0.5$. As the excess pore-pressures dissipate, Poisson's ratio decreases and finally attains its fully drained value at the end of consolidation. This change in Poisson's ratio does not have much effect on the vertical stress but the horizontal stress decreases by an amount δ to its new value $K_o p + \Delta\sigma_{h2}$, i.e.,

$$\delta = \Delta\sigma_{h1} - \Delta\sigma_{h2} \tag{7.9}$$

So during consolidation, the element follows the effective stress path BD while it would have moved from B to C had the total stress remained unchanged. After full consolidation, therefore, the stresses are:

$$\left.\begin{aligned} \sigma'_{vf} &= \Delta\sigma_v \\ \sigma'_{hf} &= K_o p + (\Delta\sigma_{h1} - \delta) \\ &= K_o p + \Delta\sigma_{h2} \end{aligned}\right\} \tag{7.10}$$

An ideal settlement analysis should take into account the complete pattern of stresses an element of soil will be subjected to in the field. In order to determine the relevant vertical strains, laboratory tests should be performed under identical stress conditions and an integration of all such vertical strains beneath a loaded area would give the settlement of the foundation (Lambe 1964, Davis and Poulos 1967).

The methods most widely used in practice, however, are the ones based on the oedometer test. Figure 7.6 summarizes all these methods indicating the effective stress-paths associated with each of them. In plotting the stress-path AF in Fig. 7.6, it has been assumed that an undisturbed sample when subjected to the in-situ vertical stress also restores the in-situ horizontal stress so that subsequent stress-path for loading starts from the point A.

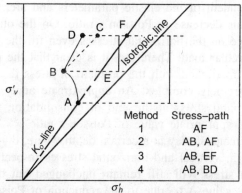

Method	Stress–path
1	AF
2	AB, AF
3	AB, EF
4	AB, BD

Fig. 7.6 Methods of settlement analysis and implied stress–path.

Method 1, described by Skempton and McDonald (1955) as the conventional method, was first proposed by Terzaghi (1929) and later used by Taylor (1948). It assumes that all settlement occurs from one-dimensional compression and that the excess pore-pressure is equal to the increase in vertical stress. Therefore, the corresponding stress–path is AF.

Method 2 recognizes that the soil undergoes shear deformation during undrained loading (path AB), and this causes the immediate settlement, but still assumes that the excess pore-pressure, which should be a function of the induced shear stress, is equal to the vertical stress increment. Consolidation settlement, therefore, occurs along the stress–path AF. This inconsistency in Method 2 has been overcome by Skempton and Bjerrum (1957) in Method 3. Here, the immediate settlement is a function of the stress–path AB while the consolidation settlement occurs along the path EF. The latter, therefore, is only a part of the total strain along the path AF depending on the magnitude of the excess pore-pressure set up during loading.

While the method of Skempton and Bjerrum introduces for the first time, the concept of stress-path in settlement analysis, the assumption is still implicit that during consolidation all strain is one-dimensional which requires the horizontal stresses to adjust accordingly. Therefore, there is a discrepancy between the field stress–path BD and the path EF used in the analysis. The condition of no lateral strain may be approximately true in cases like that of a loaded area which is very large compared to the thickness of the clay layer. But in the majority of field problems, this condition may be far from true. Moreover, while computing the immediate settlement, Method 3 obviously accepts that the clay undergoes lateral deformation during loading (with constant volume, there would be no settlement otherwise) but this is neglected in estimating the consolidation settlement. Consequently, of course, the effect of the horizontal stress is completely ignored except in determining the excess pore–water pressure.

7.2.1 Stresses During Loading and Consolidation in the Field

It is evident that during load application in the field, most of the deformation takes place under condition of no volume change, implying a value of Poisson's ratio of the soil equal to 0.5. As the excess pore-pressures dissipate, Poisson's ratio also decreases and finally drops to its fully drained value. It is known that for an isotropic homogeneous elastic medium, the vertical stresses are independent of the elastic parameters and are, therefore, unlikely to be significantly affected by this decrease in Poisson's ratio. On the other hand, Poisson's ratio has a considerable influence on the horizontal stresses even for the Boussinesq case beneath the centre of a uniform circular load. Therefore, it is clear that the problem of consolidation in the field is very much interlinked with the problem of stress distribution, whose rigorous analytical treatment is extremely complex. An approximate analysis may be made on the assumption that the horizontal stress at the end of consolidation will be the same as the Boussinesq stresses for the appropriate value of Poisson's ratio.

If we consider an element of clay at a certain depth where, during load application, $\Delta\sigma$ and $\Delta\sigma_{h1}$ are the increase in vertical and horizontal stresses respectively then, at the end of consolidation, the vertical stress will still remain unchanged but the horizontal stress will have decreased to the new value $\Delta\sigma_{h2}$ due to the reduction of Poisson's ratio.

So, the changes in effective stress during consolidation are given as

Horizontal: $$[\Delta\sigma_h]_c = \Delta u - \delta \qquad (7.11)$$

Vertical: $$[\Delta\sigma_v]_c = \Delta u$$

where $\delta = \Delta\sigma_{h1} - \Delta\sigma_{h2}$.

The ratio of the effective stress changes during consolidation, K' is then given by

$$K' = \frac{[\Delta\sigma_h]_c}{[\Delta\sigma_v]_c} = 1 - \frac{\delta}{\Delta u} \qquad (7.12)$$

Now, for the Boussinesq problem, the general expressions for stresses beneath the centre of a uniform circular load (diameter $2b$ and load intensity, q) are (Wu 1966),

$$\left.\begin{array}{l} \sigma_z = q\left[1 - \dfrac{(z/b)^3}{(1 + z^2/b^2)^{3/2}}\right] \\[4mm] \sigma_h = \dfrac{q}{2}\left[(1 + 2v) - \dfrac{2(1 + v)(z/b)}{(1 + z^2/b^2)} + \dfrac{(z/b)^3}{(1 + z^3/b^3)^{3/2}}\right] \end{array}\right\} \qquad (7.13)$$

which can be more conveniently written as,

$$\sigma_z = q(1 - \eta^3)$$

$$\sigma_h = \frac{q}{2}[(1 + 2v) - 2(1 + v)\eta + \eta^3] \qquad (7.14)$$

where,

$$\eta = \frac{z/b}{(1 + z^2/b^2)^{1/2}} \qquad (7.15)$$

Using these expressions,

$$\left.\begin{array}{l} \Delta\sigma_v = (\sigma_v)_{v=1/2} = q(1 - \eta^3) \\[3mm] \Delta\sigma_{h1} = (\sigma_h)_{v=1/2} = \dfrac{q}{2}[2 - 3\eta + \eta^3] \\[3mm] \Delta\sigma_{h2} = (\sigma_h)_{v=v'} = \dfrac{q}{2}[(1 + 2v') - 2(1 + v')\eta + \eta^3] \end{array}\right\} \qquad (7.16)$$

Combining Eqs. (7.13) to (7.16), we have

$$K' = 1 - \left[\frac{1 - 2v}{1 + \eta(1 + \eta)(3A - 1)}\right] \qquad (7.17)$$

Eq. (7.17) can then be used to determine the ratio of the stress increase during consolidation at any depth beneath the centre of a uniform circular load.

7.2.2 Influence of Stress-path on the Drained Deformation of Clay

It has been shown in the foregoing that the stress changes that occur in the field during consolidation are primarily dependent on the drained Poisson's ratio and the pore pressure parameter A of the soil. Therefore, it is of considerable practical significance to study the influence of stress-path, such as BD (see Fig. 7.5) on the axial and volumetric strain during consolidation. Experimental investigation with undisturbed London clay (Som, 1968), normally consolidated Kaolinite (Das and Som et al., 1975) and undisturbed Calcutta clay (Guha, 1979) show that the stress increment ratio during consolidation has important effect on the deformation characteristics of the soil. Two things need to be considered here:

(a) the influence of stress increment ratio on the volumetric strain of the soil and
(b) the influence of stress increment ratio on the strain ratio (that is, axial strain/ volumetric strain) of the soil.

There is conflicting evidence on the effect of stress ratio on the volumetric compressibility of clay. Some investigators have found that the volumetric compressibility is independent of the stress increment ratio and can be expressed as function of vertical effective stress only (Som, 1968). This means that the results of one-dimensional consolidation tests would be applicable irrespective of the changes in lateral pressure during consolidation. Others have found that the volumetric compressibility is not independent of the lateral pressure and can best be expressed as functions of mean effective stress during consolidation (Das, 1975). It seems that the appropriate relationship has to be determined for the particular soil under investigation.

The major effect of lateral stresses during consolidation is, however, manifested in the axial strain associated with volumetric deformation. Figure 7.7 shows some typical results of stress-path controlled triaxial tests on different clays. The results are plotted as relationships between the stress increment ratio, K' and the strain ratio, λ during consolidation or drained deformation. It is to be noted that one-dimensional consolidation implies a strain ratio, $\lambda = 1.0$ and that occurs at a particular stress increment ratio only. For any other stress ratio, the strain ratio would also the difference and assumption of one-dimensional strain (i.e. $\varepsilon_1/\varepsilon_v = 1.0$) would lead to significant error in settlement analysis.

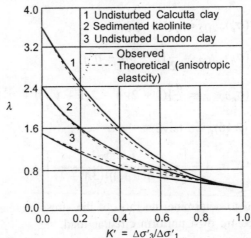

Fig. 7.7 Strain ratio, λ versus stress increment ratio, K' for drained deformation.

The relationships shown in Fig. 7.7 can best be determined from stress controlled triaxial tests. However, Som (1968), Simons and Som (1969) have demonstrated that this relationship can also be established theoretically from consideration of anisotropic elasticity as:

$$\frac{\varepsilon_1}{\varepsilon_v} = \frac{1 - 2\eta v_2' K'}{(1 - 2v_3') + 2\eta(1 - v_1' - v_2')K'} \tag{7.18}$$

where,

v_1' = effect of one horizontal strain in the other horizontal strain;

v_2' = effect of horizontal strain on vertical strain;

v_3' = effect of vertical strain on horizontal strain;

η = ratio of vertical drained deformation modulus to the horizontal drained deformation modulus.

Equation (7.18) can also be expressed as

$$\frac{\varepsilon_1}{\varepsilon_v} = \frac{1 - 2\eta v_2' K'}{(1 - 2v_3')[K'/K_o - 1] - 2\eta v_2' K'} \tag{7.19}$$

The parameters v_3' and $\eta v_2'$ can be determined from two tests with different values of K' and measuring $\varepsilon_1/\varepsilon_v$ in each case. The most convenient values of K' to choose are 0 and 1.0, that is, those from the standard drained compression test and the isotropic consolidation test.

7.2.3 Settlement Analysis by Stress-path Method

The methods of settlement analysis in current use have been described earlier. These methods have given satisfactory results in many field problems, particularly where the condition of zero lateral strain is satisfied. However, more often than not appreciable lateral deformation occurs in the field and the currently used methods of settlement analysis become inadequate to predict the settlement. A method of settlement prediction by the stress-path method has, therefore, been introduced for the case of axi-symmetric loading (Simons and Som, 1969, Som et al., 1975).

Immediate settlement

The common practice in estimating the immediate (elastic) settlement of foundation on clay is to assume the soil to be a homogeneous, isotropic, elastic medium, thus, permitting the application of Boussinesq analysis. However, in practice, the elastic modulus, E of the soil varies with depth as a consequence not only of the inherent non-homogeneity of the soil but also of varying applied stress levels at different depths. Stress–strain relationship of soft clay is essentially non-linear and the deformation modulus varies considerably with the stress level. While it is possible to make theoretical analysis of undrained deformation using the finite element technique and taking the non-linear stress versus strain behaviour of the soil into consideration, they are not yet available in readily usable form. A simple expression for immediate settlement following the stress-path method may be given as,

$$\rho_i = \int_0^z \frac{\Delta\sigma_v - \Delta\sigma_h}{E(z)} \qquad (7.20)$$

where $\Delta\sigma_v$ and $\Delta\sigma_h$ are the increase in total vertical and horizontal stresses at any depth respectively and $E(z)$ is the corresponding deformation modulus, taking due account of the vertical and horizontal effective stresses prior to and at the end of loading.

It has been shown by many investigators that the stresses in the subsoil due to a foundation loading are not significantly dependent upon the variation of E (Huang 1968, Som 1968). Therefore, in the absence of more rigorous analysis, the Boussinesq stresses may be used in evaluating the settlement by Eq. (7.21), the integration being done numerically by dividing the compressible stratum into a number of layers and taking the appropriate value of E for the layer into account.

Consolidation settlement

In the conventional method, the consolidation settlement is given by,

$$\rho_c = \int_0^z (m_v)_1 \Delta\sigma_z \, dz \qquad (7.21)$$

where $(m_v)_1$ is the coefficient of volume compressibility determined from standard consolidation test and $\Delta\sigma_z$ is the increase of vertical pressure at the centre of layer, Following the Skempton and Bjerrum (1957) method, the settlement is,

$$\rho_c = \mu \int_0^z (m_v)_1 \Delta\sigma_z \, dz \qquad (7.22)$$

where μ, the pore-pressure correction factor, is a function of the soil type and the geometry of the foundation, (refer Fig. 7.4).

For the stress-path method, the settlement is given by,

$$\rho_c' = \int_0^z \lambda (m_v)_3 \Delta u \, dz \qquad (7.23)$$

$$= \lambda \rho_c$$

where,

Δu = increase of pore–water pressure under undrained loading which dissipates during the consolidation.

$(m_v)_3$ = coefficient of volume compressibility for three-dimensional strain.

λ = the ratio of vertical strain to volumetric strain which is a function of the stress increment ratio during consolidation.

If Eq. (7.23) is to be applied in settlement analysis, K' during consolidation needs to be known in order to determine the value of λ. Accurate determination of K' for actual foundation may be difficult because it depends on many factors such as Poisson's ratio, pressure parameter A, stress–strain relationship of the soil, and so on. As an approximation, however, its value can be determined from Eq. (7.17), for a footing of diameter $2a$. According to Eq. (7.17),

$$K' = 1 - \left[\frac{1 - 2v}{1 + \eta(1 + \eta)(3A - 1)} \right]$$

where,

$$\eta = \frac{z}{a} \left(1 + \frac{z^2}{a^2} \right)^{-1/2},$$

v = Poisson's ratio

A = Skempton's pore-pressure parameter

Knowing the value of K', λ can be obtained from experimental relationships as shown in Fig. 7.7. The value of $(m_v)_3$ should, of course, be obtained from appropriate stress-path test, but, for practical purposes, can be determined from properly conducted oedometer test. Then, $(m_v)_3$ may be taken as approximately equal to $(m_v)_1$, the oedometer compressibility in terms of vertical effective stress.

It may be noted here that evaluation of the K' versus λ relationship for any problem requires determination of Poisson's ratio and pore-pressure parameter, A of the soil. But for most soils, Poisson's ratio varies between 0.1 and 0.3 and pore-pressure parameter, A between 0 and 1. Accordingly, K' varies within a narrow range of 0.6–0.9 and λ in the range 0.5–0.8. It would, therefore, be convenient to select a value of λ within this range for most practical problems (Som, 1968).

7.3 RATE OF SETTLEMENT

The rate of settlement of foundations on clay is generally determined from Terzaghi's theory of one-dimensional consolidation although field-consolidation is often three-dimensional (See Chapter 1, Section 1.9.5).

According to this, the excess pore-pressure at any point at depth z within a soil mass after a time from load application, is governed by the equation

$$C_v \frac{\delta^2 u}{\delta z^2} = \frac{\delta u}{\delta t} \tag{7.24}$$

where, u is the excess pore-pressure and

C_v coefficient of consolidation.

The degree of consolidation at any time, t (defined as the percentage dissipation of pore-pressure) may be obtained by solving Eq. (7.24) for appropriate boundary conditions. Table 7.2 gives the relationship between time factor, $T_v = C_v t / H^2$ and degree of consolidation, U for different boundary conditions.

Table 7.2 U versus T_v relationship ($T_v = C_v t/H^2$)

$U\%$	$(u_1/u_2) = 0$	0.5	1.0	5.0	T_v
0	0	0	0	0	0
20	0.01	0.05	0.031	0.014	0.01
40	0.21	0.17	0.126	0.081	0.049
60	0.39	0.31	0.287	0.195	0.155
80	0.64	0.58	0.567	0.460	0.420
100	∞	∞	∞	∞	∞

For double drainage (any distribution of pressure, that is, $u_1/u_2 =$ any value), take $H = 1/2$ times thickness of clay and T_v values for $(u_1/u_2) = 1$.

7.4 FOUNDATION ON SAND

Foundations on sand do not present the same degree of problem with regard to settlement as foundations on clay. Firstly, except for loose sand, settlement is usually small and secondly, because of high permeability the settlement is usually over during the period of construction. In this case, there is no time dependent settlement similar to the consolidation settlement in clayey soils, although large foundations may undergo minor settlement caused by fluctuations of load due to wind, machinery vibrations, and storage loads.

Prediction of settlement of foundations on sand is, by and lange, empirical although elastic theory has sometimes been used. The common methods used for this purpose are:

7.4.1 Elastic Theory

The problem is similar to calculation of immediate settlement of foundations on clay as given in Eq. (7.3). For a homogeneous, isotropic, elastic medium

$$\delta = \frac{q_n B}{E} (1 - v)^2 I_\rho \tag{7.25}$$

where the different terms have the same meaning as in Eq. (7.3). However, it is to be considered that because of volume change during load application, Poisson's ratio, v can no longer be taken as 0.5. The value of E and v should be determined from the laboratory triaxial tests for the appropriate relative density and confining pressure. The Poisson's ratio normally varies between 0.15 for coarse sand to 0.3 for fine to silty sand. The value of E can also be determined from the standard penetration resistance of the soil (refer Chapter 2, Table 2.7).

Elastic theory, however, gives only a very approximate estimate of settlement, because the assumption of homogeneity would be far from reality.

7.4.2 Semi-empirical Method (Buisman, 1948)

Buisman (1948) has given an empirical method of estimating the settlement of foundations on sand by numerical integration of the strains occurring in different strata, as shown in Fig. 7.8. Accordingly,

$$\delta = \sum \frac{2.3 p_o}{E} \log_{10} \frac{p_o + \Delta p}{p_o} \, dz \qquad (7.26)$$

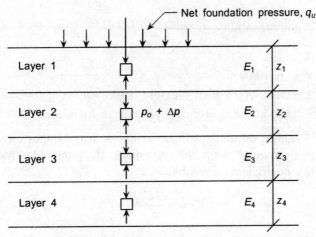

Fig. 7.8 Settlement calculation for foundations on sand (after Buisman, 1948).

where,

p_o = initial effective overburden pressure before loading,

Δp = increase of vertical stress at the centre of layer due to foundation loading, and

E = modulus of elasticity.

The sand layer is subdivided into a number of sublayers and the E values are taken from empirical relationship for the average N value for the layer, as described in Section 2.7.1 (Chapter 2).

7.4.3 Plate Load Test

Terzaghi and Peck (1948) gave an empirical relationship between the settlement, S_p of a loaded plate of diameter, D_p for a given intensity of pressure and the settlement, S_f of a foundation of diameter, D_f at the same pressure. Bjerrum and Eggestad (1963) modified this relationship as follows:

$$\frac{S_f}{S_p} = \frac{4}{(1 + D_p/D_f)^2} \qquad (7.27)$$

The use of this equation is based on the assumption that the density of the soil within the zone affected by the foundation loading is similar to that beneath the test plate. This may be approximately true if the plate diameter is of the same size as the foundation, but for large foundations, there is likely to be considerable difference in density.

7.4.4 Static Cone Test

If C_{kd} is the point resistance measured by the static cone test, the settlement of the foundation is given by the empirical relationship (Beer and Martens 1957),

$$\delta = \frac{2.3H}{C} \sum \log \frac{p_o + \Delta p}{p_o} \tag{7.28}$$

where, coefficient of compressibility, $C = \frac{3}{2} \frac{C_{kd}}{p_o}$

where,

C_{kd} = static cone resistance,

p_o = initial effective overburden pressure before applying foundation loading, and

Δp = vertical stress at the centre of layer due to foundation loading.

De Beer (1965) studied the settlement behaviour of fifty bridges on sand and concluded that Eq. (7.28) normally overestimates the settlement. He proposed a modified relationship for the compressibility coefficient, given by

$$C = \frac{1.9 C_{kd}}{p_o} \tag{7.29}$$

The cone penetration data are plotted with depth and the soil is divided into a number of layers and each layer is assigned an average C_{kd} value for obtaining the coefficient of compressibility.

Example 7.1

Figure 7.9 shows an isolated column in a building frame with a column grid of 4 m × 4 m which carries a vertical load of 400 kN and is supported on a footing, 2 m × 2 m, placed 1 m below G. L. The subsoil consists of 5 m of firm desiccated silty clay (c_u = 50 kN/m², $C_c/1 + e_0$ = 0.06) followed by medium sand (N = 20). Calculate the settlement of the footing.

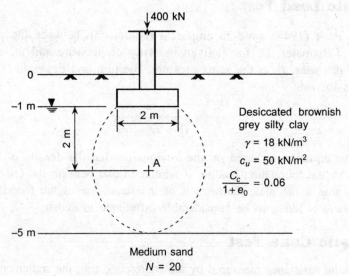

Fig. 7.9 Isolated footing on clay.

Solution

(a) *Immediate settlement*

$$\rho_i = \frac{q_n B}{E}(1 - v^2)I_\rho \quad \text{and} \quad q_n = \frac{400}{4} = 100 \text{ kN/m}^2$$

Also,

$$B = 2 \text{ m}$$
$$v = 0.5$$
$$E = 600 \, c_u$$
$$= 30,000 \text{ kN/m}^2$$
$$(I_\rho)_{\text{centre}} = 1.12$$

∴ At the centre of footing,

$$\rho_i = \frac{100 \times 2}{30,000} \times 0.75 \times 1.12$$

$$= 0.0056 \text{ m} = 5.6 \text{ mm}$$

$$\frac{L}{B} = 1.0 \quad \text{and} \quad \frac{D_f}{B} = 0.5$$

⇒ depth correction, $\alpha = 0.86$

∴ $(\rho_i)_{\text{corr}} = 0.86 \times 5.6 = 4.8$ mm

(b) *Consolidation settlement*

Take influence zone extending to twice the width of footing, that is, 4 m, below footing. Consider single layer within the zone of influence.

$$\therefore \quad \rho_{\text{oed}} = \frac{C_c}{1 + e_0} H \log \frac{p_0 + \Delta p}{p_0}$$

At A: $p_0 = 1 \times 18 + 8 \times 2 = 34 \text{ kN/m}^2$

$\Delta p = 0.34 \times 100 = 34 \text{ kN/m}^2$

$$\therefore \quad \rho_{\text{oed}} = 0.06 \times 4 \times \log \frac{34 + 34}{34}$$

$$= 0.072 \text{ m} = 72 \text{ mm}$$

Depth correction, $\alpha = 0.86$

∴ $(\rho_{\text{oed}})_{\text{corr}} = 0.86 \times 72 = 62$ mm

Skempton and Bjerrum's method

$$\rho_c = \mu (\rho_{\text{oed}})_{\text{corr}}$$

Pore-pressure correction factor, $\mu = 0.7$ (refer Fig. 7.4)

$$\rho_c = 0.7 \times 62 = 43.4 \text{ mm}$$

∴   Final settlement $= 4.8 + 43.4$
$$= 48.2 \text{ mm}$$

Stress-path method

$$\rho'_c = \lambda \rho_c$$

From Eq. (7.17),

$$K' = 1 - \left[\frac{1 - 2v}{1 + \eta(1 + \eta)(3A - 1)}\right]$$

Substituting $v = 0.3$ and $A = 0.7$,

$$K' = 1 - \left[\frac{1 - 2 \times 0.3}{1 + 0.3(1.3)(3 \times 0.7 - 1)}\right]$$

$$= 0.84$$

From Fig. 7.7, for undisturbed Calcutta soil

$$\lambda = 0.6$$

$$\therefore \quad \rho'_c = 0.6 \times 43.4 = 26 \text{ mm}$$

$$\therefore \quad \rho_f = 4.8 + 26 = 30.8 \text{ mm}$$

Example 7.2

A raft foundation, 8 m × 12 m in plan is to be placed 2 m below G. L. in the subsoil shown in Fig. 7.10. The net foundation pressure is 50 kN/m². Calculate the total settlement of the foundation

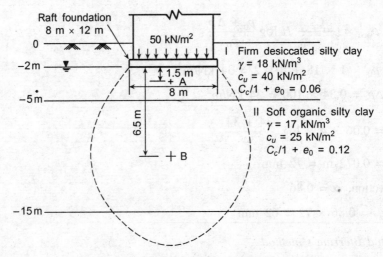

Fig. 7.10 Raft foundation on clay.

Take influence zone extending to twice the width of foundation, that is, 16 m below foundation. Neglect the effect of settlement due to sand layer.

Solution

(a) *Immediate settlement*

$$\rho_i = \frac{q_n B}{E} (1 - v^2) I_\rho$$

For the given problem,

$$q_n = 50 \text{ kN/m}^2$$
$$B = 8 \text{ m}$$
$$v = 0.5$$

I_ρ (for $L/B = 12/8 = 1.5$) = 1.36 (refer Table 7.1)

for, Stratum I: $E_1 = 600 \times 40 = 24{,}000 \text{ kN/m}^2$

Stratum II: $E_2 = 600 \times 25 = 15{,}000 \text{ kN/m}^2$

Weighted average, E

$$E = \frac{24{,}000 \times 3 + 15{,}000 \times 10}{13}$$

$$= 17{,}000 \text{ kN/m}^2$$

$$\therefore \qquad \rho_i = \frac{50 \times 8}{17{,}000} \times 0.75 \times 1.36$$

$$= 0.024 \text{ m} = 24 \text{ mm}$$

Depth correction = 1.0

Rigidity correction = 0.8 (Considering raft with interconnected beams providing rigidity to the foundation)

$$\therefore \quad (\rho_i)_{\text{corr}} = 0.8 \times 24 = 19 \text{ mm}$$

(b) *Consolidation settlement*

$$\rho_{\text{oed}} = \sum \frac{C_c}{1 + e_0} H \log \frac{p_o + \Delta p}{p_o}$$

At A, $p_o = 18 \times 2 + 8 \times 1.5 = 48 \text{ kN/m}^2$

$\Delta p = 0.82 \times 50 = 41 \text{ kN/m}^2$

At B, $p_o = 18 \times 2 + 8 \times 3 + 7 \times 5 = 95 \text{ kN/m}^2$

$\Delta p = 0.42 \times 50 = 21 \text{ kN/m}^2$

$$\therefore \qquad \rho_{\text{oed}} = 0.06 \times 3 \times \log \frac{48 + 41}{48} + 0.12 \times 10 \log \frac{95 + 21}{95}$$

$$= 0.048 + 0.104$$

$$= 0.152 \text{ m} = 152 \text{ mm}$$

Using, depth correction = 1.0

Rigidity correction = 0.8

∴ $(\rho_{oed})_{corr} = 0.8 \times 152 = 121$ mm

Skempton and Bjerrum method

$$\rho_c = \mu(\rho_{oed})_{corr}$$

Take $\mu = 0.8$ (settlement being predominantly in soft normally consolidated soil of stratum II)

∴ $\rho_c = 0.8 \times 121 = 97$ mm

∴ $\rho_f = 19 + 97 = 116$ mm

Stress-path method

$$\rho_c' = \lambda\rho_c$$

From Eq. (7.17)

$$K' = 1 - \left[\frac{1 - 2v}{1 + \eta(1 + \eta)(3A - 1)}\right]$$

Substituting $v = 0.3$ and $A = 0.8$

$$K' = 1 - \left[\frac{1 - 2 \times 0.3}{1 + 0.3(1.3)(3 \times 0.8 - 1)}\right]$$

$$= 0.84$$

From Fig. 7.7,

$$\lambda = 0.6$$

∴ $\rho_c' = 0.6 \times 97 = 58$ mm

∴ $\rho_f = 19 + 58 = 77$ mm

Example 7.3

A 5 m × 5 m foundation is placed 1 m below G. L. in the stratified sandy deposit as depicted in Fig. 7.11. Calculate the settlement of the foundation.

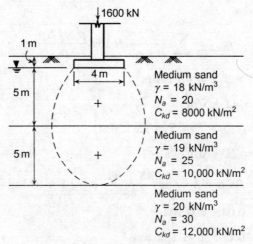

Fig. 7.11 Foundation on sand.

Solution

Net foundation pressure,

$$q_n = \frac{1600}{4 \times 4} = 100 \text{ kN/m}^2$$

(a) *Elastic theory*

$$\delta = \frac{q_n B}{E}(1 - v^2)I_\rho$$

Also,

$q_n = 100 \text{ kN/m}^2$
$B = 4 \text{ m}$
$v = 0.3 \text{ (assumed)}$
$I_\rho = 1.12$
$E = C_1 + C_2 N$ (Take average N value of the two layers having almost equal depth within the influence zone)
$= 39 + 4.5 \times 22.5$
$= 140 \text{ kg/cm}^2$
$= 14,000 \text{ kN/m}^2$

$$\therefore \quad \delta = \frac{100 \times 4}{14,000} \times 0.91 \times 1.12$$

$$= 0.029 \text{ m}$$

$$= 29 \text{ mm}$$

$$\frac{L}{B} = 1 \quad \text{and} \quad \frac{D_f}{B} = 0.25$$

$\therefore$ Depth correction, $\alpha = 0.97$

(b) *Buisman (1948) method*

Consider two layers within the influence zone

$$\delta = \sum \frac{2.3 p_o}{E} \log_{10} \frac{p_o + \Delta p}{p_o} \, dz$$

At A, $p_o = 1 \times 18 + 2.5 \times 9 = 40.5 \text{ kN/m}^2$
$\Delta p = 0.56 \times 100 = 56 \text{ kN/m}^2$
$E_1 = 39 + 4.5 \times 20 = 129 \text{ kg/cm}^2 = 12,900 \text{ kN/m}^2$

At B, $p_o = 1 \times 18 + 5.0 \times 8 + 2.5 \times 9 = 80.5 \text{ kN/m}^2$
$\Delta p = 0.11 \times 100 = 11 \text{ kN/m}^2$
$E_1 = 39 + 4.5 \times 25 = 151.5 \text{ kg/cm}^2 = 15,150 \text{ kN/m}^2$

$$\therefore \quad \delta = \frac{2.3 \times 40.5}{12,900} \times 5 \log_{10} \frac{40.5 + 56}{40.5} + \frac{2.3 \times 80.5}{15,150} \times 5 \log_{10} \frac{80.5 + 11}{80.5}$$

$$= 0.012 + 0.0034 = 0.0154 \text{ m} = 15.4 \text{ mm}$$

Depth correction = 0.97

∴ $(\delta)_{corr} = 0.97 \times 15.4 = 15$ mm

(c) *Beer and Martens (1957)*

$$\delta = \frac{2.3H}{C} \log_{10} \frac{p_o + \Delta p}{p_o}$$

Layer 1:

$H = 5$ m; $p_o = 40.5$ kN/m²; $\Delta p = 56$ kN/m²

$$\therefore \quad C = \frac{1.9 C_{kd}}{p_o}$$

$$= \frac{1.9 \times 8000}{40.5} = 375$$

Layer 2:

$H = 5$ m; $p_o = 80.5$ kN/m²; $\Delta p = 11$ kN/m²

$$\therefore \quad C = \frac{1.9 \times 10{,}000}{80.5} = 236$$

$$\therefore \quad \delta = \frac{2.3 \times 5}{12{,}900} \times 5 \log_{10} \frac{40.5 + 56}{40.5} + \frac{2.3 \times 5}{15{,}150} \times 5 \log_{10} \frac{80.5 + 11}{80.5}$$

$$= 0.017 + 0.002 = 0.019 \text{ m} = 19 \text{ mm}$$

Depth correction = 0.97

∴ $(\delta)_{corr} = 19 \times 0.97 \approx 18$ mm

Example 7.4

The results of a plate load test on a sandy stratum are shown in Fig. 7.12. The size of the plate used is 30 cm × 30 cm. Determine the size of a square footing for a column carrying a load of 1800 kN with a maximum permissible settlement of 50 mm.

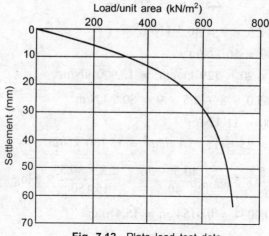

Fig. 7.12 Plate load test data.

Solution

The problem has to be solved by trial and error.

First trial

Size of footing = 3 m × 3 m

$$q_n = 1800/9 = 200 \text{ kN/m}^2$$

From Eq. (7.27)

$$\frac{S_f}{S_p} = \frac{4}{\left(1 + \dfrac{D_p}{D_f}\right)^2}$$

$$\therefore \quad S_p = \frac{S_f}{4}\left(1 + \frac{D_p}{D_f}\right)^2$$

$$= \frac{50}{4}\left(1 + \frac{300}{3000}\right)^2$$

$$= 15 \text{ mm}$$

∴ Allowable pressure (From curve) = 440 kN/m²

Second trial

Size of footing = 2 m × 2 m

$$q_n = \frac{1800}{2 \times 2} = 450 \text{ kN/m}^2$$

Again, $S_p = \dfrac{50}{4}\left(1 + \dfrac{300}{2000}\right)^2$

$$= 16.5 \text{ mm}$$

∴ Allowable pressure = 480 kN/m²

REFERENCES

Bjerrum, L. (1964), *Relasjon Mellom Malte og Beregnede Setninger a v Byggverk pa Leire og Sand*, N.G.F. Foredraget, 1964. Norwegian Geotechnical Institute, Oslo 92 pp.

Bjerrum L. (1963), *Discussion on Section 6. European Conference S.M.F.E.*, Wiesbaden, Vol. 2, pp. 135–137.

Butler, F.G. (1974), *General Report and State-of-the-Art Review*, Session 3: Heavily Overconsolidated Clays, Proceedings of Conference on Settlement of Structures, Cambridge.

Das, S.C. (1975), *Predicted and Measured Values of Stress and Displacements Downloading and Construction under Circular Footings Resting on Saturated Clay Medium*, Ph.D. Thesis, Jadavpur University, Calcutta.

Das, S.C. and N.N. Som (1976), *Settlement of Footings on Soft Clay*, Proc. Symp. on Foundations and Excavation in Weak Soil, Calcutta, Vol. 1.

Davis, E.H. and H.G. Poulos (1968), The Use of Elastic Theory for Settlement Prediction under Three-dimensional Conditions, *Geotechnique,* Vol. 18.

De Beer, E.E. (1965), *Bearing Capacity and Settlement of Shallow Foundations on Sand*, Proc. Symposium on Bearing Capacity and Settlement of Foundations, Duke University, pp. 15–33.

De Beer, E. and A. Martens (1957), *Method of Computation of an Upper Limit for the Influence of Heterogeneity of Sand Layers in Settlement of Bridges*, 4th I.C.S.M.F.E., London, Vol. 1, pp. 275–281.

Fox, E.N. (1948), *The Mean Elastic Settlement of Uniformly Loaded Area at a Depth below Ground Surface*, 2nd I.C.S.M.F.E., Rotterdam, Vol. 1, p. 129.

Guha, S. (1979), *Effect of Initial and Incremental Strain Ratios or Drained Deformation of Calcutta Soil*, MCE Thesis, Jadavpur University, Calcutta.

Ladd, C.C. (1969), *The Prediction of in-situ Stress–Strain Behaviour of Soft Saturated Clays during Undrained Shear*, Bolkesj φ Symposium on Shear Strength and Consolidation of Normally Consolidated Clays, Norwegian Geotechnical Institute, Oslo, pp.14–19.

Lambe, T.W. (1964), Methods of Estimating Settlement, *Journal of Soil Mechanics and Foundation Division,* A.S.C.E., 90, No. SM5, pp. 43–67.

Raymond, G.P., D.L. Townsend, and M.J. Lojkasek (1971), The Effect of Sampling on the Undrained Soil Properties of a Leda Soil, *Canadian Geotechnical Journal,* Vol. 8, pp. 546–557.

Simons, N.E. (1957), *Settlement Studies on Two Structures in Norway*, 4th I.C.S.M.F.E., London, Vol. 1, pp. 431–436.

Simons, N.E. and N.N. Som (1970), *Settlement of Structures on Clay with Particular Emphasis on London Clay*, Construction Industry Research and Information Association Report 22, p. 51.

Skempton, A.W. (1954), The Pore-pressure Coefficients A and B, *Geotechnique,* Vol. 4, pp. 143–147.

Skempton, A.W., R.B. Peck, and D.H. McDonald (1955), *Settlement Analyses of Six Structures in Chicago and London*, Proc. I.C.E., Part 1, Vol. 4, No. 4, p. 525.

Skempton, A.W. and L. Bjerrum (1957), A Contribution to the Settlement Analysis of Foundations on Clay, *Geotechnique,* Vol. 7, No. 4, pp. 168–178.

Som, N.N. (1968), *The Effect of Stress Path on the Deformation and Consolidation of London Clay*, Ph.D. Thesis, University of London.

Som, N.N. (1975), *Regional Deposits: Contribution for Panel Discussion*, Proc. 5th Asian Regional Conf. in SMFE, Vol. 2, p. 20.

Taylor, D.W. (1948), *Fundamentals of Soil Mechanics*, John Wiley and Sons, Inc, New York.

Terzaghi, K. (1943), *Theoretical Soil Mechanics*, John Wiley and Sons Ltd., p. 510.

Terzaghi, K. and R.B. Peck (1948), *Soil Mechanics in Engineering Practice*, 1st ed., Wiley, New York.

8 Footings and Raft Design

8.1 INTRODUCTION

Footings and rafts are the common shallow foundations wherein the subsoil immediately beneath the ground surface is required to support the foundation load. The superstructure load from a building is transmitted to the ground through columns and load-bearing walls and suitable structural members are to be provided to transfer the load into the subsoil. These structural members which are placed at some depth beneath the ground surface serve to 'spread' the load from high stress intensity of the superstructure material (concrete, steel, or brick) to the low allowable stresses in the soil mass.

The different types footings and rafts have been illustrated in Chapter 4. They are generally made of reinforced concrete. The reinforcements are provided to resist the bending moment on the footing caused by the upward soil reaction either in one direction as in the case of strip footings or in two directions as in case of isolated or combined footings, refer Fig. 8.1. The section of the footings may be sloped to economize on the volume of concrete. Heavily loaded structural steel columns are sometimes provided with grillage foundations.

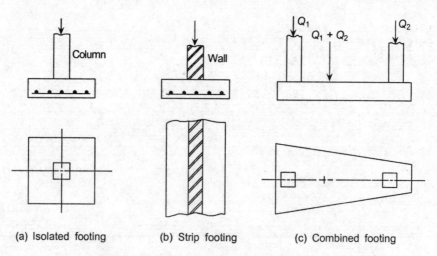

Fig. 8.1 Isolated, strip, and combined footings.

The design of shallow foundation involves determination of:

1. depth of foundation,
2. allowable bearing pressure and size of foundation, and
3. settlement of the foundation.

8.2 DESIGN OF FOOTINGS

8.2.1 Depth of Footing

The minimum depth at which a foundation should be placed depends on the soil profile, structural requirement, ground water condition, and so on. The following factors should generally be taken into consideration in determining the depth of foundations.

(a) Depth of top soil, rubbish fill, if any.
(b) Depth of poor surface deposit such as peat, muck, or sanitary land fill.
(c) Location of ground water table and its seasonal fluctuation.
(d) Depth to poor or better underlying strata.
(e) Depth of adjacent footings if any.

If the subsoil near the ground surface consists of a heterogeneous fill of uncertain properties or compressible soil like peat, muck etc. the foundation should, preferably, be taken below the fill. Figure 8.2 is a clear representation of the same. It is also desirable that the foundation be taken below the zone of seasonal fluctuation of water table where the soil is not subjected to seasonal volume changes due to alternate wetting and drying. For this, the foundation is generally placed 1–2 m below ground surface. However, if the water table fluctuates over a great depth, it may not be economical to take the foundation to the desired level. The foundation may then be placed at some convenient depth and the design done for the highest position of the water table. The depth of foundation is also influenced by the location of underground facilities below the building. In case where a stiff clay is underlain by a softer material, the foundation should be placed as high above as possible so that the pressure bulb may be restricted within the stiff clay.

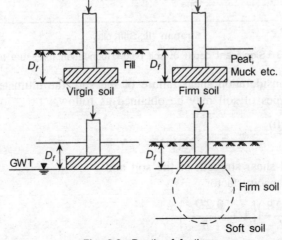

Fig. 8.2 Depth of footing.

In sandy strata, if there is marked increase in density of sand with increasing depth, it may be tempting to take the foundation deeper than normal in order to take advantage of the higher bearing capacity. This procedure may not be economical if it involves deep excavation below the water table involving large scale dewatering. Further, there may be problems with the stability of side slopes and adjacent structures.

8.2.2 Allowable Bearing Capacity

The net allowable bearing pressure on footings should be determined from considerations of bearing capacity and settlement, so as to satisfy the required design criteria. The methods of bearing capacity and settlement analysis have already been described in Chapters 6 and 7.

First, it is necessary to establish the depth of soil which is significantly affected by the foundation loading. As a general rule, this depth may be taken as twice the width of footing. All the strata contained within the significant depth of soil are to be considered in the design. The relevant soil parameters for each stratum need to be evaluated from appropriate soil tests. Figure 8.3 shows typical examples of the significant depth of soil for different size of footing.

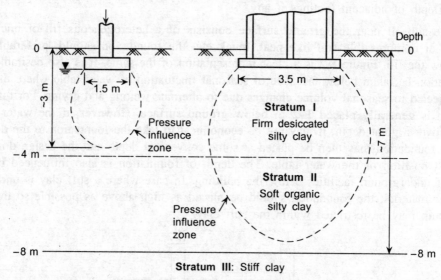

Fig. 8.3 Significant depth of soil below foundation (pressure bulb).

Keeping under consideration shear failure of the soil, the ultimate bearing capacity of footings on different types of soil may be obtained as follows:

(a) *Cohesive soil* ($\phi = 0$)

$$q_{ult}(n) = c_u N_c \tag{8.1}$$

where, c_u = undrained shear strength of the soil and
N_c = bearing capacity factor

$$= 5\left(1 + \frac{0.2B}{L}\right)\left(1 + \frac{0.2D_f}{B}\right)$$

(B and L are the width and length of footing)

The shear strength of the soil within the failure zone below the footing, that is, within a depth approximately equal to the width of footing, should be used in design. In view of the natural variation of soil properties at a building site, there is usually a wide spectrum of test results and the designer is required to exercise his judgment in selecting the design shear strength. For a stratified deposit, a weighted average shear strength may be used in design. However, it is to be realized that if the major portion of the failure surface is contained in stratum I, the bearing capacity of the foundation would be determined primarily by the undrained shear strength of stratum I. Figure 8.4 depicts a failure surface in stratified soil.

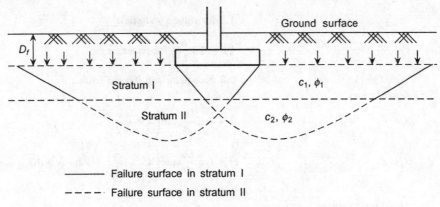

Fig. 8.4 Failure surface in stratified soil.

(b) *Granular soils* ($c = 0$)

$$q_{ult}(n) = q(N_q - 1) + 0.5s\gamma'BN_\gamma \qquad (8.2)$$

where,

q = effective surcharge at the foundation level,

γ' = effective unit weight of the soil below the foundation,

B = width of foundation,

N_q, N_γ = bearing capacity factors (see Chapter 6), and

s = shape factor (0.5–1.0).

(c) *Soil having both c and φ*

The general bearing capacity equation given in I.S. 6403—1981 may be used to determine the bearing capacity of soils having both c and ϕ'.

$$q_n = cN_cs_cd_ci_c + q(N_q - 1)s_qd_qi_q + 0.5\gamma BN_\gamma s_\gamma d_\gamma i_\gamma \qquad (8.3)$$

The shape, depth, and inclination factors to be used in Eq. (8.3) are given in Table 8.1. The effect of water level should be taken into account in determining the values of q and γ'. N_c, N_q, and N_γ are functions of the angle of shearing resistance of the soil obtained from laboratory tests or from empirical correlation with the field standard penetration resistance, N (for details, see Chapter 6).

Table 8.1 Shape, depth, and inclination factors as per IS:6403—1981

Factors	Value
Shape factors	
s_c	$\left(1 + 0.2\dfrac{B}{L}\right)$ for rectangle
	1.3 for square and circle
s_q	$\left(1 + 0.2\dfrac{B}{L}\right)$ for rectangle
	1.2 for square and circle
s_γ	$\left(1 - 0.4\dfrac{B}{L}\right)$ for rectangle
	0.8 for square and 0.6 for circle
depth factors	
d_c	$1 + 0.2\dfrac{D_f}{B}\tan\left(45° + \dfrac{\phi}{2}\right)$
$d_q = d_\gamma$	$1 + 0.1\dfrac{D_f}{B}\tan\left(45° + \dfrac{\phi}{2}\right)$, for $\phi > 10°$
	1 for $\phi < 10°$
Inclination factors	
$i_c = i_q$	$\left(1 - \dfrac{\alpha}{90}\right)$
i_γ	$\left(1 - \dfrac{\alpha}{90}\right)$ α in degrees

From the ultimate bearing capacity, the net allowable bearing capacity of a footing, $q_{all}(n)$, is obtained as

$$q_{all}(n) = \frac{q_{ult}(n)}{F} \tag{8.4}$$

where, F = factor of safety (normally 2.5–3.0).

8.2.3 Effect of Ground Water Table

The effect of ground water table on the bearing capacity of shallow foundations has been discussed in Chapter 6.

The effective surcharge, q at the foundation level is to be determined by using the appropriate density (bulk or submerged) of the soil above the foundation level. For ground water table coinciding with the ground surface, submerged density is to be used for the full depth of soil above foundation level. γ' on the other hand, is the effective unit weight of the soil below the foundation. As the failure zone extends to a depth approximately equal to the width of footing, the effective unit weight within this zone is to be considered. For the water table at or above the foundation level, submerged density should be used while for the water table at a depth greater than B below the footing, bulk density is to be used. For any intermediate position of the water table, the effective unit weight may be obtained, from interpolation between γ and γ', as shown in Fig. 8.5.

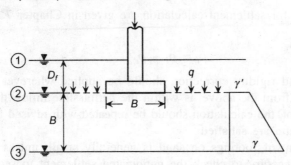

Fig. 8.5 Effect of water table on bearing capacity.

8.2.4 Settlement

Trial dimensions are chosen to accommodate the required footing as given by the allowable bearing capacity of the footing. The settlement of the footing is then estimated for the actual net foundation pressure, q_n given by,

$$q_n = \frac{Q}{\text{Area of footing}} \tag{8.5}$$

where Q = net load on the footing.

For footings on clay,

(a) *Immediate settlement*

$$\rho_i = \frac{q_n B}{E} (1 - v^2) I_\rho \tag{8.6}$$

where, q_n = net foundation pressure,
$\quad B$ = width of foundation,
$\quad E$ = modulus of elasticity of the soil within the zone of influence,
$\quad I_\rho$ = influence coefficient, and
$\quad v$ = Poisson's ratio.

(b) *Consolidation settlement*

$$\rho_c = \mu \sum \frac{C_c}{1 + e_0} H \log \frac{p_o + \Delta p}{p_o} \tag{8.7}$$

where, $C_c/(1 + e_0)$ = compressibility index of the clay,
$\quad H$ = thickness of compressible stratum,
$\quad p_o$ = in-situ effective overburden pressure at centre of stratum,
$\quad \Delta p$ = increase of vertical stress at the centre of stratum due to the foundation loading, and
$\quad \mu$ = pore-pressure correction factor.

If the coefficient of volume compressibility, m_v is used, the consolidation settlement is given by

$$\rho_c = \mu \sum m_v \Delta pH \tag{8.8}$$

The detailed procedures for settlement calculation are given in Chapter 7.

(c) *Total settlement*

$$\rho_f = \rho_i + \rho_c \tag{8.9}$$

The depth correction and rigidity correction should be applied wherever applicable. If ·the settlement, as estimated from the above, is within the permissible limits, the foundation may be considered final. If not, the calculation should be repeated with revised footing dimensions until the settlement criteria are satisfied.

Although settlement of footings on sand is generally small, unless the sand is loose ($N < 10$)—it may be necessary to check the anticipated settlement to see if it exceeds the tolerable limits. As has been discussed in Chapter 7, the semi-empirical methods would be preferable to any theoretical analysis (See subsection 7.4.2).

Some complications arise when two different kinds of soil (say, sand and clay or vice versa) fall within the influence zone. In such cases, the settlement of the sand layer can be obtained separately and added to the settlement caused by the clay strata. There is usually no problem in determining the consolidation settlement of the relevant depth of clay strata using appropriate expressions. However, if the clay stratum is of lesser thickness than the depth of pressure bulb, the immediate settlement due to the clay layer may be overestimated. The designer may apply his judgment in evaluating the immediate settlement of individual stratum giving due importance to the relative contribution of each stratum.

8.2.5 Dimensioning Footing Foundations

(a) Isolated footings

As a general rule, square footings give the most economic structural design for a given column load. In case of space restriction, it may be necessary to go for rectangular footings, as shown in Fig. 8.6.

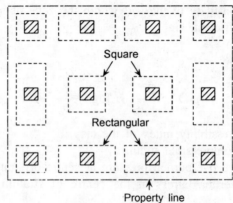

Fig. 8.6 Square and rectangular footings.

(b) Combined footings

Two or more columns in a building frame may be supported on a single footing, either rectangular or trapezoidal when isolated footings tend to overlap or extend beyond the property line. If the allowable bearing pressure is known, the size of the footing may be determined for a permissible settlement. In order to get uniform bearing pressure, the footing should be so proportioned that the resultant column loads pass through the centroid of the footing area. Figure 8.7 shows typical combined footings of rectangular and trapezoidal shape.

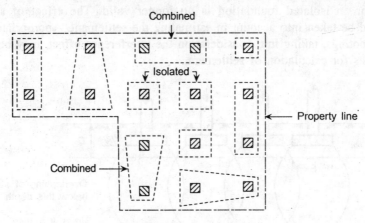

Fig. 8.7 Combined footings.

(c) Strip footings

Strip footings are provided below load bearing walls or a row of columns in building frame. Even if the column loads vary a little, the longitudinal extent of the footing normally ensures a uniformly distributed load below the footing. A pedestal may be provided if a singly reinforced strip footing along the width is desired below closely spaced columns. This is depicted in Fig. 8.8.

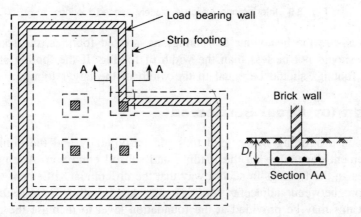

Fig. 8.8 Strip footings.

8.2.6 Interference effect

When a number of closely spaced columns are required to be supported on isolated footings, the pressure bulbs under individual footings tend to overlap, as shown in Fig. 8.9. According to Boussinesq stress analysis, the pressure bulb below a footing extends to a distance of $B/2$ on either side of the footing (where B is the width of footing). If two adjacent footings come to within a distance B from each other, the pressure bulbs overlap and the soil below the depth, D behaves as if it is loaded by a combined footing. The settlement of the footing as determined for an isolated foundation is no longer valid. The effect of superposition of stresses should be taken into account in estimating the settlement. Appropriate stress analysis for adjacent footings, taking into consideration the interference effect, may be done to obtain the soil stresses for calculation of settlement.

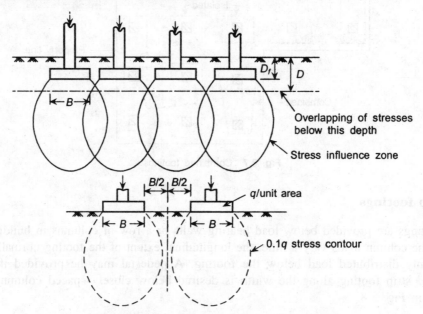

Fig. 8.9 Interference effect in closely spaced footings.

In order to ensure the behaviour of footings as isolated footings, the spacing between adjacent footings should not be less than the width of footing. If the footing size varies, the spacing between footings should be equal to the width of the larger footing.

8.2.7 Design for Equal Settlement

Isolated footings in a building frame tend to settle individually and the columns undergo differential settlement depending on the loading and subsoil condition. Attempts should be made either to design the footings in such a way that the differential settlement and hence, the angular distortion between adjacent footings are kept within permissible limits or interconnected beams may be provided at the foundation level to increase the rigidity of the foundation.

The basic factors which affect the settlement of footings on a given soil are the intensity of loading and the size of the loaded area. The settlement increases in almost direct proportion to these parameters. It is a common experience that even though the foundation pressure is the same under all footings, a bowl shaped deformation trough is obtained with greater settlement at the centre than along the edges of the building mainly due to greater load in the central columns and the overlapping of stresses from adjacent footings. Hence, to produce uniform settlement, it may be necessary to adjust the pressure with size of footing, that is, to impose greater pressure under smaller footings than under larger ones and also to use larger pressure under the footings along the edge of the building. The principle was utilized in the CB1 Esplanada Building, Sao Paulo which was founded on moderately compact fine to medium sand. A pressure of 550 kN/m² was used under the outer footings and 400 kN/m² under the inner footings, with the result that the maximum differential settlement did not exceed 8 mm (Skempton, 1955). So, the bearing pressure of footings in a building frame may be varied though it is always kept within the safe bearing capacity so as to have the maximum and differential settlement within permissible limits. This may require a number of trials before the final design is arrived at. A simple calculation for achieving uniform settlement of adjacent footings is shown in Example 8.4.

A satisfactory method of reducing the differential settlement of footings in a building frame is to provide interconnected beams between columns at the foundation level. A framed structure is tied laterally, from first floor onwards, by beams and slabs but no such connection is generally provided at the foundation level. Consequently, the individual footings can settle differentially. Providing interconnected beams increases the rigidity of the foundation and differential settlement is substantially reduced. An approximate way of analysis would be to estimate the settlement of individual footings as isolated/strip footings and then to design interconnected beams to resist the bending moment at the joints due to the differential settlement. For more accurate design, theoretical analysis of the building frame may be done with predetermined sections of the interconnected beams and footings. Figure 8.10 shows a typical strip foundation with interconnected beams.

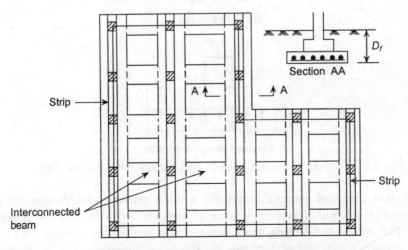

Fig. 8.10 Strip footings with interconnected beams.

8.2.8 Structural Design of Footings

Structural designing of footings is mostly done by the conventional 'rigid' method. Here, the footing is assumed to be infinitely rigid so that the displacement of the footing does not affect the pressure distribution. However, in actual practice, the pressure distribution below a footing depends on the footing rigidity and the soil type. In loose sand, the soil near the edges of the footing tends to displace laterally whereas the soil towards the inside remains confined. Cohesive soil has high shear stress concentration near the edges which leads to local yielding even at high factor of safety against bearing capacity failure. Bowles (1988) indicates the probable pressure distribution beneath rigid footings for different soil types, refer Fig. 8.11.

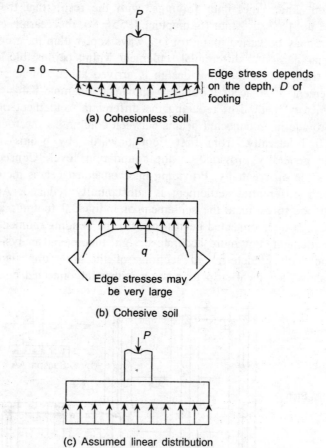

Fig. 8.11   Pressure distribution beneath rigid footings (after Bowles 1988).

However, it is common practice to assume a linear pressure distribution for practical design. The centroid of soil pressure is made to concide with the line of action of the resultant load. For an isolated footing, the column centre line passes through the centroid of the footing area while for combined footings, the shape and dimensions of the footing are to be determined to achieve this condition. Once the pressure distribution has been obtained for the chosen geometry of the footing, the structural design of the footing is made to resist the shear forces and bending movement for the appropriate support conditions. For isolated footings, the cantilever bending moment at the column face will normally determine the amount of reinforcement to be provided. In case of combined footings, the design is a simply supported beam with overhang on both sides while a strip footing is designed as a continuous beam with support at the column locations.

8.3 DESIGN OF RAFT FOUNDATION

A raft foundation usually covers the entire area of the building, thereby distributing the total load to a larger area than a footing foundation and reduces the bearing pressure to a minimum. The choice between a raft and a footing foundation depends on the soil properties and the weight of the building. If a preliminary design with footing foundations reveals that the sum of the footing areas required to support the structure exceeds 60% of the total building area, a raft foundation covering the entire area of the building should be preferred. Moreover, where the soil properties vary largely throughout the site, the amount of differential settlement may be excessive for a footing foundation, but with a raft foundation the effect of weak zones scattered at random tend to even out. Therefore, the settlement pattern is less erratic and the differential settlement is also reduced considerably. Also, a raft foundation provides increased rigidity which reduces the differential movement of the superstructure.

8.3.1 Types of Raft Foundation

Two types of raft foundation are in common use which form a subject matter of discussion in this subsection.

(a) Conventional raft

In conventional raft (refer Chapter 4, Fig. 4.7) foundation, a flat concrete slab of uniform thickness covering the entire area of a building is placed at some depth beneath the ground surface and the columns are built directly on the raft. Then backfilling is done upto the ground level. This is followed by further filling, called the plinth filling, on which the ground floor slab is placed. The ground floor load, therefore, goes to the raft by direct bearing through the plinth and backfill.

Figure 8.12 shows some typical raft foundations. The columns may be placed directly on the slab or the slab may be thickened under large column loads to give sufficient strength against shear and support moment. Rigidity of the raft foundation may be increased by cellular construction or rigid beam and slab construction. Figure 8.13 shows a simple raft foundation for the fourteen-storeyed block of flats at Golden Lane, London (Skempton, 1955).

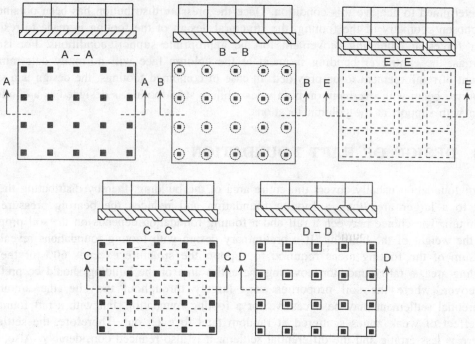

Fig. 8.12 Typical raft foundations.

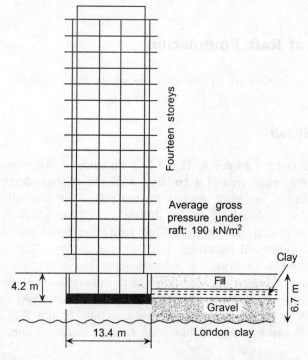

Fig. 8.13 Raft foundation for Golden Lane flats, London (Skempton, 1955).

(b) Buoyancy raft foundation

With the increased number of tall buildings now being built for various commercial and residential purposes, raft foundations have been coveniently combined with the principle of floatation. Buoyancy rafts (refer Chapter 4, Fig. 4.8), with or without basement, are placed at some depth beneath the ground surface, but no backfilling is done on the raft. A structural slab is made, usually a little above the ground level to serve as the ground floor of the building. The space between the ground floor slab and the raft is kept void such that the net foundation pressure gets reduced and remains within the permissible bearing pressure on the foundation.

That is,
$$q_{net} = q_{gross} - \gamma D_f \tag{8.10}$$
$$< q_{all}(n)$$

where, q_{gross} = gross foundation pressure,
γD_f = total overburden pressure at the foundation level, and
$q_{all}(n)$ = net permissible bearing pressure.

When a buoyancy raft has sufficient head room below the ground floor slab, the space may be utilized as a basement floor below the ground. Although a basement is, in effect, a buoyancy raft, it is not necessarily designed for the purpose. The main function of a basement is to provide additional floor space in the building and the fact that it reduces the net bearing pressure may be quite incidental. A buoyancy raft may, however, be designed solely for the purpose of providing support to the structure and to reduce the total and differential settlements within permissible limits.

8.3.2 Bearing Capacity

The mechanism of bearing capacity failure of raft foundation is similar to that of footings, except that the former involves a much bigger and deeper zone of failure.

The net ultimate bearing capacity of a raft foundation $q_{ult}(n)$ is, therefore, given by the same expressions as or footings, as illustrated by Eqs. (8.1)–(8.3). However, raft foundations being much larger than footings, the pressure bulb as well as the failure surface extend to considerable depth below the foundation. A typical raft foundation on normal Calcutta soil is shown in Fig. 8.14. As seen, more than one stratum with different shear parameters are involved. A weighted average strength or the strength or the most predominant stratum within the failure zone may be used to determine the bearing capacity.

8.3.3 Settlement

The immediate and consolidation settlements of raft foundations on clay are obtained in the same manner as for footings, see Eqs. (8.5) and (8.6). The compression of all the strata within the significant depth below the foundation are to be calculated separately and the summation of all these settlements will give the total settlement of the foundation. Use of Eq. (8.6) to calculate the immediate settlement presumes that the subsoil behaves as homogeneous, isotropic, elastic medium. In case of non-homogeneous/stratified deposit, this is somewhat approximate, but is often accepted for practical purposes. More rigorous analysis, taking into account the variation of subsoil properties, can also be done. Depth correction is not significant for raft foundations because of small D_f/B ratio but rigidity correction is generally applicable. Pore-pressure correction is to be applied giving due weightage to the contribution of the most significant layers in the settlement calculation.

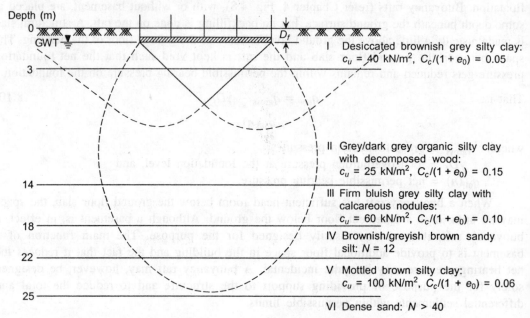

Fig. 8.14 Bearing capacity of raft foundation.

8.3.4 Floating Foundation

A *floating foundation* is a particular case of buoyancy raft foundation where the net foundation pressure is zero, that is

$$q_{net} = q_{gross} - \gamma D_f \qquad (8.11)$$
$$= 0$$

This condition can be achieved by excavating the soil to such a depth that the weight of soil removed is equal to the weight of the building, including that of the substructure. This principle can be adopted to construct even multistorey buildings on very soft clay by taking the foundation to the desired depth and providing one or more basement floors below the ground surface.

Though examples of fully floating foundation (i.e., zero net pressure) are often found, it is only in exceptional cases that such a foundation is necessary. An excellent example of almost a fully buoyant foundation is the New England Mutual Life Insurance Building in Boston, Massachussets (Casagrande and Fadum, 1942). The building covers an area of 104 m × 61 m and consists of one 10-storey section with two 2-storey and two 4-storey sections to be supported on a soil containing 21 m of very soft Boston blue clay. The construction of a 10.7 m basement has been found to be extremely successful even with such differential loading. Out of a total settlement of 62.5 mm (including the resettlement of the heaved soil), more than 35 mm occurred during construction of foundation. The long-term settlements due to consolidation of the clay were reduced as can be seen from the very flat slope of the settlement curve in Fig. 8.15.

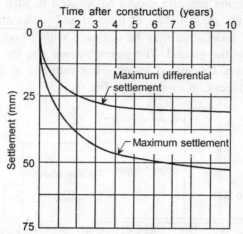

Fig. 8.15 Settlement of Liberty Mutual Life Insurance Building, Boston (Casagrande and Fadum, 1942).

In the high Paddington project, a thirty four-storey building was built on overconsolidated London clay by providing a 18 m basement which reduced the foundation pressure to 100 kN/m^2 while the gross pressure remained 425 kN/m^2. This is depicted through Fig. 8.16. The construction of a box type foundation provides extra stiffness to the structure which reduces the chance of excessive differential settlement (Skempton, 1955).

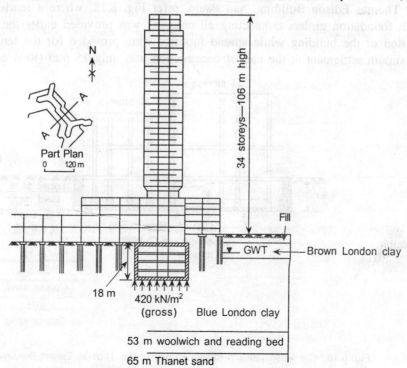

Fig. 8.16 Buoyancy raft foundation for High Paddington Project (Skempton, 1955).

Glossop (1972) indicated two types of floating foundations depending on whether floatation is required to reduce the settlement or to prevent shear failure of the subsoil. In either case, the net foundation pressure should be reduced to such an extent that the design criteria with respect to both, shear failure of the soil and settlement, are satisfied.

Increased rigidity of foundation may be achieved with a buoyancy raft by providing a cellular construction below the ground. The superstructure and the cellular raft provided for the Cossipore Power Station, Calcutta is an excellent example of a near floating foundation on soft normal Calcutta deposit. It is shown in Fig. 8.17.

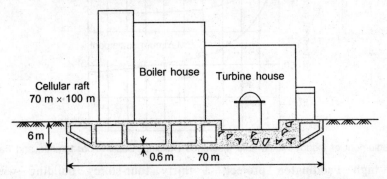

Fig. 8.17 Cellular raft foundation for Cossipore Power Station, Calcutta (Skempton, 1955).

When the height of a building varies, it is possible to combine rafts under the heavier part with footings under the lighter part of the building. A good example of this is found in the Thomas Edison Building, Sao Paulo, refer Fig. 8.18, where a reinforced concrete mat with foundation girders connecting all columns was provided under the twenty one-storey section of the building while spread footings were provided for the ten-storey block. The maximum settlement at the end of construction was only 15 mm (Rios and P. Silva, 1948).

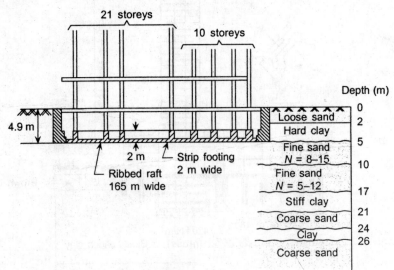

Fig. 8.18 Combined raft and footing foundation for Thomas Edison Building, Sao Paulo (Rios and P. Silva, 1948).

8.3.5 Basement Raft

Buoyancy raft foundation can be conveniently used to provide basements for multistorey buildings in soft clay. This gives additional floor space to a building but requires no additional ground coverage. Depending on the depth of foundation multiple basements may be accommodated. Figure 8.19 depicts Albany telephone building in New York which was founded on a 7 m deep buoyancy raft with two basement floors. The subsoil consisted of sensitive plastic clay. But the net foundation pressure of 40 kN/m² restricted the settlement of the building to only 60 mm (Casagrande 1932, Glossop 1972).

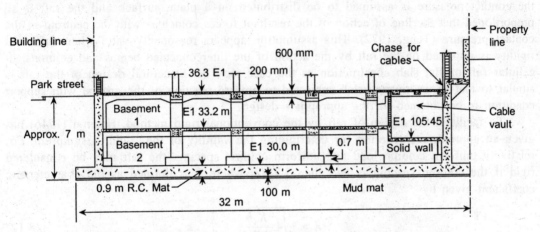

Fig. 8.19 Multiple basement for Albany telephone building (Glossop, 1972).

Analysis of buoyancy raft foundation in normal Calcutta soil gives some interesting observations. For a typical 10 m × 20 m raft, the net foundation pressure is to be restricted to 40 kN/m² to keep the settlement within the permissible limit of 125 mm (IS 1904 : 1966). Accordingly, the depth of foundation for different gross foundation pressures can be calculated. Table 8.2 gives the relevant data. It is clear that one, two, or three levels of basement may be provided within the depth of foundation to restrict the net foundation pressure to 40 kN/m² for eight, twelve, and sixteen-storey buildings respectively. That means, four upper storeys can be built for each basement floor for buildings beyond four storeys high.

Table 8.2 Depth of basement in normal Calcutta soil

q_{gross} (t/m²)	D_f (m)	q_{net} (t/m²)	No. of storeys	No. of basement floors
4.0	—	4.0	4	—
10.0	3.3	4.1	8	1
16.0	6.6	4.1	12	2
22.0	10.0	4.0	16	3

$$q_{net} = q_{gross} - \gamma D_f$$
$$= q_{gross} - 1.8 D_f$$

8.3.6 Structural Design of Raft Foundation

A raft foundation can be structurally designed in the same way as the reinforced concrete floors of a framed building. However, in a foundation raft, the floor is inverted so that its dead weight can be subtracted from the upward soil reaction.

(a) Rigid method

Raft foundations are mostly designed by the rigid method. This assumes the raft to be fully rigid and ensures that the deflection of the raft does not affect the soil contact pressure. Also, the contact pressure is assumed to be distributed on a plane surface and the raft is so proportioned that the line of action of the resultant forces coincides with the centroid of the contact pressure (Teng, 1972). This assumption appears reasonably valid when sufficient rigidity is imparted on the raft by the action of the interconnected beams and columns, or cellular raft or flat slab construction as the case may be. Structural design of the raft is similar to that of an inverted slab subjected to upward reaction of the ground with support condition to be obtained for the appropriate design.

To facilitate the design of raft by the conventional rigid method, Hetenyi (1946) has given an approximate criterion for determining the validity of the 'rigid' assumption. For relatively uniform column load and uniform column spacing, the raft may be considered rigid if the column spacing is less than $1.75/\lambda$, where λ is defined as the characteristic coefficient given by

$$\lambda = \sqrt{4 \frac{K_s b}{4 E_c I}}$$

(8.12)

where,

K_s = coefficient of subgrade reaction (pressure required to cause unit displacement),

b = width of a strip of raft between centres of adjacent bays,

E_c = modulus of elasticity of concrete, and

I = moment of inertia of the strip of width b.

(b) Elastic method

In the simplified elastic method, the soil is assumed to consist of infinite number of elastic springs each acting independently. The Winkler model is considered valid. The springs are assumed to be able to resist both tension and compression while the soil deformation is considered proportional to the pressure. The elastic constant of the springs is determined by the coefficient of subgrade reaction of the soil (defined as the unit pressure required to produce unit settlement). The analysis is carried out following the concept of beam on elastic foundation, the rigorous analysis for which was done by Hetenyi (1946). The raft is considered as a plate and the column loads are distributed in the surrounding areas in the zone of influence. The bending moment and shear force are calculated for each column point and the data are superimposed to obtain the moment and shear for the raft foundation. American Concrete Institute has recommended a procedure for design (ACI 1966, 1988). Computer programs are now available to carry out the complex mathematical calculations for practical use. Finite element formulation can also be used to work out a computer program. Given the scope of this book, the details are not discussed here.

Example 8.1

An isolated column in a building frame, with a column grid of 3.5 m × 3.5 m carries a superimposed load of 450 kN. The subsoil condition is shown in Fig. 8.20. Design a suitable square footing for the column.

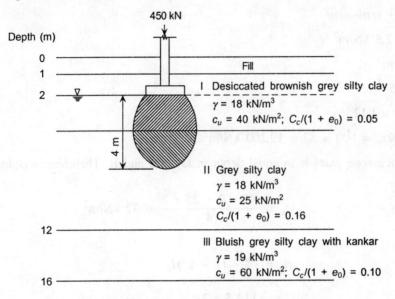

Fig. 8.20 Subsoil condition for Example 8.1.

Given superimposed load = 450 kN

Take footing size = 2 m × 2 m

$$q_{net} = \frac{450}{2 \times 2} = 112.5 \text{ kN/m}^2$$

Place foundation below fill and at depth of G.W.T. i.e.

$$D_f = 2 \text{ m}$$

Solution

(a) *Bearing capacity*

$$N_c = 5\left(1 + 0.2\frac{B}{L}\right)\left(1 + 0.2\frac{D_f}{B}\right)$$

$$= 7.2$$

Failure surface lies in stratum I. Hence, take $c_u = 40$ kN/m².

Cohesive soil: $q_{ult}(n) = c_u N_c$

$$= 40 \times 7.2$$

$$= 288 \text{ kN/m}^2$$

$$Fs = \frac{288}{112.5} = 2.56 > 2.5$$

(b) *Immediate settlement*

$q_u = 112.5 \text{ kN/m}^2$

$B = 2 \text{ m}$

$v = 0.5$

$I_{\rho(centre)} = 1.12$

$E = 600 c_u = 600 \times 32 = 19,200 \text{ kN/m}^2$

Stress influence zone extends to equal depth in layers I and II. Therefore, weighted average is

$$\therefore \quad c_u = \frac{40 \times 2 + 25 \times 2}{4} = 32 \text{ kN/m}^2$$

and

$$\rho_i = \frac{q_n B}{E} (1 - v^2) I_\rho$$

$$= \frac{112.5 \times 2}{19,200} \times 0.75 \times 1.12$$

$$= 0.0098 \text{ m}$$

Depth correction:

$$\frac{D}{\sqrt{LB}} = \frac{2}{\sqrt{2 \times 2}} = 1$$

and

$$\frac{L}{B} = 1$$

∴ Correction factor = 0.73

∴ $(\rho_i)_{corr} = 0.73 \times 0.0098 = 0.007 \text{ m} = 7 \text{ mm}$

(c) *Consolidation settlement*

$$\rho_c = \sum \frac{C_c}{1 + e_0} H \log \frac{p_o + \Delta p}{p_o}$$

Influence zone extends to depth of 4 m below footing, that is, 2 m in stratum I and 2 m in stratum II. Hence, consider settlement due to both.

At A: $p_o = 18 \times 2.0 + 8 \times 1.0 = 44$ kN/m^2

Δp (for $Z/B = 1/2$) $= 0.6 \times 112.5 = 67.5$ kN/m^2

At B: $p_o = 18 \times 2.0 + 8 \times 2.0 + 8 \times 1.0 = 60$ kN/m^2

Δp (for $Z/B = 3/2$) $= 0.15 \times 112.5 = 16.9$ kN/m^2

$\therefore \quad \rho_c = 0.05 \times 2 \times \log \left(\dfrac{44 + 67.5}{44} \right) + 0.16 \times 2 \log \left(\dfrac{60 + 16.9}{60} \right)$

$= (0.041 + 0.034)$ m

$= 0.075$ m

Depth correction $= 0.73$

Pore-pressure correction, $\mu = 0.7$

(Contribution of layers I and II are nearly equal. Hence, take $\mu = 0.7$)

$\therefore \quad (\rho_c)_{corr} = 0.73 \times 0.7 \times 0.075 = 0.038$ m $= 38$ mm

$\therefore \quad$ Total settlement, $\rho_f = 7 + 38 = 45$ mm

The foundation is therefore adequate

Example 8.2

Design a strip foundation for the column row C in the building frame shown in Fig. 8.21. The soil data are also given in the figure.

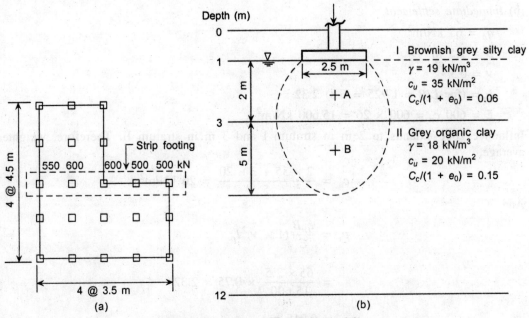

Fig. 8.21 Building frame (Example 8.2).

Total load, Q_{net} = 550 + 600 + 600 + 500 + 500 = 2750 kN

Take footing of size = 17 m × 2.5 m

(that is, extend length of footing by 1.5 m beyond outermost columns)

$$\therefore \quad q_{net} = \frac{2750}{17 \times 2.5} = 65 \ kN/m^2$$

Solution

(a) *Bearing capacity*

$$N_c = 5\left(1 + 0.2\frac{D_f}{B}\right)\left(1 + 0.2\frac{B}{L}\right)$$

$$= 5\left(1 + 0.2\frac{1}{2.5}\right)\left(1 + 0.2\frac{2.5}{17}\right)$$

$$= 5.6$$

$$\therefore \quad q_{ult}(n) = c_u N_c$$

$$= 35 \times 5.6$$

$$= 196 \ kN/m^2$$

Failure zone is restricted in stratum I. Hence, take c_u of stratum I.

$$Fs = \frac{196}{65} = 3.0 > 2.5$$

(b) *Immediate settlement*

q_n = 65 kN/m²

B = 2.5 m

v = 0.5

I_ρ = (for L/B = 17/25 = 6.8) 2.32

E = 600 c_u = 600 × 26 = 15,600 kN/m²

Influence zone extends to 2 m in stratum I and 3 m in stratum II. Therefore, weighted average,

$$c_u = \frac{2 \times 35 + 3 \times 20}{5} = 26 \ kN/m^2$$

And

$$\rho_i = \frac{q_n B}{E}(1 - v)^2 I_\rho$$

$$= \frac{65 \times 2.5}{15,600} \times 0.75 \times 2.32$$

$$= 0.018 \ m$$

Depth correction:

$$\frac{L}{B} = \frac{17}{2.5} = 6.8$$

$$\frac{D}{\sqrt{LB}} = \frac{1.5}{\sqrt{2.5 \times 1.7}} = 0.23$$

∴   Correction factor = 0.95

∴ $(\rho_i)_{corr} = 0.95 \times 0.018 = 0.017$ m = 17 mm

(c) *Consolidation settlement*

$$\rho_c = \sum \frac{C_c}{1 + e_0} H \log \frac{p_o + \Delta p}{p_o}$$

Two layers (I and II) are involved in the pressure influence zone.

At   A: $p_o = 19 \times 1.0 + 9 \times 1.0 = 28$ kN/m²

Δp [for $Z/B = 1.6/2.5 = 0.4$] = $0.85 \times 65 = 55.2$ kN/m²

At   B: $p_o = 19 \times 1.0 + 9 \times 2.0 + 8 \times 1.5 = 49$ kN/m²

Δp (for $Z/B = 3.5/2.5 = 1.4$) = $0.42 \times 65 = 27.3$ kN/m²

∴ $\rho_c = 0.04 \times 2 \log \dfrac{28 + 55.2}{28} + 0.15 \times 2 \log \dfrac{49 + 27.3}{49}$

$= 0.038 + 0.058$ m

$= 0.096$ m

Depth correction = 0.95

Rigidity correction = 0.8 (Interconnected strip provides rigidity to the foundation)

Pore-pressure correction, $\mu = 0.75$

(Major contribution to settlement comes from strata I and II. Hence, μ correction for N.C. clay is considered.)

∴ $(\rho_c)_{corr} = 0.95 \times 0.8 \times 0.75 \times 0.096 = 0.055$ m = 55 mm

∴   Total settlement, $\rho_f = 17 + 55 = 72$ mm < 75 mm

Example 8.3

A column carrying a superimposed load of 1500 kN is to be founded in a medium sand as shown in Fig. 8.22. Design a suitable isolated footing.

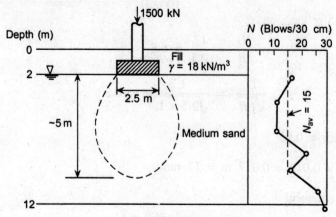

Fig. 8.22 Example 8.3.

Average N value, $N_{av} = 15$

$\phi' = 34°$ (Eq. (2.5))

$N_q = 30$

$N_\gamma = 41$

$Q_{net} = 1500$ kN/m^2

Assume footing size = 2.5 m × 2.5 m

$$\therefore \quad q_{net} = \frac{1500}{2.5 \times 2.5} = 240 \text{ kN/m}^2$$

Solution

(a) *Bearing capacity*

$s_q = 1.2$; $s_\gamma = 0.8$; $d_q = d_\gamma = 1.15$; $i_q = i_\gamma = 1.0$

$q_{ult}(n) = q(N_q - 1)s_q d_q i_q + 0.5\gamma B N_\gamma s_\gamma d_\gamma i_\gamma$

$= [18 \times 2(30 - 1) \times 1.2 \times 1.15 \times 1.0] + (0.5 \times 9 \times 2.5 \times 41 \times 0.8 \times 1.15 \times 1.0)$

$= 1440 + 424$

$= 1864$ kN/m^2

$$Fs = \frac{1864}{240} = 7.8 > 2.5$$

(b) *Settlement*

Considering single layer of 5 m depth

$$\delta = \frac{2.3 p_o}{E} H \log \frac{p_o + \Delta p}{p_o}$$

$$p_o = 18 \times 2 + 9 \times 2.5 = 58.5 \text{ kN/m}^2$$

$$\Delta p = (\text{for } Z/B = 1.0) = 0.35 \times 240 = 84 \text{ kN/m}^2$$

$$E = 39 + 4.5 \times 15 = 1065 \text{ kg/cm}^2 = 10,650 \text{ kN/m}^2 \text{ (refer Eq. (2.6))}$$

$$\therefore \quad \delta = \frac{2.3 \times 58.5}{10,650} \times 5 \log \frac{58.5 + 84}{58.5}$$

$$= 0.024 \text{ m}$$

$$= 24 \text{ mm}$$

Example 8.4

Three columns in a building frame spaced 4 m apart carry vertical superimposed loads of 500 kN, 720 kN and 600 kN. Design suitable isolated footings for the columns for equal settlement. Subsoil condition is shown in Fig. 8.23.

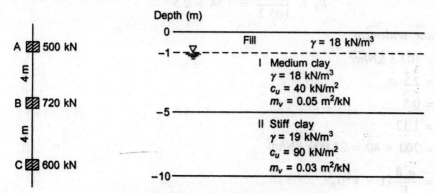

Fig. 8.23 Subsoil condition (Example 8.4).

Solution

Approximate $q_{ult}(n)$ of stratum I

$$= c_u N_c$$

$$= 40 \times 6.0 = 240 \text{ kN/m}^2$$

$$q_{all}(n) = \frac{240}{2.5} = 96 \text{ kN/m}^2 \text{ (for } Fs = 2.5)$$

Choosing approximate footing size accordingly and carrying out detailed analysis,

Column A

Take footing size $= 2.2 \text{ m} \times 2.2 \text{ m}$

$$\text{Area} = 4.84 \text{ m}^2$$

$$q_{net} = \frac{500}{4.84} = 103.3 \text{ kN/m}^2$$

(a) *Bearing capacity*

Failure surface in layer I: Take $c_u = 40$ kN/m²

$$N_c = 5\left(1 + 0.2\frac{D_f}{B}\right)\left(1 + 0.2\frac{B}{L}\right)$$

$$= 5\left(1 + 0.2 \times \frac{1}{2.2}\right)(1 + 0.2 \times 1)$$

$$= 6.55$$

$$q_{ult}(n) = c_u N_c$$

$$= 40 \times 6.55$$

$$= 262 \text{ kN/m}^2$$

$$Fs = \frac{262}{103.3} = 2.54 > 2.5$$

(b) *Immediate settlement*

$q_n = 103.3$ kN/m²

$B = 2.2$ m

$v = 0.5$

$I_\rho = 1.12$

$E = 700 \times 40 = 28,000$ kN/m²

$$\rho_i = \frac{q_n B}{E}(1 - v^2)I_\rho$$

$$= \frac{103.3 \times 2.2}{28,000} \times 0.75 \times 1.12$$

$$= 0.007 \text{ m}$$

Depth correction:

$$\frac{D}{\sqrt{LB}} = \frac{1}{2} = 0.5$$

$$\frac{L}{B} = 1.0$$

∴   Correction factor = 0.85

∴ $(\rho_i)_{corr} = 0.007 \times 0.85 = 0.006$ m = 6 mm

(b) *Consolidation settlement*

$$\rho_c = \Sigma m_v \Delta p H$$

At A: $p_o = 18 \times 1.0 + 8 \times 2.2 = 35.6$ kN/m²

Δp (for $Z/B = 1$) = $0.35 \times 103.3 = 36$ kN/m²

$$\therefore \quad \rho_c = m_v \Delta pH$$

$$= 0.05 \times 36.5 \times 4.4 = 0.080 \text{ m}$$

Depth correction = 0.85

Pore-pressure correction = 0.7

$$\therefore \quad (\rho_c)_{\text{corr}} = 0.85 \times 0.7 \times 0.08 = 0.048 \text{ m} = 48 \text{ mm}$$

$$\therefore \quad \rho_f = 6 + 48 = 54 \text{ mm}$$

Column B

Take footing size of 3 m × 3 m

$$\text{Area} = 9 \text{ m}^2$$

$$\therefore \quad q_{\text{net}} = \frac{720}{9} = 80 \text{ kN/m}^2$$

(a) *Bearing capacity*

$$N_c = 5\left(1 + 0.2\frac{D_f}{B}\right)\left(1 + 0.2\frac{B}{L}\right)$$

$$= 5\left(1 + 0.2 \times \frac{1}{2.9}\right)(1 + 0.2)$$

$$= 6.4$$

$$q_{\text{ult}}(n) = c_u N_c$$

$$= 40 \times 6.4$$

$$= 256 \text{ kN/m}^2$$

$$Fs = \frac{256}{80} = 3.2$$

(b) *Immediate settlement*

$$q_n = 89 \text{ kN/m}^2$$

$$B = 28 \text{ m}$$

$$v = 0.5$$

$$I_{\rho(\text{centre})} = 1.12$$

$$c_u = \frac{40 \times 4 + 2 \times 80}{6}$$

$$= 53 \text{ kN/m}^2$$

Influence zone extends approximately to 4 m in stratum I and 2 m in stratum II. Therefore, weighted average taken is

$$E = 700 \times 53 = 37,000 \text{ kN/m}^2$$

$$\rho_i = \frac{q_n B}{E}(1 - v^2)I_\rho$$

$$= \frac{80 \times 3}{37,000} \times 0.75 \times 1.12$$

$$= 0.0054$$

Depth correction:

$$\frac{L}{B} = 1$$

$$\frac{D}{\sqrt{LB}} = 0.33$$

∴ Correction factor = 0.9

∴ $(\rho_i)_{\text{corr}} = 0.9 \times 0.0054 = 0.005 \text{ m} = 5 \text{ mm}$

Consolidation settlement

$$\rho_c = \sum m_v \Delta p H$$

$$= (0.05 \times 4 \times 36.8) + (0.03 \times 2 \times 9.6)$$

for point A $(Z/B = 0.66)$, $\Delta p = 0.46 \times 80 = 36.8 \text{ kN/m}^2$

$$= 0.074 + 0.006$$

for point B $(Z/B = 1.67)$, $\Delta p = 0.12 \times 80 = 96 \text{ kN/m}^2$

$$= 0.080 \text{ m}$$

Depth correction = 0.9

Pore-pressure correction = 0.7

∴ $(\rho_i)_{\text{corr}} = 0.9 \times 0.7 \times 0.086 = 0.050 \text{ m} = 50 \text{ mm}$

∴ $\rho_f = 5 + 50 = 55 \text{ mm}$

Column C

Take footing size of 2.5 m × 2.5 m, and

$$\text{area} = 6.25 \text{ m}^2$$

∴ $q_{\text{net}} = \frac{600}{6.25} = 96 \text{ kN/m}^2$

Settlement

By similar calculation as in the previous parts,

$$\rho_i = 5 \text{ mm}$$

$$\rho_c = 50 \text{ mm}$$

$$\rho_f = 5 + 50 = 55 \text{ mm}$$

Column	Footing size (m × m)	q_{net} (t/m²)	Settlement (mm)
A	2.2 × 2.2	10.3	54
B	3.0 × 3.0	8.3	55
C	2.5 × 2.5	9.6	55

Example 8.5

Design a raft foundation for the buildings shown in Fig. 8.24.

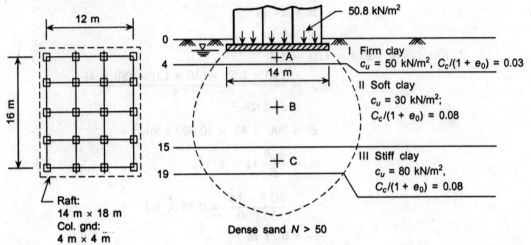

Fig. 8.24 Example 8.5.

Solution

Total load = 720 × 10 + 560 × 10
= 7200 + 5600 = 12,800 kN

Consider raft of size = 14 m × 18 m

$$\therefore \quad q = \frac{12,800}{14 \times 18} = 50.8 \text{ kN/m}^2$$

Depth of foundation, $D_f = 1.5$ m

(a) *Bearing capacity*

$$N_c = 5\left(1 + 0.2\frac{B}{L}\right)\left(1 + 0.2\frac{D_f}{B}\right)$$

$$= 5\left(1 + 0.2 \times \frac{14}{18}\right)\left(1 + 0.2 \times \frac{1.5}{14}\right)$$

$$= 5.8$$

$$q_{ult}(n) = c_u N_c$$

$$= 30 \times 5.8$$
$$= 174 \text{ kN/m}^2$$
$$Fs = \frac{174}{58} = 3.0$$

Major portion of failure zone lies in stratum II, hence, we have taken $c_u = 30$ kN/m^2

(b) *Immediate settlement*

$$q_n = 50.8 \text{ kN/m}^2$$
$$B = 14 \text{ m}$$
$$v = 0.5$$
$$I_\rho \text{ (for } L/B = 18/14 = 1.3) = 1.2$$
$$(c_u)_{av} = \frac{(50 \times 1.5) + (30 \times 12) + (80 \times 4)}{17.5}$$
$$= 43 \text{ kN/m}^2$$
$$E = 700 \times 43 = 30,000 \text{ kN/m}^2$$
$$\therefore \quad \rho_i = \frac{q_n B}{E} (1 - v^2) I_\rho$$
$$= \frac{50.8 \times 14}{30,000} \times 0.75 \times 1.2$$
$$= 0.21 \text{ m}$$

Depth correction = 1.0
Rigidity correction = 0.8

$$\therefore \quad (\rho_i)_{corr} = 0.8 \times 0.021 = 0.017$$

(c) *Consolidation settlement*

$$\rho_c = \sum \frac{C_c}{1 + e_0} H \log \frac{p_o + \Delta p}{p_o}$$

At A: $p_o = 1.5 \times 18 + 8 \times 0.75 = 33$ kN/m^2

Δp (for $Z/B = 0.75/14 = 0.05$) $= 0.98 \times 50.8 = 49.8$ kN/m^2

At B: $p_o = 18 \times 1.5 + 8 \times 1.5 + 7 \times 6$

$$= 27 + 12 + 42$$
$$= 81 \text{ kN/m}^2$$

Δp (for $Z/B = 7.5/14 = 0.53$) $= 0.74 \times 50.8 = 37.5$ kN/m^2

At C: $p_o = 18 \times 1.5 + 8 \times 1.5 + 7 \times 12 + 8 \times 2$

$$= 27 + 12 + 84 + 16$$
$$= 139 \text{ kN/m}^2$$

$$\Delta p = \left(\text{for } Z/B = \frac{15.5}{14} \right) \ 0.35 \times 50.8 = 17.8 \text{ kN/m}^2$$

$$\frac{\Delta p}{p_o} = \frac{12.9}{139} = 0.09 < 0.1 \quad \text{Hence, soil below stratum III is not significant.}$$

$$\rho_c = 0.03 \times 1.5 \times \log \frac{33 + 49.8}{33} + 0.08 \times 12 \log \frac{81 + 37.5}{81}$$

$$+ \ 0.06 \times 4 \log \frac{139 + 17.8}{139}$$

$$= 0.018 + 0.158 + 0.012$$

$$= 0.188 \text{ m}$$

Depth correction = 1.0

Rigidity correction = 0.8

Pore-pressure correction = 0.7

$$\therefore \quad (\rho_c)_{\text{corr}} = 0.8 \times 0.7 \times 0.188 = 0.105 \text{ m}$$

$$\rho_f = 0.017 + 0.105$$

$$= 0.122 \text{ m} = 122 \text{ mm} < 125 \text{ mm}$$

Example 8.6

Design of a buoyancy raft foundation for a 4-storey dormitory building with two columns in each bay, depicted in Fig. 8.25.

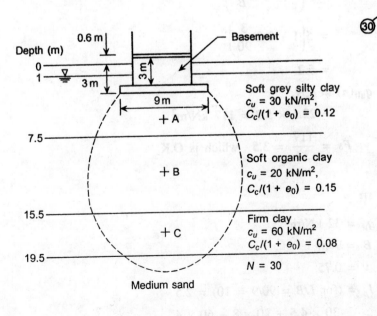

Fig. 8.25 Example 8.6.

Solution

Size of raft = 9 m × 90 m
Area of each column bay = 3 m × 9 m = 27 m²
Load in each bay = 2 × 900 = 1800 kN

Gross foundation pressure

for superstructure $q = \dfrac{1800}{27} = 67$ kN/m²

Assume basement to be 25%. Thus, $q_{(basement)} = 17$ kN/m²

∴ $q_{gross} = 84$ kN/m²

Provide depth of foundation, $D_f = 3$ m.

$$q_{net} = q_{gross} - \gamma D_f$$
$$= 84 - 18 \times 3$$
$$= 30 \text{ kN/m}^2$$

(a) Bearing capacity

Consider 9 m depth below raft

$$\therefore \ c_u = \frac{25 \times 4.5 + 20 \times 4.5}{9} = 22.5 \text{ kN/m}^2$$

and

$$N_c = 5\left(1 + 0.2\frac{D_f}{B}\right)$$

$$= 5\left(1 + 0.2\frac{3}{90}\right)$$

$$= 5.2$$

$$q_{ult}(n) = c_u N_c$$

$$= 22.5 \times 5.2 = 117 \text{ kN/m}^2$$

$$Fs = \frac{117}{33} = 3.5 \quad \text{which is O.K.}$$

(b) Immediate settlement

Consider layers I, II, and III.

$$q_u = 32 \text{ kN/m}^2$$
$$B = 9 \text{ m}$$
$$v = 0.75$$
$$I_\rho = (\text{for } L/B = 90/9 = 10) = 2.5$$
$$(c_u)_{av} = \frac{30 \times 4.5 + 20 \times 8 + 60 \times 4}{16.5} = 32 \text{ kN/m}^2$$

$$E = 600 \times 32 = 19{,}200 \text{ kN/m}^2$$

$$\rho_i = \frac{q_n B}{E}(1 - v) I_\rho$$

$$= \frac{30 \times 9 \times 0.75 \times 2.5}{19{,}200}$$

$$= 0.026 \text{ m}$$

Depth correction = 1.0
Rigidity correction = 0.8

$$(\rho_i)_{corr} = 0.08 \times 0.026 = 0.02 \text{ m} = 20 \text{ mm}$$

(c) *Consolidation settlement*

Consider layers I, II, and III.

At A: $p_o = 18 \times 1.0 + 8 \times 4.25 = 52 \text{ kN/m}^2$
$\Delta p = 0.96 \times 30 = 28.8 \text{ kN/m}^2$

At B: $p_o = 18 \times 1 + 8 \times 6.5 + 8 \times 4.0$
$= 102 \text{ kN/m}^2$
$\Delta p = \quad 0.55 \times 30 = 16.5 \text{ kN/m}^2$

At C: $p_o = 18 \times 1 + 8 \times 6.5 + 8 \times 8 + 8 \times 2$
$= 150 \text{ kN/m}^2$
$\Delta p = 0.36 \times 33 = 10.8 \text{ kN/m}^2$

$$\therefore \quad \rho_c = \sum \frac{C_c}{1 + e_0} H \log \frac{p_o + \Delta p}{p_o}$$

$$= 0.10 \times 4.5 \log \frac{52 + 28.8}{52} + 0.15 \times 8 \log \frac{102 + 16.5}{102}$$

$$+ \ 0.08 \times 4 \log \frac{150 + 12}{150}$$

$$= 0.086 + 0.078 + 0.01$$

$$= 0.174 \text{ m}$$

Depth correction = 1.0
Rigidity correction = 0.8
Pore-pressure correction = 0.8

$$(\rho_c)_{corr} = 0.8 \times 0.8 \times 0.174 = 0.110 \text{ m} = 110 \text{ mm}$$

Total settlement,

$$\rho_f = 20 + 110 = 130 \text{ mm}$$

Marginal extra settlement beyond the permissible limit of 125 mm may be allowed in view of the rigidity of the raft foundation.

========================= **REFERENCES** =========================

Bowles, J.E. (1988), *Foundation Analysis and Design*, McGraw-Hill Book Company, International ed., Singapore.

Casagrande, A. (1936), *The Determination of Preconsolidation Load and its Practical Significance*, Proc. Ist Int. Conference on Soil Mechanics and Foundation Engineering, Cambridge, Mass, Vol. 3.

Casagrande, A. and R.E. Fadum, (1942), *Application of Soil Mechanics in Designing Building Foundations*, Proc. American Society of Civil Engineers, Nov. 1942.

Glossop, R. (1972), *Floating Foundations*, *Foundation Engineering Handbook* Ist ed., Van Nostrand Reinhold Co., New York.

Hetenyi, M. (1946), *Beams on Elastic Foundation*, The University of Michigan Press, Ann Arban, Mi, USA.

IS 1904 (1966), *Code of Practice for Design of Shallow Foundations*, Bureau of Indian Standards, New Delhi.

Rios, L. and P. Silva, *Foundations in Down Town, Sao Paulo*, Proc. 2nd Int. Conf. Soil Mech (Rotterdam) Vol. 4, p. 69.

Skempton, A.W. (1955), *Foundations for High Buildings*, Proc. Institution of Civil Engineers, London. Pt. 3, Vol. 4, p. 246.

Terzaghi, K. (1938), *Settlement of Structures in Europe and Methods of Observation*, Proc. ASCE, Vol. 103, p. 1432.

Pile Foundations

9.1 INTRODUCTION

Piles are relatively long and slender structural members used to transmit foundation loads through soil of low bearing capacity to deeper strata (soil or rock) having high bearing capacity. They are also used in normal ground conditions to resist uplift and lateral forces.

The principal uses of piles are:

1. To carry vertical compression load from buildings, bridges, and so on.
2. To resist horizontal or inclined loads by retaining wall, bridge pier, water front structures and structures subjected to wind or seismic loads.
3. To resist uplift forces in transmission towers and underground structures below water table.

Piles are described as end bearing piles and friction piles depending on the manner in which the load is transmitted into the surrounding soil. These can be defined as follows:

If the pile rests in a hard and relatively incompressible stratum, for example, rock or dense sand/gravel, the pile derives most of its carrying capacity from end bearing at the pile tip. Such piles are called *end bearing* or *point-bearing* piles. Figure 9.1(a) depicts end-bearing piles. The soft compressible layer through which the pile passes may not carry any significant load by side friction.

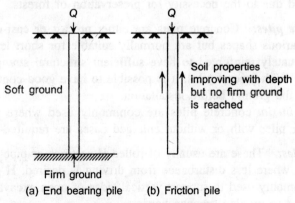

Fig. 9.1 Piles: methods of load transfer.

If the pile does not reach an incompressible stratum but is driven for some depth into a penetrable soil, the carrying capacity of the pile is derived primarily from skin friction or adhesion between the embedded surface of the pile and the surrounding soil. Such piles are called *friction piles*, as shown in Fig. 9.1(b).

Although, in common terminology, piles are often referred to as friction piles or end-bearing piles—in reality, there is no pile that transmits the load to the surrounding soil solely by friction or solely by end bearing. The distinction only serves to indicate the relative magnitude of the load that is transmitted by friction and by end bearing. For example, a straight pile embedded in homogeneous clay will mostly transfer the load by friction and a pile with its tip resting in dense sand underlying soft clay can be considered an end bearing pile.

The relative magnitude of the skin friction and the base resistance of a pile, however, depends on various factors such as,

1. Geometry of the pile shaft—its shape, length, and diameter and whether it is with or without enlarged base.
2. Subsoil stratification along the pile shaft and the properties of soil in which the pile is embedded and where the tip of the pile rests.
3. Method of construction of the pile, that is, driven pile, bored pile, and so on.

9.2 CLASSIFICATION

Piles are classified according to their composition or method of installation.

9.2.1 Classification Based on Composition

(a) *Timber piles:* In India, timber piles are mostly made up of sal tree trunks and are called salballah piles. These are commonly available in length between 4–6 m, with diameter ranging from 15–25 cm. These may be suitable where good bearing stratum is available at a relatively shallow depth. Now a days, use of timber piles is restricted due to the necessity for preservation of forests.

(b) *Concrete piles:* Concrete piles are either precast or cast-in-situ. *Precast* piles may be of various shapes but are normally suitable for short lengths. These piles should be adequately reinforced to have sufficient structural strength to withstand handling stresses. With precast piles, it is possible to have good control on quality as they are cast on the ground before installation.

   Cast-in-situ concrete piles are commonly used where relatively long and large diameter piles with or without enlarged bases are required to support heavy loads.

(c) *Steel piles:* These are usually of rolled H sections or pipe sections. These piles may be used where less disturbance from driving is desired. H-piles and steel sheet piles are commonly used to support vertical sides of open excavation. Steel sheet piles are also used to provide seepage barrier.

9.2.2 Classification Based on Method of Installation

(a) *Driven piles:* These may comprise of timber, steel, or precast concrete. The piles are driven by the impact of a hammer or by vibrations induced by a vibratory hammer.

When a pile is driven into the soil, it displaces a volume of the soil equal to the volume of the pile. So, these piles are also called *displacement piles*. In granular soils, the driving operation densifies the soil and increases strength of the soil in the vicinity. When piles are driven in saturated clay, the soil instead of being compacted gets remoulded often with reduction of strength. The soil, however, regains strength with time due to consolidation and thixotropic hardening. Because of the displacement of soil by the piles, there may be ground heaving around the piles. Also, driving of piles imparts vibration to surrounding soil, which in some cases, may be detrimental to structures located very close to the site.

(b) *Driven cast-in-situ-piles:* These piles are also a kind of driven pile. Steel casing is driven into the ground with a shoe at the bottom. The hole is then filled up with concrete, and the casing is gradually lifted as the concrete is poured.

(c) *Bored piles:* These piles are formed in prebored holes in the ground either using a casing or by circulation of a drilling fluid, such as bentonite slurry. Concrete is poured into the hole by displacing bentonite and then gradually lifting the casing.

Bored piles may be of the following types:

(i) Small diameter bored piles—generally upto 600 mm diameter.
(ii) Large diameter bored piles—diameter generally greater than 600 mm. They are advantageous where heavy structural loads are to be supported.
(iii) Under-reamed piles—one or more bulbs of larger diameter than that of the shaft are formed by suitably enlarging the borehole with special tools to increase the end resistance. These piles are suitable when a good bearing stratum is available at a relatively shallow depth. The uplift capacity of these piles is high. In India, these piles are widely used to provide suitable foundations in expansive soils.

Bored piles are non-displacement piles and may be used when pile driving is detrimental to adjoining structures.

9.3 PILE BEHAVIOUR UNDER AXIAL LOAD

Figure 9.2 shows a typical load settlement diagram for a pile under gradually increasing load. At small deformations, say upto the point A in the figure, the pile–soil system behaves elastically, as indicated by the linear load versus settlement relationship. Upto this point, the entire load is carried by skin friction and there is virtually no transfer of load to the pile tip. With more deformation, the pile shaft carries more frictional forces while a part of the load is transferred to the pile tip also. At B, the pile tip carries the maximum skin friction that can be mobilized in the soil. Therefore, further increase of load on the pile is carried out at the pile tip and the pile is said to have failed, at C, when the tip load has also reached its maximum value. Thus, the load carrying capacity of a pile may be determined from separate evaluation of the ultimate skin friction and the ultimate base resistance.

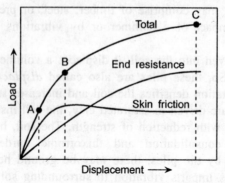

Fig. 9.2 Load settlement relationship of piles.

Thus,

$$Q_p = Q_f + Q_b - W_p - W_e \qquad (9.1)$$

where Q_p is the load carrying capacity of the pile,
Q_f is the maximum skin friction on the pile shaft,
Q_b is the ultimate tip resistance,
W_p is the weight of the pile, and
W_e is the weight of soil displaced by the pile.

In general, W_p and W_e are small in relation to Q_p and their differences even smaller. Thus, for practical purposes

$$Q_p = Q_f + Q_b \qquad (9.2)$$

Therefore, it is apparent, that full mobilization of Q_f and Q_p are primarily dependent on the movement of the pile in the soil (refer Fig. 9.2). Cooke (1974), and Cooke and Price (1973) have shown that, for an elastic soil, full mobilization of skin friction in a cylindrical pile occurs at vertical displacement of 0.5–1% of the pile diameter. On the other hand, full mobilization of base resistance requires a much greater deformation—even upto 20% of the base diameter (Whitaker, 1976). This suggests that skin friction is mobilized almost fully before any appreciable base resistance develops in the soil. Therefore, the entire problem of load distribution through piles has to be viewed as a pile–soil interaction where the response of the soil to a given deformation in terms of skin friction and end bearing has to be taken into account.

An evaluation of the zone of soil that is influenced by piling is important to determine the soil properties that come into play in resisting the movement of the pile in the soil. Meyerhof (1959) and Kerisel (1961) considered the influence zone as shown in Fig. 9.3. According to Meyerhof (1959), the shear zone at the base causing bearing capacity failure of the soil extends to some distance above the pile tip. The zone of skin friction may extend laterally to a distance of three to four times the diameter of the pile while compaction of the pile may occur upto a distance of five to six times the diameter. Cooke and Price (1973) have, however, observed that displacement of the soil may extend to a radial distance of ten diameters around the pile. Pile driving causes significant changes in the soil within the zone of influence. Therefore, it is important to study the various factors that affect the soil during piling.

Fig. 9.3 Stress influence zone around piles.

9.4 PILE CAPACITY TO RESIST AXIAL FORCES

The load carrying capacity of a pile is controlled by its structural strength and the supporting strength of the soil. The smaller of the two is considered for design purposes. There are three apporaches to the computation of pile capacity based on soil support. These are:

    (a) Static analysis
    (b) Dynamic analysis
    (c) Load test

9.4.1 Structural Capacity of Piles

The structural capacity of a pile is its strength as a column. When the pile is completely embedded in soil, the restraint offered by the soil is generally sufficient to consider the pile as a short column (except for the case of a long pile in very soft clay). Pre-cast concrete piles are adequately reinforced to withstand handling and driving stresses. Cast-in-situ piles are also reinforced to increase column strength and also to resist moment that may have developed due to horizontal load or eccentricity of vertical loads. Reinforcements are also helpful in resisting tensile stresses that may develop due to heave resulting from driving of adjacent piles in clay.

9.4.2 Pile Capacity from Static Analysis

The static analysis relates the shear strength of the soil to the skin friction along the pile shaft and end bearing at the pile tip. The carrying capacity of a pile is given by the sum of the ultimate bearing capacity of the soil at the pile tip and the ultimate adhesion between the pile and the surrounding soil, (refer Fig. 9.2).

$$Q_{ult} = \sum A_s f_s + A_p q_p \tag{9.3}$$

where

f_s = unit skin friction in each layer,
A_s = pile area providing skin friction in each layer,
A_p = pile area providing end bearing, and
q_p = ultimate bearing capacity of soil at pile tip.

Piles in cohesive soil

A total stress approach is generally used to determine the shaft and tip resistances of piles in cohesive soil considering that the load transfer from the pile to the soil occurs under undrained condition.

Driven piles: When a pile is driven into a cohesive soil, two effects of pile driving become significant.

The clay is displaced laterally and in an upward direction, resulting in a ground heave. The soil close to the pile shaft gets disturbed to cause remoulding of the clay. This may lead to considerable loss in strength of the soil. This has been discussed by Flaate (1972) who observed appreciable increase in water content and corresponding reduction of shear strength of the soil between two adjacent driven piles in clay. With time, soft clay regains this strength either completely or partially by thixotropic hardening (Skempton and Northey, 1952). A corresponding increase in load carrying capacity of the pile with time is shown in Fig. 9.4 (Tomlinson, 1994). On the other hand, in stiff clays, extensive cracking of the soil occurs during pile driving. Clay in the upper part of the pile *breaks away from the shaft and may never regain contact with it* (Tomlinson, 1977). Thus, the effect of pile driving on the shear strength of the clay differs with the soil consistency. Considerable gain in strength may be observed with time in soft clays while no appreciable effect may be noted in stiff clays.

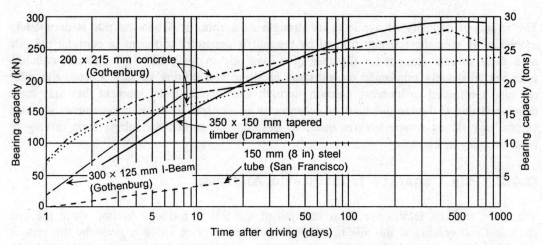

Fig. 9.4 Gain in load carrying capacity with time of driven piles in soft clay (Tomlinson, 1994).

High pore-pressure is developed in the clay due to pile driving forces. This pore-pressure often takes time to dissipate. When reconsolidation of the clay occurs, the soft clay tends to stick to the pile shaft. In the case of stiff clays, the high pore-pressure dissipates quickly through the cracks and fissures and often a negative pore-pressure is set up in the surrounding clay. Tomlinson carried out extensive research on the behaviour of driven piles in stiff clays. Steel tubular piles were driven into stiff London clay and were subjected to load test 1 month, 3 months, and 1 year after driving. Examination of the soil around the pile showed that gaps had been created near the top of the pile during driving which had not closed even after one year. Tomlinson observed three zones of adhesion of varying magnitude in a driven pile. Below a length of 8 × diameter at the top where the gap occurred, there was a length of partial adhesion followed by a length of very close adhesion where most of the frictional resistance of the pile developed.

Bored piles: In bored piles, the in-situ soil is removed from the hole by drilling and the void is later filled with concrete. Water and/or drilling fluid is used to stabilize the borehole before concreting. Boring causes release of lateral stresses on the walls of the borehole. The soil in contact with the borehole undergoes swelling and migration of pore water occurs towards the exposed clay face. Use of drilling fluid such as bentonite, may considerably minimize this migration of pore water.

When concrete is poured in the borehole to make the pile, there is further migration of water from the green concrete into the surrounding soil which causes further softening of the clay. Meyerhof and Murdock (1953) measured 4–8% increase of water content in stiff fissured clay close to the interface of the soil with concrete.

Pile in cohesionless soil

Drained condition generally prevails in cohesionless soil. Accordingly, pile capacity is determined from effective stress analysis.

Driven piles: The effect of pile driving in cohesionless soil is to compact the soil and, thereby, increase its shearing resistance. The question of loosening of the soil, as in the case of clays, does not normally arise. Whatever loosening may occur during withdrawal of casing, is more than compensated by the vibration caused by ramming of the concrete and the density of the soil is not likely to come down below its original value.

Bored piles: Boring a hole in granular soil generally causes a loosening of the soil, thereby reducing its relative density. This causes a marked reduction in both the skin friction and the end bearing. Further, if bentonite slurry is used to stabilize the hole, there often remains a screen of bentonite between the hole and the pile. It is believed that this may result in a reduction of the frictional resistance.

Sliwinski and Fleming (1974), and Broms and Hill (1973) have observed some reduction of skin friction of bored piles made with bentonite slurry as compared with those made without bentonite slurry. On the other hand, Touma and Reese (1973), did not find any significant difference in behaviour between the piles installed with or without bentonite slurry. However, if chances of reduction of skin friction is anticipated, measures to expell the bentonite screen (say, by grouting or other means) may be adopted (Tomlinson, 1977).

9.5 FRICTIONAL RESISTANCE

We have, so far discussed the various factors that affect the soil during pile driving. The quantitative evaluation of frictional resistance should take into account the effect of all these factors on the shear behaviour of the soil. Although the frictional resistance is basically a function of the shear strength of the soil, this friction develops as an adhesion between the pile and the soil and not as pure friction between soil particles. The complex nature of stresses developed around a pile, the effects of remoulding and/or stress release of the soil, and a somewhat indeterminate nature of the pore-pressure dissipation and pile soil interaction make it difficult to analyze the problem of skin friction evaluation.

Paton (1895) applied Rankine's earth pressure theory to obtain the bearing capacity of piles (see Whitaker, 1976). The frictional resistance along the pile shaft was expressed as,

$$F_s = \mu\gamma \frac{L}{2} \frac{1 - \sin\phi}{1 + \sin\phi} A_s \tag{9.4}$$

where μ is the coefficient of friction between the pile and the soil,
   γ is the density of the soil,
   L is the length of the pile,
   ϕ is angle of internal friction of the soil, and
   A_s is surface area of the pile.

The major uncertainty in using Eq. (9.4) to evaluate the skin friction lies in the estimation of μ and ϕ of the soil. In particular, no distinction is made between clay and sand, and the choice of these parameters for actual conditions that exist around a pile during and after driving leads to difficulties.

Burland (1973) developed a method of calculation of shaft friction from the effective stress principle. It is assumed that the zone of distortion around the pile being small, the pore-pressure dissipates quickly and loading takes place under drained condition. It is further assumed that there is no effective cohesion in the soil due to remoulding of the soil around the pile. The unit skin friction at any depth is then given by,

$$f_s = kp \tan\delta \tag{9.5}$$

where k is the lateral earth pressure coefficient,
   p is the effective overburden pressure at the depth considered, and
   δ is the angle of friction between the pile and the soil.

It poses extreme difficulties to select appropriate values of k and p. p is initially small due to high pore-pressure developed during driving but its value soon increases with dissipation of pore water pressure. In driven piles, k is also likely to be high initially because of the driving forces whereas in bored piles, k is observed to have small value due to swelling of the soil. These values may ultimately attain some stability after installation but uncertainties still remain regarding the time required for drainage of the soil around the pile and about values of k and p that should be applicable in a field situation. Burland made a further simplifying assumption that $k = k_0$ (k_0 is the coefficient of earth pressure at rest) and obtained the following expression for the total shaft friction:

$$F_s = \pi B \sum_0^L p k_0 \tan\delta (\Delta L) \tag{9.6}$$

Owing to the uncertainties involved in selecting field parameters for analytical solutions, attempts for evaluation of frictional resistance of piles have been based on semi-empirical approaches.

9.5.1 Frictional Resistance in Cohesive Soil

When a pile is installed in a cohesive soil, either by methods of driving or boring, the immediate deformation of the soil takes place essentially under undrained condition. In this case, $\phi = 0$ analysis (Skempton, 1948) would be valid. The unit skin friction between the soil and the pile may then be considered to be a function of the undrained shear strength of the clay. Notwithstanding the fact that this shear strength may undergo change with time due to thixotropic hardening or due to dissipation of excess pore-pressure, the unit ultimate skin friction may be expressed by the equation,

$$f_s = \alpha\, c_u \tag{9.7}$$

where c_u is the undrained shear strength of the clay, and
α is an adhesion factor.

The total frictional resistance of the piles would then be,

$$F_s = \sum A_f\, \alpha c_u \tag{9.8}$$

where A_f is the surface area of the pile shaft in contact with the soil in different layers.

While the undrained shear strength of the clay, c_u can be determined from field vane shear test or from laboratory triaxial test, a proper assessment of the adhesion factor is required for the evaluation of the unit skin friction. This adhesion factor gives a measure of the part of the undrained shear strength of the soil that is mobilized as skin friction.

Driven piles in cohesive soil

Extensive research by Tomlinson (1965, 1977) has shown that the adhesion factor depends on the consistency of the soil, as represented by its cohesive strength. In addition, the penetration depth of the pile appears to have some effect on the adhesion factor. Figure 9.5 illustrates following three different cases.

(i) In the case of short piles in uniform clay, the gap formed near the pile during driving may occupy a large part of the penetration depth. The average adhesion factor would, accordingly, be low. For large penetration depth, the gap would be small compared to the pile length and a higher average adhesion factor may be adopted. Figure 9.5(a) shows the variation of adhesion factor with undrained shear strength for piles in uniform clay.

(ii) If the pile is driven through a soft clay with an underlying stiff clay, a skin of soft clay would be dragged into the gap formed near the top of the stiff clay. This would reduce the adhesion over a certain length of penetration depth into the stiff clay, the effect being more predominant in the case of short penetration lengths. The corresponding adhesion factors for different values of c_u are shown in Fig. 9.5(b).

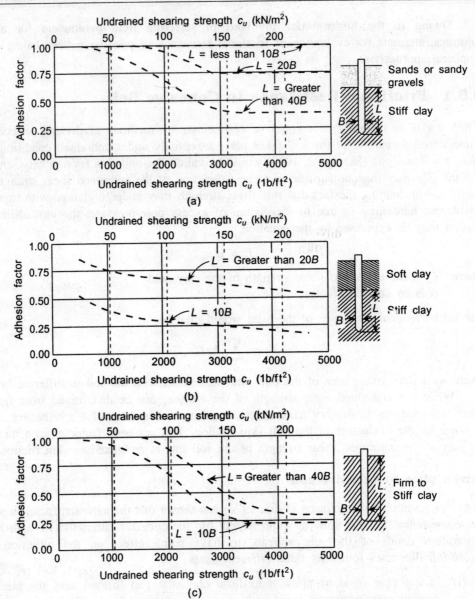

Fig. 9.5 Skin friction in cohesive soil and adhesion factor.

(iii) If the pile is driven through a granular soil into a stiff clay, the gap near the top of the stiff clay explained previously will be filled with a sand skin and the corresponding adhesion factors would be higher. The design curves for adhesion factor are shown in Fig. 9.5(c). In general, deep piles will show higher adhesion factor for all values of shear strength.

Tomlinson (1977) points out that the adhesion factors given in Fig. 9.5 are applicable to driven piles of uniform cross-section. The adhesion factors for driven and cast-in-situ piles may be slightly less than the precast piles because of the migration of water from the green concrete into the surrounding soil, although the extent of reduction cannot be precisely determined.

Vijayvergiya and Focht (1972) employed a different approach to the determination of adhesion factor for driven piles in cohesive soils. They suggested that pile driving forces would mobilize passive resistance in the soil and this would influence the development of skin friction in pile shaft. The unit skin friction could then be expressed as a function of the passive earth pressure.

$$f_s = \lambda\,(p_0 + 2c_u) \tag{9.9}$$

where p_0 is the average vertical effective stress between the ground surface and the pile tip,
c_u is the average undrained shear strength along the pile shaft, and
λ is a dimensionless coefficient.

The values of λ for different penetration depths of piles in clay as proposed by Vijayvergiya and Focht are shown in Fig. 9.6. When calculating p_0, due account of the ground water table should be taken. McClelland (1974) compared the results of 47 pile tests, mostly from U.S.A. with the predictions based on Tomlinson's and Vijayvergiya and Focht's adhesion factor. Better correlation was obtained with the latter.

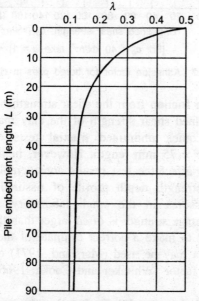

Fig. 9.6 Values of λ (Vijayvergiya and Focht, 1972).

Bored piles in cohesive soil

In bored piles, as discussed earlier, softening of the clay occurs in the vicinity of the borehole. This results in a reduction of the adhesion factor which should be duly taken into account in design. This effect is more pronounced in stiff clay than in soft clay. Skempton (1959) observed an adhesion factor as low as 0.3 for bored piles in stiff London Clay. This factor is likely to vary with time that is allowed between boring and casting of the concrete. Greater time gap would allow more swelling of the surrounding clay and thereby reduce the adhesion factor. For bored piles, the adhesion factor may best be obtained from correlation with undrained strength of the clay as shown in Fig. 9.7.

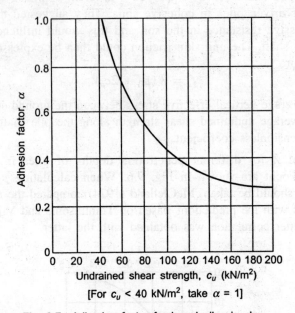

[For $c_u < 40$ kN/m², take $\alpha = 1$]

Fig. 9.7 Adhesion factor for bored piles in clay.

The determination of skin friction from the shear strength of the clay requires a proper evaluation of the in-situ undrained shear strength of the clay mass. This strength is usually determined from laboratory quick undrained triaxial tests on cylindrical samples of dimensions 37.5 mm diameter × 75 mm length. However, in practice, anisotropy, rate of strain, sample size, and so on affect the test results considerably (Bishop, 1966). There is often a large scatter in the strength–depth profile of fissured clays due to the random presence or otherwise of fissures in the small laboratory specimens (Skempton and La Rochelle, 1955) and the average strength is often larger than the representative strength of the clay mass in-situ. In order to make a correct estimate of the in-situ strength, either the lower limit of the strength data may be used (Marsland, 1971) or the average strength may be multiplied by an empirical factor (Whitaker and Cooke, 1966).

9.5.2 Frictional Resistance in Cohesionless Soil

The classical formula for calculation of the unit skin friction at any depth on a pile shaft in a cohesionless soil is given by the equation

$$f_s = K_s p \tan \delta \qquad (9.10)$$

where,

K_s is the coefficient of lateral earth pressure which depends on the relative density of the soil,

p is the effective overburden pressure at the depth considered, and

δ is the angle of friction between the pile and soil.

The unit skin friction, according to Eq. (9.9). Varies linearly with depth along the pile shaft. This is shown in Fig. 9.8. The total frictional resistance for a pile of depth D_f can be obtained as

$$Q_f = \frac{1}{2}K_s\gamma'D_f \tan \delta A_f \qquad (9.11)$$

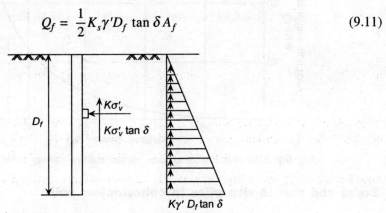

Fig. 9.8 Skin friction in piles within sand.

where,

A_f is the area of the pile shaft in contact with the soil and
$\gamma' D_f$ is the effective overburden pressure at the pile tip.

The values of K_s and δ as given by Broms (1966) are shown in Table 9.1. While K_s is a function of the relative density of the soil, δ is related to the angle of shearing resistance of the soil and the nature of pile soil contact. In piles with liner, the skin friction is even less than direct soil–concrete friction and a reduced value of δ should be used.

Table 9.1 Values of K_s and δ (Broms 1966)

Installation method	K_s	Pile–soil interface	δ
Driven piles, large displacement	1.5–2.0	Steel/sand	$0.5\phi'$–$0.7\phi'$
Driven piles, small displacement	1.0–2.0	Precast concrete/sand	$0.8\phi'$–$1.0\phi'$
Bored and cast-in-situ piles	0.7–1.0	Cast-in-situ concrete/sand	$1.0\phi'$

Driven piles in cohesionless soil

In driven piles, Poulos and Davis (1980) have recommended the use of the relationship

$$\delta = \frac{3}{4}\phi' + 10° \qquad (9.12)$$

to determine the angle of pile soil friction accounting for the effect of pile driving, ϕ' being the angle of shearing resistance prior to installation of the pile.

Tomlinson (1977) has proposed a relationship between average unit skin friction and relative density, based on the results of actual pile load tests, as depicted by Fig. 9.9. According to this relationship, the skin friction attains a peak value of 107 kN/m^2 (10 t/m^2) for high relative density.

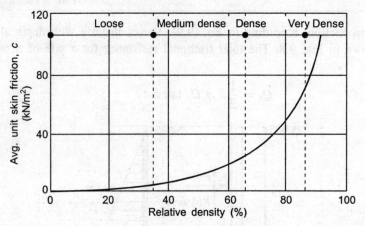

Fig. 9.9 Unit skin friction in sand versus relative density (after Tomlinson, 1977).

Bored and cast-in-situ piles in cohesionless soil

For bored and cast-in-situ piles in cohesionless soil, there may be a marked reduction in the relative density of the soil around the walls of the borehole after drilling. This may lead to a reduction of skin friction on the pile shaft. If appreciable loosening of the soil is anticipated a low angle of shearing resistance should be used in calculating skin friction by Eq. (9.9). The presence of a bentonite skin on the face of the borehole may also cause a reduction of skin friction. Poulos and Davis (1980) suggest reduction in ϕ of sand in bored piles by 3°. That is,

$$\phi = \phi' - 3° \tag{9.13}$$

Touma and Reese (1974) suggest a value of 0.7 for the earth pressure coefficient, K_s for calculating skin friction of bored piles in sand.

The skin friction mobilized in cohesionless soil may be best understood by considering the value of $K_s \tan \delta$ in Eq. (9.10). Vesic (1967) and Meyerhof (1976) have done extensive work on model piles. They gave the relationships between $K_s \tan \delta$ and ϕ' as shown in Fig. 9.10. The effect of pile installation on the skin friction coefficient is clearly seen.

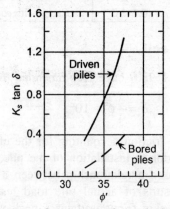

Fig. 9.10 Relationship between $K_s \tan \delta$ and angle of shearing resistance, ϕ' (Poulos and Davis, 1980).

9.6 END BEARING

The ultimate bearing capacity of shallow foundation has been discussed in Chapter 5. The general equation for bearing capacity of a foundation is given by

$$q_{ult} = cN_c + qN_q + 0.5B\gamma'N_\gamma \tag{9.14}$$

where c = cohesion of the soil,
  q = effective overburden pressure at the level of foundation,
  B = width of footing,
  γ' = effective unit weight of soil, and
  N'_c, N'_q, and N'_γ = bearing capacity factors suitably corrected for shape and depth.

For a pile having depth D_f and diameter D, the ultimate bearing capacity at the pile tip becomes

$$q_p = cN'_c + qN'_q + 0.5\gamma'DN'_\gamma \tag{9.15}$$

Now the pile diameter, D is generally small compared to the depth D_f. So, the term $\gamma DN'_\gamma$ is small and often neglected. End bearing at the pile tip is then obtained as

$$q = cN'_c + qN'_q \tag{9.16}$$

9.6.1 End Bearing in Cohesive Soil

For cohesive soil having $\phi = 0$, under undrained condition

$$q_p = c_uN_c \tag{9.17}$$

where, N_c = bearing capacity factor.

For $\phi = 0$ and for deep foundations, which a pile may be treated as, N_c may be taken to be equal to 9 on the basis of analytical and experimental evidences (Skempton, 1951). However, for deep piles in homogeneous clay, contribution of end resistance is small in comparison to the shaft resistance.

9.6.2 End Bearing in Cohesionless Soil

For a cohesionless soil having $c = 0$, the end bearing is

$$q_p = qN'_q \tag{9.18}$$

where

  q = effective vertical stress at the pile tip and
  N'_q = bearing capacity factor which is a function of ϕ'.

Different investigators have obtained different relations between N_q and ϕ' depending on the shape of failure surface considered. Figure 9.11 shows some assumed failure surfaces and corresponding relationships between N_q and ϕ.

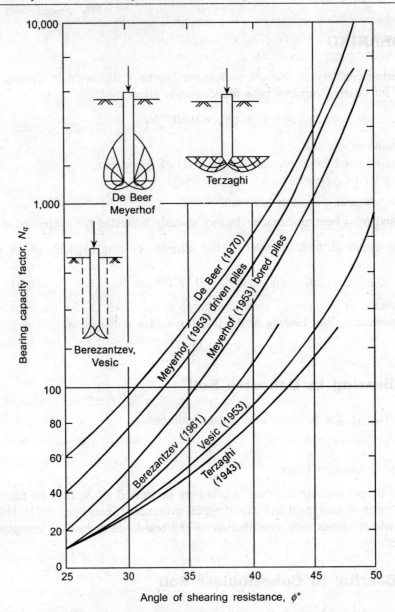

Fig. 9.11 Variation of N_q with ϕ' as obtained by different investigators.

It can be seen that Terzaghi (1943) gives the lower bound and De Beer (1970) gives the upper bound values of N_q. Comparison of observed point resistances of piles in sand by Nordlund (1969) and Vesic (1967) reveal that N_q values established by Berezantzev et al. (1961) give better estimates of point resistance. Although N_q values of Berezantzev are dependent on the relative embedment depth D_f/B, the variation is small. An average Berezantzev curve has been given in IS 2911 (part I, 1979), Fig. 9.12.

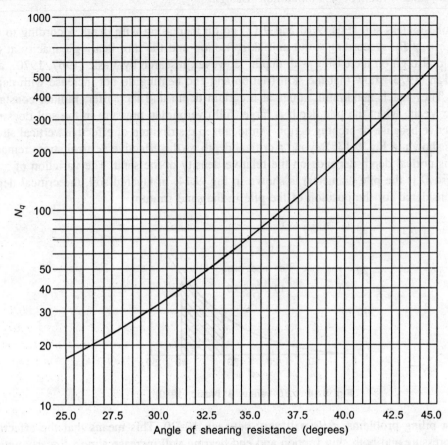

Fig. 9.12 Variation of N_q with ϕ' (Berezantzev, 1961).

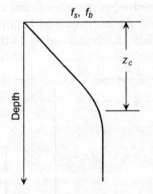

Fig. 9.13 Concept of critical depth in cohesionless soil (after Vesic, 1970).

9.7 CRITICAL DEPTH

The effective overburden pressure required for calculation of frictional resistance and end bearing in cohesionless soil (Eqs. 9.10 and 9.17) go on increasing with depth according to the relationship $q = \gamma' D_f$. Accordingly the unit skin friction and the unit bearing capacity at the pile tip should also go on increasing with depth. However, research by Vesic (1967, 1970), and Hanna and Tan (1973) show that the skin friction and end bearing do not increase with depth indefinitely. They reach maximum values at a certain depth and thereafter, remain constant. According to Vesic, when the pile depth is large, full overburden pressure in the soil does not become effective because of arching action. An idealized distribution of effective vertical stress with depth is shown in Fig. 9.13. Beyond a critical depth z_c, the effective vertical stress remains constant. This critical depth depends on the relative density of the sand. The variation of z_c/D with ϕ' (where D is the pile diameter) is shown in Fig. 9.14. In layered soil, the critical depth should be considered for the position of the pile in the sand stratum.

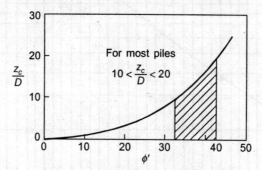

Fig. 9.14 z_c/D versus ϕ' (Vesic, 1970).

In most piling problems, z_c/D will vary between 10–20. This means that the effective overburden pressure and both skin friction and end bearing will increase almost linearly upto a depth of 10–20 times the diameter of the pile and thereafter remain constant. The ultimate skin friction should therefore be obtained from the trapezoidal distribution of skin friction with depth, as shown in Fig. 9.15, while the ultimate end bearing should be determined for the maximum overburden pressure mobilized, corresponding to the critical depth z_c.

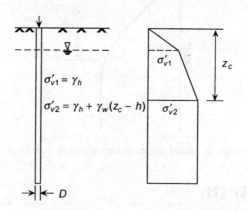

Fig. 9.15 Distribution of skin friction with depth.

9.8 PILE CAPACITY FROM IN-SITU SOIL TESTS

The methods of determination of axial load capacity of driven and bored piles as described earlier in the Chapter, make use of the shear strength parameters of the soil as determined from laboratory tests. For cohesive soils, the undrained shear strength, c_u can be determined from the field vane shear test also. For sensitive clays, in particular, sampling disturbances may cause reduction of the cohesive strength of the soil. Vane test results may be conveniently used in such clays.

For cohesionless soil, the angle of shearing resistance of the soil is best determined from the standard penetration test. This test is performed at different depths within boreholes and the average N value for a layer is related to the angle of shearing resistance, ϕ'.

A number of semi-empirical relationships have been proposed to obtain the skin friction and end resistance directly from the standard penetration test and static cone penetration test. Thorburn and MacVicar (1970), based on their experiences in Scotland, proposed an empirical formula for the evaluation of pile capacity directly from the standard penetration resistance of the sand as

$$Q_u = \frac{\bar{N}}{6} A_f + 4NA_p \quad \text{tonnes} \tag{9.19}$$

where A_f is area of pile strength (m^2),

A_p is area of pile tip (m^2),

$\bar{N}$ is the average N value over the pile shaft, and

N is the SPT value at pile base.

Tomlinson has given the relationship for ultimate bearing capacity of displacement piles from static cone penetration test.

$$Q_u = q_c A_p + \frac{\bar{q}_c}{200} A_s \tag{9.20}$$

where

q_c = cone resistance in t/m^2 at pile tip. It is taken as the average value of cone resistance over a depth equal to three times the pile diameter above the tip and one pile diameter below the tip,

A_p = area of pile tip in sq. m.,

$\bar{q}_c$ = average cone resistance over the embedded length of the pile shaft, and

A_s = surface area of pile shaft in sq. m.

For taking unit base resistance as q_c, the pile should penetrate at least 8 diameter into the bearing stratum and sand should be present to a depth of at least 3 diameter beneath the base.

9.9 UNDER-REAMED PILES

Under-reamed piles are often used as load bearing and anchor piles in expansive clays. A single under-reamed pile is suitable for anchor pile while double under-reamed piles are used to increase the load bearing capacity of the pile, this is shown in Fig. 9.16. The ultimate capacity of an under-reamed pile is given by

$$Q_u = \sum \alpha c_{u1} f_s + \frac{\pi d_0^2}{4} c_{u2} N_c \tag{9.21}$$

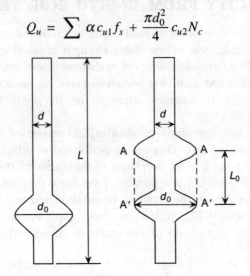

Fig. 9.16 Under-reamed pile.

where c_{u1} = undrained shear strength of the clay along pile shaft,
  α = adhesion factor,
  d_0 = diameter of under-reamed,
  c_{u2} = undrained shear strength of the clay at pile tip, and
  N_c = bearing capacity factor.

For double and multi-under-reamed piles, the soil between the bulbs acts as part of the pile and the frictional resistance for the length, L_0 may be calculated for the diameter, d_0 while for rest of the pile length above the under-ream, the skin friction develops over the diameter d. Mohan et al. (1967) suggest an optimum spacing of $1.25d_0 - 1.5d_0$ between bulbs and under-reamed diameter, d_0 approximately equal to $2.5d$.

Under-reamed piles are not normally recommended for cohesionless soil owing to doubts about formation of bulb in soils having no cohesion.

9.10 ALLOWABLE LOAD ON PILES FROM STATIC ANALYSIS

The load carrying capacity of a pile is determined from separate evaluation of the ultimate skin friction and the ultimate base resistance. The two are added together to arrive at the ultimate capacity of the pile, refer Eq. (9.2). However, this would imply a total settlement of the order of 20–30% of the pile diameter if the ultimate base resistance is to be fully mobilized, the skin friction having already mobilized at a settlement of only 0.5–1% of the pile diameter. For large diameter piles (diameter > 600 mm) that are now being increasingly used, this means a total settlement which may go beyond the permissible limit for an engineering structure. Therefore, the allowable load on a pile has to be restricted to the extent by which the settlement is to be restricted for the pile. A factor of safety of 2.5 generally ensures that the settlement is well within the permissible limits, that is,

$$Q_a = \frac{Q_u}{2.5} \qquad (9.22)$$

where, Q_a = allowable load on pile.

However, it may be noted that a concept of overall factor of safety does not necessarily imply identical factors of safety with respect to both frictional resistance and base resistance of a pile. Tomlinson (1977) has observed that for a large diameter bored pile in stiff clay, an overall factor of safety equal to 2.0 will result in the mobilization of full frictional resistance (i.e., $FS = 1$) while only 22% of the ultimate base resistance will come into play (i.e $FS = 5$). Settlement of a pile foundation will, therefore, be primarily governed by the compression of the soil around the pile shaft where the load is transmitted to the soil by skin friction.

Design of a pile foundation involves not only the determination of the allowable load on a pile from consideration of ultimate resistance of single pile but also ensuring the stability of the pile group where a large number of closely spaced piles are provided to support the column load. The entire soil including the pile group will be stressed as a block and the settlement of the pile group will have to be evaluated under the working load and the design should ensure that the settlement is acceptable. Group action of piles is discussed in Section 9.12.

9.11 DYNAMICS OF PILE DRIVING (DYNAMIC ANALYSIS)

The resistance to penetration of a pile during driving may be related to its static bearing capacity by applying the principal of conservation of energy as

$$WH = RS \qquad (9.23)$$

where W = weight of driving hammer,
   H = height of fall of hammer,
   R = resistance to penetration, and
   S = pile penetration under each blow.

The left hand side of Eq. (9.23) is the energy supplied per blow, and the right hand side is the corresponding work done to facilitate the penetration of pile. The basic elements involved in the driving analysis is shown in Fig. 9.17.

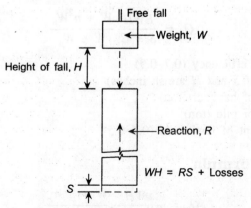

Fig. 9.17 Dynamics of pile driving.

Equation (9.23) is, to some extent, a simplification of a complex field problem. In reality, following uncertainties tend to influence the measured response of the pile to the driving energy:

1. The soil resistance is not constant during pile driving because of elasticity and damping characteristics of soil.
2. The elastic compression of the cap block, cushion, pile, and soil absorb some energy and does not contribute to the penetration of pile.
3. Some energy is lost through impact, and the sound and heat generated during the hammer blow.
4. A pile is a long, slender member and at any instant, different lengths of the piles experience different kinds of motion in the soil.

Notwithstanding the above uncertainties a number of pile driving formulae have been developed to determine the ultimate capacity of the pile during driving. They differ mainly in the manner of accounting for energy losses involved in the driving operation. Some of the commonly used formulae are discussed in the subsequent parts of this section.

9.11.1 ENR Formula

The earliest pile driving formula assumes that for a given hammer blow, the resistance increases in an elastic manner as the pile is displaced, remains constant for further displacement, and finally falls to zero as the pile rebounds. Equating the energy supplied to the work done, the following formula was obtained.

$$WH = Q_u(S + C) \tag{9.24}$$

where W = weight of hammer (ton),
H = fall of hammer (ft),
S = penetration per blow (in),
R = pile resistance (ton), and
C = constant which accounts for elastic settlement of pile-soil system (1.0 in for drop hammer and 0.1 in for single acting steam hammer)

Equation (9.24) has subsequently been revised to a more generalized form,

$$Q_u = \frac{EWH}{S + C} \frac{W + n^2 W_p}{W + W_p} \tag{9.25}$$

where E = hammer efficiency (0.7–0.9)
C = 0.1 in (if S and H are in inches)
W = weight of hammer (ton)
W_p = weight of pile (ton)
n = coefficient of restitution (0.4–0.5)

9.11.2 Hiley Formula

$$Q_u = \frac{\eta WH}{S + C/2} \frac{W + n^2 W_p}{W + W_p}$$

Here, Q_u, W, H, n, S, and W_p have the same meaning as in Eq. (9.25).

η = efficiency of hammer blow (0.75–1.0)

C = a factor which accounts for energy losses due to elastic compression of pile, C_1, elastic compression of the head assembly, C_2, and elastic compression of the soil, C_3, that is,

$$C = C_1 + C_2 + C_3$$

The approximate values of C_1, C_2, and C_3 to be used in Hiley formula for concrete piles are given in Table 9.2.

Table 9.2   Values of C_1, C_2, C_3 in Hiley formula

C_1 = 0.075–0.10 in, for hard driving
$C_2 = R_u L/AE_p$,   where R_u = ultimate test load,
L = Length of pile,
A = Cross sectional area of pile, and
E_p = Elastic modulus of pile material.
C_3 = 0.1 in (0 for hard soil and 0.2 for resilient soil)

9.11.3 Simplex Formula

This formula has been found to give quite good prediction of load carrying capacity for driven piles in alluvial deposits. The ultimate capacity is given by:

$$Q_u = NWH \frac{\sqrt{(L/50)}}{L(1 + S)} \qquad (9.26)$$

where Q_u = ultimate load in tons,

L = embedded length of pile in ft.,

W = weight of hammer in tons,

N = total number of blows,

S = penetration for last blow in inches, and

H = drop of hammer in ft.

9.11.4 Janbu's Formula (Janbu, 1953)

$$Q_u = \frac{EH}{K_u S} \qquad (9.27)$$

where

$$K_u = C_d \left(1 + \sqrt{1 + \frac{\lambda}{C_d}} \right);$$

$$C_d = 0.75 + 0.14 \left(\frac{W_p}{W} \right), \text{ and}$$

$$\lambda = \frac{EHL}{A_p E_p S^2}$$

The ultimate pile capacity as determined from the pile driving formulae may be used to determine the safe load capacity of a pile by using a factor of safety. In view of the uncertainties involved in the calculation, a high factor of safety, not less than 3, should be used. The early ENR formula even recommended a factor of safety equal to 6.

Sorensen and Hansen (1957), Housel (1966), and Olsen and Flaate (1967) made comprehensive studies on the use of different pile driving formulae in predicting the pile capacity. It appears from their studies, "if driving formulae are to be used, those which involve the least uncertainty are the Hiley and Janbu formulae while the most uncertain is the ENR formula", (Poulos and Davis, 1980).

9.11.5 Wave Equation

It has been long recognized that the phenomenon of pile driving involves transmission of compression waves down the pile. Smith (1960) gave a practical method of solving the wave equation using a digital computer for studying the dynamic behaviour of a pile during driving. The analysis involves dividing the pile into a number of segments, each beam represented by a weight joined to the adjacent weights by springs. The hammer, pile cap, cap block, and cushion block are also represented by weights and springs. The soil resistance is represented by a Kelvin rheological model consisting of spring and dash pot. A set of equations can be set up considering dynamic equilibrium of each element. The pile resistance corresponding to a given set observed in the field can be obtained by solving the set of simultaneous equations. The procedure has been described by Bowles (1968, 1974).

The success of this method is handicapped by the lack of knowledge of essential parameters involved in the equation. Nowadays, with the advent of sophisticated equipment such as Pile Driving Analyzer (PDA) and suitable transducers which can be attached to piles to measure the relevant parameters, for example, velocity, acceleration, and so on, the method is becoming increasingly popular. A number of computer program have been developed to get direct output of the pile capacity from the PDA.

9.11.6 Limitations of Dynamic Analysis

Most of the dynamic formulae are based on assumptions which have little regard for the driving of an actual pile. For instance, Newtonian Impact Theory is not valid for pile driving because of the presence of cushioning material. Also it is difficult to correctly estimate the various energy losses involved in pile driving. Although Smith's wave equation is theoretically more sound, it's practical usefulness is limited by the lack of precise information about quake, damping factor, and distribution of resistance through the shaft and the base.

One major limitation of dynamic analysis is that the dynamic resistance can hardly be equal to the static bearing capacity. Even if this may be assumed to be true at the time of driving or immediately afterwards, the situation is likely to change with time, and, as a result, there may be significant difference between the actual pile capacity and the value obtained from dynamic analysis. Nevertheless, dynamic analysis may be used in pile driving to obtain assessment of the variation of soil consistency with depth. It gives a good indication if the pile has reached good bearing stratum. Also, through proper correlation with load test data, a dynamic formula can be used to have a very good control over the pile installation.

9.12 PILE GROUPS

Piles are commonly used in groups with a foundation slab or pile cap cast over the pile heads to distribute the load to the piles. Although the piles in a group are generally identical, the group capacity is not necessarily the same as the capacity of single pile multiplied by the number of piles in the group. Also, the settlement of a pile group may be different from the settlement of a single pile subjected to the same average load per pile. This is because of interaction among piles which is known as *group action*.

Figure 9.18 shows the pressure bulb of a single pile and that of a group of piles installed in a homogeneous medium. In case of a closely spaced group, the pressure bulb of individual piles overlap and a much larger bulb is formed. This is likely to result in a larger settlement. Also, due to overlapping of stresses, the capacity of individual piles may decrease and result in a lower capacity of the group.

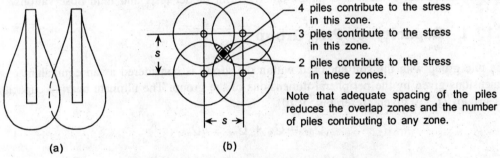

Fig. 9.18 Pressure bulb for single pile and pile group.

A vivid example of this effect is seen in the foundations of the Charity Hospital Building in New Orleans, (Terzaghi, 1942), as depicted in Fig. 9.19. A steel framed building with stone facing was founded on 13 m long timber piles. Soil investigation was done prior to design to a depth of 15 m and the piles were driven into the dense sand below the soft

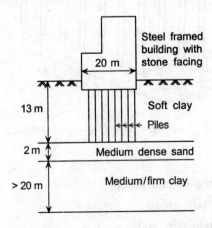

Fig. 9.19 Foundation for Charity Hospital Building, New Orleans (Terzaghi, 1942).

clay. Against an estimated working load of 15 ton, a single pile was tested for a maximum load of 30 ton and gave a settlement of only 6 mm. On this basis, a maximum settlement of 6 mm was anticipated for the building. However, two years after construction, the building had settled 270 mm with a maximum differential settlement of 200 cm. Obviously, as later investigation revealed, the pressure bulb below the pile group extended well into the compressible clay below the sand layer which caused large settlement of the building. In the Hotel Sao Paulo building which is founded on spread footings supported on RCC piles, the settlement of the pile group under full structural load was 15 mm while the test pile showed settlement of only 3 mm (Vargas, 1948). Another good example of this group effect is an apartment house in Vienna where the settlement of the building was 20 times more than that of the test pile under the same load (Terzaghi, 1955).

Theoretical analysis of pile groups considering interaction of piles is quite complicated. Load carrying capacity and settlement of pile groups are generally determined by methods based on empirical evidences obtained chiefly from model tests and field observations.

9.12.1 Capacity of Pile Group

The pile group with the soil enclosed within the group is considered as an equivalent pier of dimensions given by the peripheral dimensions of the group. The ultimate bearing capacity of the group, Q_{ug} is given (refer Fig. 9.20).

$$Q_{ug} = Q_{fg} + Q_{pg} \leq nQ_u \tag{9.28}$$

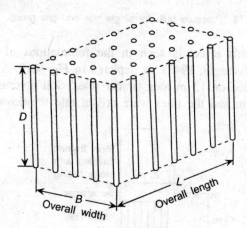

Fig. 9.20 Capacity of pile group: block action.

where

Q_{fg} = frictional resistance of pile group,

Q_{pg} = point bearing of pile group,

Q_u = capacity of a single pile, and

n = no. of piles in the group.

For friction piles in clay,

$$\left. \begin{array}{l} Q_{fg} = 2(B + L)D_f c_u \\ Q_{pg} = (B \times L)9c_u \end{array} \right\} \qquad (9.29)$$

where

B = width of pile group,

L = length of pile group,

D_f = depth of piles,

c_u = average undrained shear strength of clay along the pile depth, and

Terzaghi and Peck (1967) recommended $FS = 3$ to obtain allowable load on the pile group. This approach appears reasonable for friction piles in clay. For piles resting in sand, end bearing resistance of the pile group will be high and block failure would hardly be possible. The group capacity may then be taken as the capacity of a single pile multiplied by the number of piles in the group.

9.12.2 Pile Spacing

The spacing of piles in a group is to be selected basically keeping under consideration the factor that the influence zones around individual piles do not overlap and each pile in the group is allowed to develop its full capacity. The practical considerations important from this point of view are

(a) *The type of soil and the method of installation:* Larger spacing may be required for driven piles in saturated clay to minimize the effect of heaving.

(b) *The minimum spacing:* Should be such that with due tolerance for error in layout and verticality, reasonable gap is maintained between adjacent piles.

(c) *The efficiency of the group:* The bearing should be such that the group efficiency is close to unity, that is, each pile can develop its full capacity.

(d) *The economy:* The larger the group, the more costly is the cap.

The building codes provide minimum allowable spacing of piles in a group. I.S. 2911 (part I), 1979, suggests the following guidelines for spacing:

(a) For piles bearing on a hard stratum and deriving their strength mainly from end bearing, the minimum spacing shall be $2.5d$, d being the diameter of the pile.

(b) For friction piles in clay, the minimum spacing shall be $3d$.

(c) For piles driven through loose sand or fill, the minimum spacing may be $2d$.

However, the most commonly used spacing is $3d$. Figure 9.21 shows some typical pile groups with different number of piles.

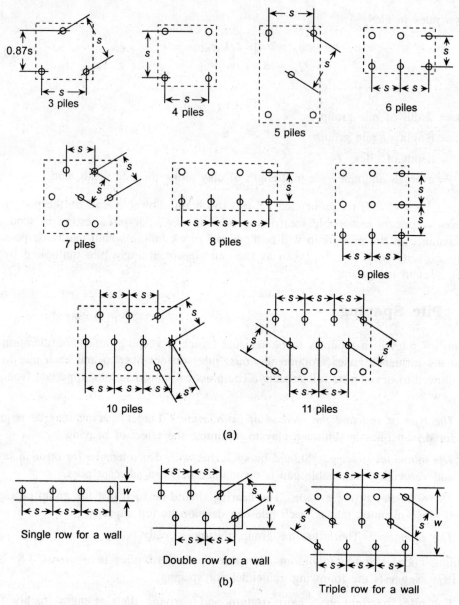

Fig. 9.21 Typical pile groups.

9.12.3 Pile Group Subjected to Vertical Load and Moment

A pile group may be subjected to moments due to horizontal load or eccentrically applied vertical load. The following approximate method is commonly used in practical design of group of identical piles subjected to vertical load and moments. The cap is assumed to be rigid, and the reaction of any pile is assumed to be proportional to the displacement of the pile head.

Figure 9.22 shows a pile group subjected to a vertical load, V at its CG = 0 (CG = centre of gravity) and moments M_{xx} and M_{yy} about XX and YY axes respectively, passing through 0. Due to vertical load, V there is uniform deflection and reaction. Therefore, pile reaction due to $V = V/N$, where N represents no. of piles.

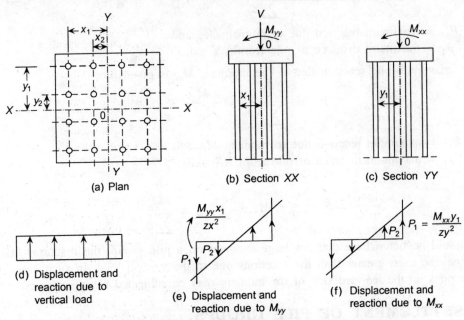

(a) Plan   (b) Section XX   (c) Section YY

(d) Displacement and reaction due to vertical load

(e) Displacement and reaction due to M_{yy}

(f) Displacement and reaction due to M_{xx}

Fig. 9.22 Pile group subjected to vertical load and moment.

Due to M_{yy}, there is tilting of the cap about 0, and the displacement of the head of each pile is proportional to the distance of the pile from 0, as shown.

For a row of N piles,

$$\frac{P_{y_1}}{x_1} = \frac{P_{y_2}}{x_2} = \cdots = \frac{P_{y_n}}{x_n}$$

This gives,

$$P_{y_1} = P_{y_2}\frac{x_1}{x_2}; \; P_{y_2} = P_{y_1}\frac{x_2}{x_1}; \cdots ; \; P_{y_n} = P_{y_1}\frac{x_n}{x_1}$$

But

$$M_{yy} = P_{y_1}x_1 + P_{y_2}x_2 + \cdots + P_{y_n}x_n$$

or

$$M_{yy} = P_{y_1}\frac{x_1^2}{x_1} + P_{y_2}\frac{x_2^2}{x_1} + \cdots + P_{y_n}\frac{x_n^2}{x_1}$$

$$= \frac{P_{y_1}}{x_1}(x_1^2 + x_2^2 + \cdots + x_n^2)$$

$$P_{y_1} = \frac{M_{yy}x_1}{\sum x_n^2} = \frac{M_{yy}}{z_{xx}}$$

On the other side of CG, the reaction is upward, that is, P_y is negative.
Thus,

$$P_{y_1} = \frac{\pm M_{yy}.x_1}{\sum x_n^2} = \frac{\pm M_{yy}}{z_{xx}}$$

where P_{y1} = maximum reaction due to moment M_{yy} and
x_1 = maximum distance of pile from YY axis.

Similarly, maximum pile reaction due to the moment, M_{xx} is given by

$$P_{x_1} = \frac{\pm M_{xx}y_1}{\sum y_n^2} = \frac{\pm M_{xx}}{z_{yy}}$$

where, P_{x1} = maximum reaction due to moment M_{xx} and
x_1 = maximum distance of pile from XX axis.

Thus,

$$\frac{\text{load}}{\text{pile}} = P = \frac{V}{N} \pm M_{yy}\, x_1 \sum x_n^2 \pm \frac{M_{xx}\, y_1}{\sum y_n^2} \qquad (9.30)$$

It should be noted here that for large moment on a pile group, the reactions due to moment may be even greater than the reactions due to the vertical load. In such cases, the outermost piles on the leeward side of the moment may be subjected to uplift forces.

9.13   SETTLEMENT OF PILE GROUPS

As yet, there is no precise method of estimating the settlement of pile groups. The problem is still approached in an approximate manner.

9.13.1   Pile Groups in Cohesive Soil

The settlement of a pile group in predominantly cohesive soil is given by the sum of the immediate or elastic settlement and the long-term consolidation settlement of the subsoil. The procedure for calculating these settlements are similar to that for raft foundations with appropriate depth corrections as described in Chapters 7 and 8. However, it is, necessary to determine the load distribution around a pile group in order to obtain the size and depth of the equivalent raft from which the settlement is to be determined.

It requires rigorous mathematical analysis, for example, finite element technique, to determine the true stress distribution around a pile group (Poulos and Davis, 1980) but such analysis is highly complex and is seldom used in practical design because pile foundations, in any case are expected to undergo small settlement. Simplified method of load distribution has been suggested by Terzahi and Peck (1962) and Tomlinson (1967, 1977). Accordingly, total load, Q is assumed to get dispersed from the foundation level at a slope of 4:1 upto the depth of an imaginary raft at depth $2D_f/3$, as shown in Fig. 9.23. Here, the size of the imaginary raft becomes $(B + D_f/6)(L + D_f/6)$. Thereafter, the load is assumed to get dispersed at a slope of 2:1 into the underlying strata. If the pile group passes through a very weak stratum to an underlying hard stratum, the load is assumed to spread at a slope 4:1 on an equivalent raft at a depth $2L/3$, where L is the embedment of the pile in the stiff stratum.

Thereafter, a spread of 2:1 may be followed. For a wholly end bearing pile, the load is assumed to get transferred to an equivalent raft at the level of pile tip without any dispersion through the upper strata. It is, therefore, necessary to determine the compressibility characteristics of the soil below the depth $2D_f/3$. The settlement of the pile group may then be calculated in the usual way.

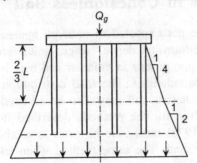

Fig. 9.23 Settlement of pile group: load dispersion.

(a) *Immediate settlement*

$$\rho_i = \alpha \frac{q_n B}{E} \left(1 - v^2\right) I_\rho \qquad (9.31)$$

where
q_n = net pressure on the equivalent raft at depth $2D_f/3$,
B = width of equivalent raft at depth $2D_f/3$,
v = Poisson's ratio of soil below the depth equivalent raft (may be taken as 0.5),
E = elastic modulus of the soil below equivalent raft,
I_ρ = influence coefficient, $f(L/B)$, and
α = depth correction factor, $f(D_f/B, L/B)$

The elastic modulus of the soil may be determined from the undrained shear strength of the soil by using the relationship $E = (500–1000) c_u$ as discussed in Chapter 7. The influence coefficient, I_ρ and the depth correction factor, α are given by the elastic analysis of Boussinesq (1985) and Fox (1948) (refer Chapter 7).

(b) *Consolidation settlement*

The consolidation settlement of the pile group is obtained from the standard equation,

$$\rho_c = \left(\sum \frac{C_c}{1 + e_0} H \log \frac{p_o + \Delta p}{p_o}\right) \alpha \mu \qquad (9.32)$$

or,

$$\rho_c = \left(\sum m_v \Delta p H\right) \alpha \mu$$

where
$C_c/(1 + e_0)$ and m_v are respectively the compressibility index and the coefficient of volume decrease for the appropriate stress level of the relevant strata,
H is the thickness of strata,
p_o and Δp are the in-situ vertical effective stress and the increase of stress in the respective strata, and
α and μ are respectively the depth and pore-pressure correction factors.

To know the effective depth of soil below the pile group, the best method is to determine the stress increment ratio $\Delta p/p_o$ at different depths and consider the depth of soil for which $\Delta p/p > 0.1$.

9.13.2 Pile Groups in Cohesionless Soil

Pile foundations in sand are not expected to undergo appreciable settlement because of the low compressibility of medium to dense, where piles are usually terminated. Still, if necessary, the settlement of pile groups in granular soil may be determined by using the same methods as given for raft foundations. The load distribution in the soil and the equivalent raft concept proposed for cohesive soil may also be adopted to obtain the geometry of the problem to be solved. Further, for the methods described in Chapter 7 for foundations on sand, Schultze and Sherif (1973) used case histories to establish a method of predicting the settlement of foundation on sand. The same method when adopted for pile groups gives

$$\delta = \frac{Sq_u}{N^{0.81}\left(1 + 0.4\dfrac{D_f}{B}\right)} \qquad (9.33)$$

where S = settlement coefficient,
  q_u = average pressure on the equivalent raft,
  N = average SPT value over a depth $2B$ below the foundation level or D_s if the depth of cohesionless soil is less than $2B$,
  D_f = depth of equivalent raft, and
  B = width of equivalent raft.

The settlement coefficient, S varies with D_f/B as shown in Fig. 9.24.

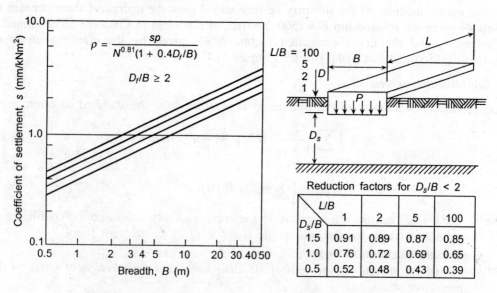

Fig. 9.24 Settlement coefficient, S versus D_f/B (Schultze and Sherif, 1973).

9.14 UPLIFT RESISTANCE OF PILES

A straight shaft pile, when subjected to uplift forces, derives its ultimate capacity from frictional resistance of the pile which can be determined in the same way as indicated for piles under compression (see Section 9.7). However, for cyclic loading, skin friction may be reduced by the degradation of soil strength at the pile–soil interface under repetitive load. In particular, for sandy soils, reduction in uplift capacity to 50% of the ultimate skin friction has been reported (St. John et al., 1983). Even for cohesive soils, Radhakrishnan and Adams (1973) have observed 30–50% reduction of uplift capacity in short augured piles. Such reduction in uplift capacity has been attributed to long term creep under sustained loading whereby the strength of the soil reaches its residual value. As a general rule, a factor of safety of 3–4 on the frictional resistance calculated for compression may be applied to determine the uplift capacity of piles. However, it should be noted that an upward movement of only 0.5–1% of the pile diameter is required to mobilize the peak frictional resistance.

In case of pile groups, the uplift resistance may be calculated by taking the resistance of the soil enclosed within the groups, as depicted in Fig. 9.25. For cohesionless soil, an assumed spread of 1:4 from the pile tip to the ground and the weight of soil block enclosed within the group give the frictional resistance. In such cases, Tomlinson suggests a factor of safety equals to unity against uplift since skin friction around the periphery of the group is ignored. The submerged weight of the soil below the ground water table should be taken.

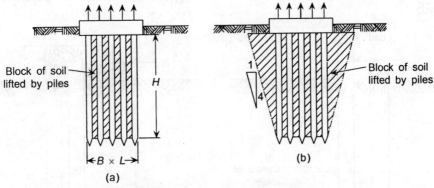

Fig. 9.25 Uplift resistance of pile group.

For cohesive soils, the uplift resistance of the block may be obtained by summing up the undrained shearing resistance around the periphery of the block and the weight of soil enclosed by the group as

$$Q_u = 2(L + B)D_f c_u + W \tag{9.34}$$

where L = length of the pile group,
B = width of the pile group,
D_f = depth of the pile group,
c_u = average undrained shear strength of the clay, and
W = weight of the soil enclosed within the block.

A safety factor of 3 should be used to determine the safe uplift capacity of the group.

9.15 PILES UNDER HORIZONTAL FORCES

Vertical piles can resist lateral forces to a certain extent depending on the strength and stiffness of the pile and the soil. According to I.S. 2911—1985, permissible lateral load of a vertical pile is 2–5% of the permissible vertical load. For greater horizontal load, additional reinforcement is to be provided in the pile or raker piles may be used.

Ultimate lateral resistance and load-deflection behaviour of a pile under lateral load is a complex problem of soil-structure interaction. The lateral load on the pile head is initially carried by the soil close to the ground surface resulting in elastic deformation of the soil. As the load increases, the soil yields and transfers the load to greater depths.

9.15.1 Failure Mechanisms

Short and long piles fail under different mechanisms. This subsection briefly discusses these failure mechanisms.

A short rigid pile, unrestrained at the head, tends to rotate or tilt as shown in Fig. 9.26, and passive resistance develops above and below the point of rotation on opposite sides of the pile. If the pile head is restrained by a cap, there will be lateral translation. In both the cases, the pile will fail when the applied load exceeds the passive resistance of the soil.

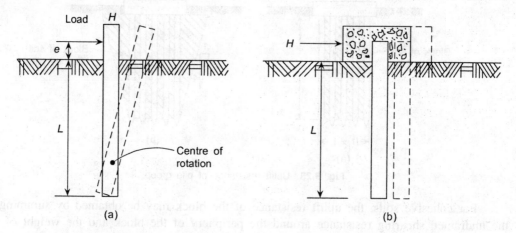

Fig. 9.26 Failure mechanism of rigid pile under horizontal load: (a) free head, (b) fixed head.

For a long pile, the passive resistance is very large and pile cannot rotate or tilt. The lower portion remains almost vertical due to fixity while the upper part deflects in flexure. The pile fails when a plastic hinge is formed at the point of maximum bending moment, Fig. 9.27.

As a consequence, a short pile fails when passive resistance of soil is exceeded (soil failure) and a long pile fails when the moment capacity is exceeded (structural failure).

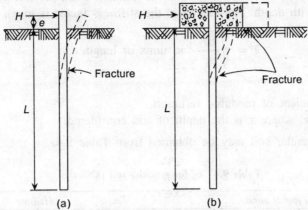

Fig. 9.27 Failure mechanism of long pile under horizontal load: (a) free head, (b) fixed head.

For development of ultimate resistance, the lateral movement is generally too large. Therefore, after calculating the ultimate resistance and dividing by a factor of safety, it is necessary to check that the permissible deflection of pile head is not exceeded.

9.15.2 Stiffness Factors and Subgrade Modulus

The stiffness factors, R and T, of the pile–soil system determine the behaviour of a pile as a short rigid pile or a long flexible one. These factors depend on the stiffness, EI of the pile and the compressibility of the soil expressed in terms of a soil modulus. The soil modulus depends on the type of soil, the width of pile, and the depth of influence area and is related to Terzaghi's modulus of subgrade reaction.

For stiff overconsolidated clay, the soil modulus is assumed to be constant with depth, and the stiffness factor is given by

$$R = \sqrt[4]{\frac{EI}{KD}} \quad \text{in units of length} \tag{9.35}$$

where $K = K_1/1.5$ in which K_1 is Terzaghi's subgrade modulus in kg/m³ or kN/m³ and
D = diameter of pile.

According to Terzaghi (1955), K_1 is related to the undrained shear strength of the clay as shown in Table 9.3.

Table 9.3 Values of K_1 for different consistencies of clay (after Terzaghi)

Consistency	Unconfined comp. strength, q_u(kN/m²)	Range of K_1 (kN/m³)	Recommended, K_1 (kN/m³)
Stiff	100–200	18–36	27
Very stiff	200–400	36–72	54
Hard	>400	>72	>108

For soft normally consolidated clays and for granular soils, the soil modulus is assumed to increase linearly with depth. For this case, the stiffness factor is given by

$$T = \sqrt[5]{\frac{EI}{n_h}} \quad \text{in units of length} \tag{9.36}$$

where

n_h = coefficient of modulus variation

= $K D/x$, where x is the depth of soil considered.

The n_h values for granular soil may be obtained from Table 9.4.

Table 9.4 n_h for granular soil (kN/m^3)

Type of sand		Loose	Medium	Dense
Dry or moist sand	(Terzaghi (1955))	2500	7500	20000
Sub merged sand		1400	5000	12000

For soft normally consolidated clay,

$$n_h = 350 \text{ to } 750 \text{ kN/m}^3$$

For soft organic silt,

$$n_h = 150 \text{ kN/m}^3$$

The criteria for behaviour of a pile as a short pile or a long pile are related to the embedded length, L as follows:

Long Pile: $L \geq 4T$ or $L \geq 3.5R$

Short Pile: $L \leq 2T$ or $L \leq 2R$

9.15.3 Ultimate Lateral Resistance

The method suggested by Broms (Broms, 1964; Poulos and Davis, 1980) appears to be most convenient for design. Broms assumed simplified distribution of soil resistance for cohesive and cohesionless soils and determined the ultimate capacity of short and long piles in terms of the flexural rigidity of the pile. The design charts prepared by Broms are given Fig. 9.28 through Fig. 9.31.

In Fig. 9.28, the variation of $H_u/c_u D^2$ versus L/D (where H_u = ultimate horizontal load, c_u = undrained shear strength of soil, D = pile dia, L = embedded length) have been plotted for short piles in cohesive soils. Curves for both restrained and unrestrained piles with different eccentricity ratio have been presented. In Fig. 9.29, the corresponding plots for short piles in cohesionless soil with design curves for $H_u/K_p D^3 \gamma$ versus L/D (K_p = coefficient of passive earth pressure and γ = unit weight of soil) are presented. The curves for long piles in cohesive soil are given in Fig. 9.30. The normalized ultimate lateral resistance, $H_u/c_u D^2$ is plotted against normalized yield moment, $M_{yield}/c_u D^3$. Figure 9.31 shows corresponding curves for cohesionless soil, where normalized ultimate lateral resistance, $H_u/K_p D^3 \gamma$ has been plotted against normalized yield moment, $M_{yield}/K_p \gamma D^4$.

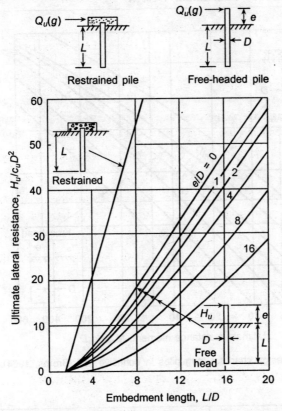

Fig. 9.28 Design charts for short piles in cohesive soil (Broms 1964).

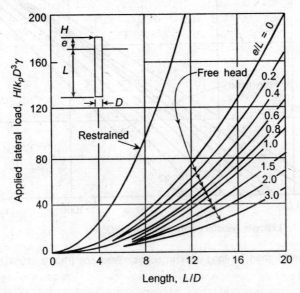

Fig. 9.29 Design charts for short piles in cohesionless soil (Broms, 1964).

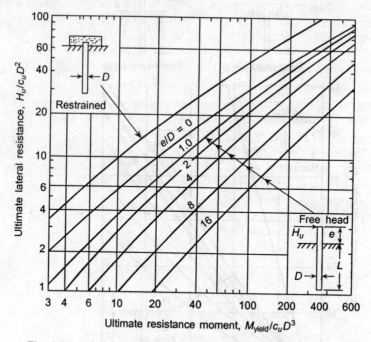

Fig. 9.30 Design chart for long piles in cohesive soil (Broms, 1964).

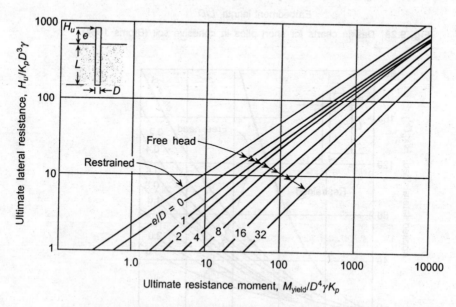

Fig. 9.31 Design chart for long piles in cohesionless soil (Broms, 1964).

9.15.4 Deflection, Moment, and Shear under Working Load

Under the working load, the pile and soil are generally assumed to behave elastically, and Winkler model is applied to determine load-deformation behaviour of the pile. The soil pressure and deflection, y at a point are assumed to be related through the horizontal modulus of subgrade reaction as follows:

$$p = Ky \tag{9.37}$$

or,

$$w = KDy \tag{9.38}$$

where

w = soil reaction per unit length of pile,
K = modulus of subgrade reaction (force/length^3), and
D = diameter or width of pile.

The behaviour of the pile is assumed to be governed by the beam equation (Heteny, 1946)

$$E_p I_p D^4 y/D^4 + KDy = 0 \tag{9.39}$$

Where

E_p = elastic modulus of pile and
I_p = moment of inertia of pile section.

Closed form solution of Eq. (9.39) can be obtained for the simple case of constant K. For the cases where K varies with depth, numerical analysis has to be applied.

A solution of Eq. (9.39) for appropriate boundary conditions gives the following results (Rees and Matlock 1956, Das 1998):

Pile deflection at any depth

$$x_z(z) = A_x \frac{Q_g T^3}{E_p I_p} + B_x \frac{M_g T^2}{E_p I_p} \tag{9.40}$$

where
Q_g = Applied lateral load and
M_g = Applied moment
$\Bigr\}$ at frow surface

Moment in pile at any depth

$$M_z(z) = A_m Q_g T + B_m M_g \tag{9.41}$$

where A_x, B_x, A_m, and B_m are coefficients and T is the characteristic length of the pile–soil system given by

$$T = \sqrt{\frac{E_p I_p}{n_h}} \tag{9.42}$$

Table 9.4 gives the representative values of n_h. Figure 9.32 shows the variation of A_x, B_x, A_m, and B_m for different values of L/T. Beyond $L/T > 5$, the coefficients remain essentially constant.

For cohesive soil, the elastic solution has been given by Davisson and Gill (1963). The pile deflection $x_z(z)$ is given by

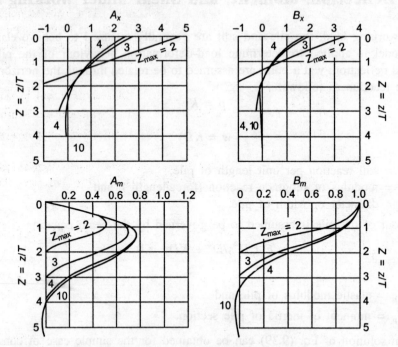

Fig. 9.32 Pile deflection and moment in cohesionless soil: Elastic analysis (Reese and Matlock (1956), Das (1998)).

$$x_z(z) = A'_x \frac{Q_g R^3}{E_p I_p} + B'_x \frac{M_g R^2}{E_p I_p} \qquad (9.43)$$

and

$$M_z(z) = A'_x Q_g R + B'_m M_g \qquad (9.44)$$

where A'_x, B'_x, A'_m, and B'_m are coefficients and $R = \sqrt{E_p I_p / K}$.

The variation of A'_x, B'_x, A'_m, and B'_m with z is given in Fig. 9.33.

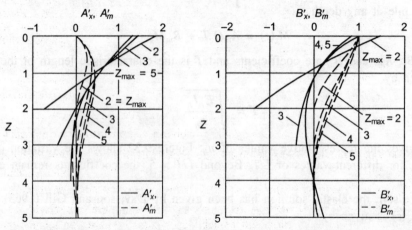

Fig. 9.33 Pile deflection and moment in cohesive soil: Elastic analysis (Davisson and Gill, 1963).

Broms (1966) worked out a procedure for obtaining the deflection of the pile head from an ultimate load analysis. He presented the relationship between dimensionless parameters.

$$\frac{x_0(E_pI_p)^{3/5}(n_h)^{2/5}}{Q_gL} \quad \text{and} \quad \eta L, \quad \text{for sand}$$

and

$$\frac{x_0(KDL)}{Q_g} \quad \text{and} \quad \beta L, \quad \text{for clay.}$$

where x_0 = deflection at pile head,
   E_p = elastic modulus of pile,
   I_p = moment of inertia of pile section,
   $n_h = \dfrac{K_z}{h}$ = constant of modulus of horizontal subgrade reaction,
   K = modulus of subgrade reaction,
   D = diameter of pile,
   L = length of pile,
   Q_g = working load,

$$\eta = \sqrt{\frac{n_h}{E_pI_p}}$$

$$\beta = \sqrt{\frac{KD}{4E_pI_p}}$$

These relationships are illustrated through Figs. 9.34 and 9.35.

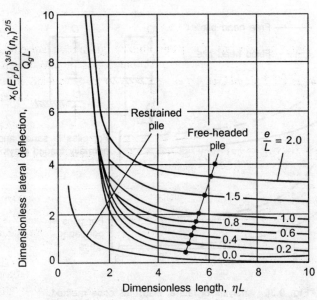

Fig. 9.34 Deflection of pile head under working load: Cohesionless soil (Broms, 1966).

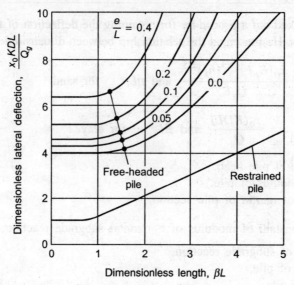

Fig. 9.35 Deflection of pile head under working load: Cohesive soil (Broms, 1966).

9.15.5 I.S. Code Method

(I.S. 2911 (Part 1/Sec. 4)—1984, Amended in 1987)

In this method, a long pile is considered as a cantilever with fixity at some depth below the ground surface. The depth of fixity depends on the relative stiffness of pile and soil, Fig. 9.36.

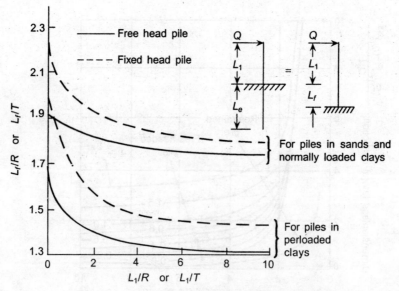

Fig. 9.36 Analysis of lateral load: IS code method.

Considering the pile as an equivalent cantilever, the pile load deflection, y is obtained as

$$y = \frac{Q(L_1 + L_f)^3}{3E_pI_p}, \quad \text{for free head pile}$$

$$= \frac{Q(L_1 + L_f)^3}{12E_pI_p}, \quad \text{for fixed head pile}$$
(9.45)

where Q = lateral load and L_1 and L_f are as shown in the figure.

The fixed end moment (M_F) of the equivalent cantilever is given by

$$M_F = Q(L_1 + L_f), \quad \text{for free head}$$

$$= Q(L_1 + L_f), \quad \text{for fixed head}$$
(9.46)

The actual moment is obtained by applying reduction factor, as depicted in Fig. 9.37, which is given by

$$M = m(M_F)$$

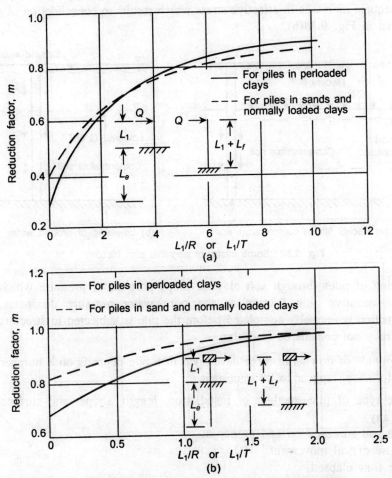

Fig. 9.37 Correction factor for long piles (IS 2911).

9.16 NEGATIVE SKIN FRICTION

When a soil surrounding the pile shaft moves downward relative to the pile shaft, the friction between the soil and the pile tends to drag the pile downwards. The skin friction in that case increases the pile load instead of resisting it. This additional load on the pile is known as *negative skin friction* or *downdrag force*.

Negative skin friction may develop under any the following situations:

1. A pile is installed through a recently placed fill overlying soft compressible clay to a stiff bearing stratum. Alternatively, the fill might have been placed after the pile is installed through the soft compressible layer to the bearing stratum. The situation is illustrated in Fig. 9.38(a). As the soft clay layer consolidates, the fill as well as the soft clay layer move downward and exert a downdrag force on the pile. The magnitude of negative skin friction can be very high and may lead to failure of piles in the extreme case.

2. Negative skin friction may also occur due to lowering of ground water table and subsequent increases in effective stress which results in consolidation of soft clay as shown in Fig. 9.38(b).

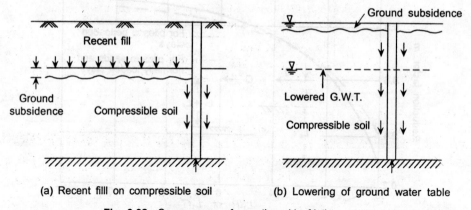

(a) Recent filll on compressible soil   (b) Lowering of ground water table

Fig. 9.38 Some cases of negative skin friction.

3. Driving of piles through soft clay develops pore water pressure which may also cause negative skin friction as the pore water pressure dissipates. But this dissipation is normally completed before the pile is subjected to structural load and generally not considered in design.

The magnitude of downdrag force developed in a pile depends on a number of factors of which the following appear to be important.

(a) The type of pile, method of installation, length, shape, and surface treatment (if any).

(b) The soil stratification and properties.

(c) Cause of soil movement.

(d) The time elapsed.

When a pile is subjected to external load in a soil which is undergoing consolidation under its own weight due to a fill, there is relative movement between the soil and the pile. If the elastic compression of the pile is neglected, the pile settles uniformly while the surrounding soil settlement is maximum at the top and minimum at the pile trip, visible in Fig. 9.39. Accordingly, a neutral point develops above which the soil moves more than the pile and below which the pile moves more than the soil. Obviously negative drag occurs on the length of the pile above the neutral point. In case of end bearing pile, the neutral point develops only at the pile trip and the negative drag occurs over the full length of the pile, as shown in Fig. 9.39.

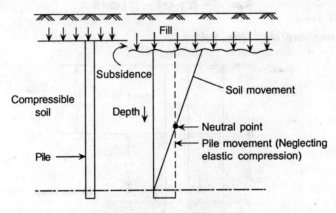

Fig. 9.39 Neutral Point.

The unit negative drag on a pile at any depth may be determined in the same way as frictional resistance is determined. Two cases are considered, Das (1998), as follows:

(a) *Clay overlying sand fill (Fig. 9.40.)*

$$f_n = K'\gamma'z \tan \delta \qquad (9.47)$$

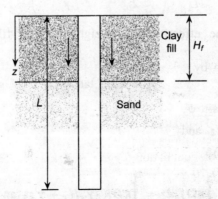

Fig. 9.40 Negative skin friction in clay overlying sand fill.

where K' = earth pressure coefficient (take $K' = K_o = 1 - \sin \phi'$),

$\gamma'z$ = effective vertical stress at depth z, and

δ = soil–pile friction angle: $0.5\phi'-0.7\phi'$.

Hence, total downdrag force on the pile

$$Q_u = \pi D \int_{z_1}^{z_2} K'\gamma'\tan\delta z\,dz$$

$$= \frac{\pi D}{2} K'\gamma'(z_2^2 - z_1^2)\tan\delta \tag{9.48}$$

(b) *Sand fill overlying clay (Fig. 9.41.)*

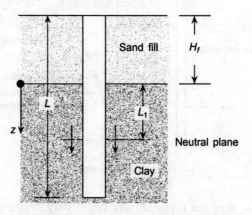

Fig. 9.41 Negative skin friction in fill overlying clay.

For a height of fill, H_f above the clay the depth of neutral point, L_1 may be obtained from the approximate relationship (Bowles, 1982)

$$L_1 = \frac{L - H_f}{L_1}\left(\frac{L - H_f}{2} + \frac{\gamma'_f - H_f}{\gamma'}\right) - \frac{2\gamma'H_f}{\gamma'} \tag{9.49}$$

where γ'_f and γ' are the effective unit weights of the fill and the underlying clay respectively.

The negative drag is given by

$$f_n = K'\sigma'_v \tan \delta$$

where $K' = K_o = 1 - \sin \phi'$,

$\sigma'_v = \gamma'_f H_f + \gamma'z$, and

$\delta = (0.5 - 0.7)\phi'$

$\therefore$
$$Q_u = \int_0^{L_1}\pi D f_n dz = \int_0^{L_1}\pi D K'(\gamma_f H_f + \gamma'z)\tan\delta\,dz$$

$$= (\pi D K'\gamma_f H_f \tan\delta)\,L_1 + \frac{\pi D L_1^2 K'\gamma'\tan\delta}{2} \tag{9.50}$$

A semi-empirical method of estimating the negative drag based on the one described by Terzaghi and Peck (1967) has been suggested by Nayak (1967). It can be undertaken through the following steps:

1. Estimate the length of pile over which negative skin friction develops.
2. Estimate the positive skin friction over that length.
3. Multiply the value obtained in (2) by a factor to obtain dragdown. This factor depends on the cause of soil movement and may be taken as a value varying over 0.6–0.8.
4. Estimate the ultimate pile capacity, Q_{up} as consisting of point bearing and shaft resistance over the length *where there is no negative skin friction*.
5. Calculate factor of safety, F from

$$F = \frac{Q_{up}}{(Q + Q_{ns})} \tag{9.51}$$

where Q = structural load on pile and
         Q_{ns} = downdrag force on the pile.

In a pile group, the total downdrag force on the group, Q_{ng} is the minimum of the following values.

$$Q_{ng} = N Q_{ns} \tag{9.52}$$

and

$$Q_{ng} = \gamma L_n P \tag{9.53}$$

the weight of soil enclosed with the pile groups,

where N = no. of piles in the group,
         γ = unit weight of soil,
         L_n = length over which negative friction develops, and
         P = perimeter of the group.

Poulos and Davis (1980) have described a method based on elastic theory, with allowance for pile–soil slip, to estimate downdrag force. In this method, the dependence of downdrag on the settlement of the soil can be included, rather than assuming that sufficient soil movement occurs to mobilize the full adhesion along the pile shaft. Based on a study of the various parameters responsible for negative skin friction, the authors have presented charts which can be used very conveniently to calculate downdrag force.

9.17 TESTING OF PILES

Recent trends in foundation design follow the use of large diameter, high capacity piles for supporting heavy structural loads. Till the early sixties, driven piles of diameter upto 500 mm were normally in use in India. These piles would have safe load capacity upto 100 tonnes depending on the depth of pile and subsoil condition. With the coming of bored cast-in-situ systems, large diameter piles are now common and typical 1500 mm diameter piles with safe capacity of 500 tonnes or more have frequently been used in recent years. This obviously necessitates a strict quality control in the installation of the pile so that each pile carries the design load with certainty.

9.17.1 Purpose of Pile Testing

In bored cast-in-situ piles, a vertical hole is made into the soil by augering and the same is stabilized by casing or drilling mud. A reinforcement cage is inserted and the hole is filled with concrete by the tremie method. The entire operation being directed from the ground, uncertainties remain as to the condition of the soil at the pile tip and the integrity of the concrete in the pile shaft. The purpose of pile testing is, therefore, to determine

(a) whether the pile tip has reached firm stratum or it rests on loose soil at the bottom of the hole,
(b) whether the concreting of the pile shaft has been done properly and without any discontinuity, and
(c) whether the load carrying capacity of the pile has been correctly assessed.

There is no readily available method of checking the condition of the soil at the pile tip prior to concreting. Normally, at the end of boring, reverse circulation is done to remove all debris from the bottom of the hole and the depth is finally obtained from the length of the tremie pipe and the depth of boring. Even then, uncertainities remain about the possible existence of any thin layer of loose soil at the pile tip.

9.17.2 Causes of Defect in Piles

Defects in pile shaft normally occur in the form of unfilled voids which cause discontinuity in the pile shaft. Major causes of such defects have been listed by Tomlinson (1981) as

(a) Encrustation of hardened concrete on the inside of the casing which may cause the concrete to be lifted as the casing is withdrawn.
(b) The falling concrete which may arch across the casing or between the casing and the reinforcement.
(c) Falling concrete may get jammed between the reinforcing bars and not move towards the borehole wall.
(d) Clay lumps may fall into the hole as the concrete is placed.
(e) Soft or loose soil may squeeze into the pile shaft from the bottom of the lining.

Most of these defects can be minimized, if not eliminated, by having the inside of the casing properly cleaned, using a high slump concrete and by avoiding conjestion of reinforcing bars. Also, proper care needs to be taken in lifting the casing while concreting, particularly in unstable soils.

9.17.3 Integrity Testing

Often excavations for pile caps show defective construction near the pile head. Similar defects may be there at greater depth also. Therefore, integrity testing is done to check the soundness of the pile shaft after installation. The following methods of integrity testing of piles are generally available (Weltman, 1980; Robertson, 1982)

(a) Excavation surrounding the pile shaft

(b) Exploratory boring through the pile shaft

(c) Acoustic tests

(d) Radiometric tests

(e) Dynamic response of pile

(f) Load test

Excavation around the pile shaft is only possible for shallow depths. It is hardly conceivable that a deep pile can be fully exposed for visual inspection. However, short piles of limited length or pile near the ground surface can be examined thoroughly.

Drilling/boring through the pile shaft is possible through large diameter piles. Cores of concrete can be examined for soundness and they can be tested to determine their compressive strength. Even a TV Camera may be lowered into the hole to look for cavities and honeycombs.

Various radiometric and acoustic tests are also done in drilled holes. Pairs of ducts are made in the pile shaft at the time of concreting and suitable scanning devices are introduced to scan the concrete between the ducts. However, these require specialized equipment. Sonic integrity testing is one such test which is gaining popularity.

Seismic and dynamic response tests are extensively done because of their simplicity and adaptability. No hole is required to be made in the pile shaft. In the seismic method, a weight is dropped on the pile head and the time for return of the seismic wave after reflection from the toe is measured. In the dynamic response method, an electrodynamic vibrator is mounted on the pile head to apply a constant amplitude stress wave at the pile top and the response of the pile is seen through an oscillograph or digital indicator. Various types of pile diagnostic systems/pile driving analyzers are now commercially available to facilitate such testing.

Load test is the most direct method of determining the capacity of the pile. While other methods of integrity testing determine primarily the soundness of the concrete, the load test gives an integrated method of determining both the soundness of the concrete and the response of the soil under load. It also permits an evaluation of the load carrying capacity of the pile on the basis of soil response. The next section discusses load test on piles in a much greater depth.

9.18 LOAD TEST ON PILES

Initial and routine load tests are commonly undertaken in all piling work in India. The basic purpose of an *initial load test* is to obtain the failure load of a pile and then to determine the design load by applying a factor of safety. However, for larger diameter piles, this means applying a very heavy load on the pile to take it to failure. *Routine tests* are carried out on working piles to serve as a proof test to ensure that the behaviour of pile is satisfactory. This test is carried out on a limited number of working piles as a measure of quality control.

9.18.1 Test Procedure

The test load is applied on the pile by jacking against a reaction frame which is either loaded with kentledge or supported on anchor piles. The hydraulic jack is provided with a remote controlled pump and the pressure gauge gives a measure of the load applied. For more accurate measurement of load, a load cell or a proving ring may be used. The distance of the reaction frame supports from the pile should be at least five times the diameter of the pile. The displacement of the pile head is measured by dial gauges (3 or 4 nos.) suspended from datum bars resting on unyielding supports, sufficiently away from the pile, and reaction frame supports. A pile test arrangement is shown in Fig. 9.42.

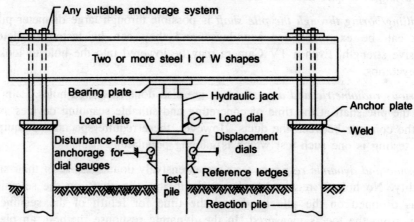

Fig. 9.42 Load test on pile.

9.18.2 Maintained Load Test

This is the usual method of test to obtain a load settlement curve, which may be used for estimating the ultimate load as well as settlement under working load. The load is applied in suitable increments (say 20% of the estimated working load) and settlement observation under each load is made as a function of time. Each load is maintained till the settlement becomes negligible. According to I.S. 2911—1985, each load should be maintained for two hours or till the rate of settlement becomes 0.2 mm/hour, whichever is earlier. For initial load test, the loading is continued till the settlement of pile head equals one tenth of the pile diameter, or the load is three times the estimated safe load, whichever is earlier. In case of routine load test, the load is continued upto one and half times the working load. The load is then released in equal steps to zero, and settlement observations are made at each stage.

The result of load test is presented in the form of a load-settlement curve for both loading and unloading, as illustrated in Fig. 9.43. The ultimate load is obtained as the one among any of the following approaches:

(a) The load corresponding to the point of intersection of the tangents of upper and lower portion of the curve.

(b) The load corresponding to a break point when the load settlement curve is drawn on a log–log plot.

(c) The load at which the settlement of the pile head is equal to 10% of the pile diameter.

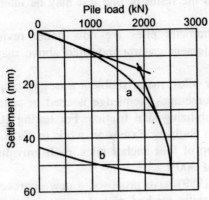

Fig. 9.43 Load versus settlement relationship from pile test.

According to IS 2911—1985, the safe load from initial load test is taken as the least of the following:

(a) Two third of the load at which gross settlement is 12 mm or
(b) 50% of the final load at which gross settlement becomes 1/10th of the pile diameter.

The ultimate load obtained by these methods for the pile test data, as depicted through Fig. 9.43, is shown in Table 9.5.

Table 9.5 Pile capacity from load test (Fig. 9.43)

Method	Ultimate load (kN)	Safe load (kN) FS = 2
Double tangent (on arithmetic plot)	1900	950
Double tangent (on log–log plot)	1320	660
Settlement 10% pile diameter	2300	1150
I. S. 1985		
2/3 gross load at 12 mm settlement	1300	870
50% of load at settlement of 10% pile diameter	2300	1150

In routine load test, the pile is considered to be safe if the gross settlement under 1.5 times the working load is not greater than 12 mm.

Maintained load test is very time consuming. Also, some amount of consolidation and creep may take place during the test. So, it may be difficult to correlate the test results with undrained shear strength data obtained from laboratory tests.

9.18.3 Constant Rate of Penetration (CRP) Test

This is a short duration test in which the load is applied at a constant rate of strain till the pile fails. The rate of strain depends on the type of soil. A common rate is 0.75 mm/min for friction piles in clay, and 1.5 mm/min for end bearing piles in sand. The duration for the test corresponds to the time required for failure in a laboratory undrained shear test. So, the two tests have a common basis and the result of one test may be interpreted on the basis of the other.

The criteria for large diameter piles are now under review by Bureau of Indian Standards. Obviously, the settlement criteria indicated above may hardly be applicable to large diameter piles.

For testing large diameter piles, major problem arises in building up a huge kentledge to take the desired reaction. Sometimes, kentledge is used in conjunction with anchor piles which provides reaction by mobilizing skin friction. For testing 1000 mm diameter × 20 m long bored piles for the Metro Railway Extension work in Calcutta with a design load of 230 t, the reaction was made up of four anchor piles, each providing a reaction load of 70 t and a dead weight kentledge of 300 t.

Birmingham and James (1989) have proposed a new approach to load testing for high capacity piles. This is a quasi-static method of test using a statnamic apparatus. A reaction mass is placed over a pressure chamber on top of the test pile. Fuel is burnt within the pressure chamber to create large pressures to drive the reaction mass upwards at high velocity. The reaction force acts equally downwards on the pile top. A schematic diagram of the system and the details of the test chamber are shown in Fig. 9.44. The force–time history of the mass allows a linearly increasing load on the pile of greater than 600 t. The duration of the pulse is upto 80 ms which is an order of magnitude higher than that associated with the conventional drop hammer blow. The acceleration imparted to the pile is less than 1 g as compared to 50–100 g for dynamic loading. This emulates the stress condition within a pile during static loading.

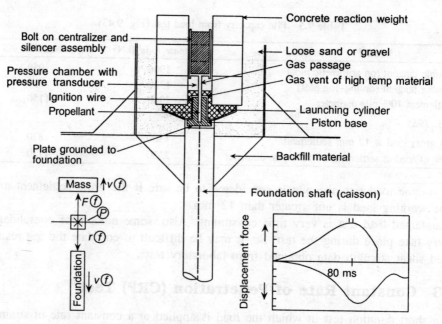

Fig. 9.44 Statnamic test apparatus.

9.18.4 Pile Driving Analyzer

Measurement of stress waves in piles has led to the development of a new and versatile equipment for testing the integrity of piles and predicting pile capacity. Different types of pile driving analyzers (PDA) are now commercially available. These are essentially based on the propagation of stress waves down and up the length of the pile while the boundary values applied at the pile top are used as inputs for the wave equation analysis. The commonly available pile driving analyzers are capable of determining the

(a) pile bearing capacity,
(b) maximum compressible and tensile stresses,
(c) maximum pile top movement, velocity, and acceleration and
(d) pile integrity.

For interpretation of data, a wave equation analysis programme is developed and fed into the system. For example, PWDWAP (Tan et al., 1987) uses a model of the pile as discrete line elements and the soil as a system of elasto-plastic springs and viscous dashpots. The programme uses an implicit Newmark integration analysis and displays the measured and predicted responses on a computer screen and presents the data in convenient worksheet format. Tan et al. (1987) present five case histories of piles in residual soil and one in alluvium formation. The pile capacities that predicted the Hiley's formula were compared with the results of PDA and load tests data (refer Table 9.6).

Table 9.6 Summary of pile capacities (Tan et al., 1987)

S.No.	Size of pile (mm)	Length (m)	Design load (m)	Ultimate Load		
				Hiley formula (t)	PDA (t)	Load test (t)
	Residual Soil					
1.	Conc. 280 × 280	25.1	93	220	272	268
2.	Conc. 325 × 325	25.6	115	355	360	380
3.	Steel 356 × 368	40.8	120	368	310	390
4.	Steel 356 × 368	29.4	120	352	250	283
5.	Steel 356 × 368	28.0	120	427	400	390
	Alluvium					
6.	Steel 356 × 368	11.0	110	252	372	330

A field test with PDA is shown in Fig. 9.45. Major defects such as cracks, soil incursions, necking, and changes in cross section are easily detected using this test. For complete pile driving analysis, two specially designed combined acceleration and strain transducers are mounted on the side of the pile near the pile top. During driving of piles, signals obtained through the mounted transducers are passed on to the signal conditioning subsystem for processing and finally to the computer. The processed signal for each blow is stored in a hard disc memory and displayed on the computer screen. After pile driving, a full record of all relevant parameters are available with respect to the number of blows as well as depth. Typical test data obtained for structurally sound piles and piles with discontinuity are shown in Fig. 9.45.

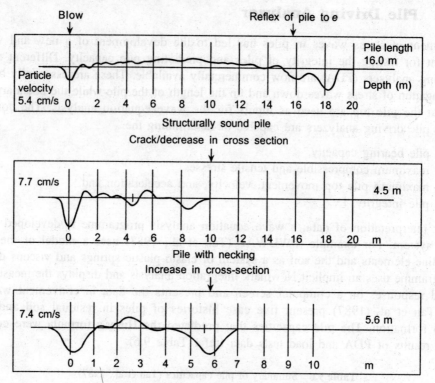

Fig. 9.45 PDA results cast-in-situ concrete piles.

Example 9.1

A closed ended steel pipe pile, 30 cm in diameter and 20 m long is driven to a sand deposit with cut off at ground level. The depth-wise variation of N and ϕ values is given in Fig. 9.46. The water table is 3 m below ground level. Compute the ultimate load capacity of the pile based on soil support, using

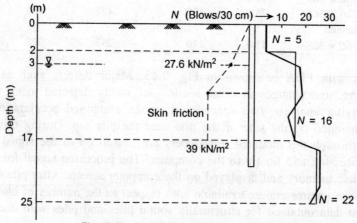

Fig. 9.46 Variation of N and ϕ with depth.

(i) N value.
(ii) Overburden pressure considering critical depth.

Take effective unit, $\gamma' = 1.73$ t/m^3 above water table, and 0.73 t/m^3 below water table.

Solution

(i) Using N value

Average N along pile shaft, $N_{av} = [(5 \times 2) + (16 \times 15) + (22 \times 3.6)]/20.6 = 15.98$
$\simeq 16$ (upto $2D$ below tip)
N value below tip $= 22$

Using Eq. (9.19)

$$Q_u = 4NA_p + \frac{N_{av}}{6} A_f$$
$$= 4 \times 22 \times 0.785 \times (0.3)^2 + (16/6)\ 3.14 \times 0.3 \times 20$$
$$= 62.22 + 50.24 = 112.41 \simeq 112\ \text{t} = 1120\ \text{kN}$$

(ii) Considering critical depth for overburden
For medium sand, $z_c/D = 20$, $z_c = 6$ m, $N_q = 45$, using Berezantzev's curve, Fig. 9.12.

For driven steel pile, take $K_s = 1.5$, $\phi_{av} = 0.7\phi'$

$$0\text{–}3\ \text{m:}\quad \phi_{av} = 0.7\phi' \times 28° = 19.50°$$
$$3\text{–}20\ \text{m:}\quad \phi_{av} = 0.7\phi' \times 32° = 22.5°$$

Also, $\quad Q_u = \sum A_s f_s + A_p q_p$

$$f_{s_1} = K_s \gamma D_f \tan 19.5° = 1.5 \times 17.3 \times 3 \times \tan 19.50° = 27.6\ \text{kN/m}^2$$

$$f_{s_2} = K_s \gamma' D_f \tan 22.5°$$
$$= 1.5(17.3 \times 3 + 7.3 \times 1.5) \tan 22.5°$$
$$= 39.0\ \text{t/m}^2$$

$$\therefore\ \sum A_s f_s = \pi \times 0.3\left[\frac{3 \times 27.6}{2} + 3\left(\frac{27.6 + 39}{2}\right) + (14 \times 39)\right]$$
$$= \pi \times 0.3(41.4 + 99.9 + 546)$$
$$= 647\ \text{kN}$$

and $\quad A_p q_p = \dfrac{\pi \times 0.3^2}{4}(\gamma' D_f N_q)$

$$= \frac{\pi \times 0.3^2}{4}(17.3 \times 3 + 7.3 \times 17) \times 45$$
$$= 559\ \text{kN}$$

$$Q_u = 647 + 559 = 1206\ \text{kN}$$

Example 9.2

Determine the safe load of a group of 9 driven cast-in-situ piles, 40 cm in diameter, 15 m long, installed in a cohesive deposit, as shown in Fig. 9.47. The cut off level is 1.5 m below ground level and the spacing of piles 1.2 m.

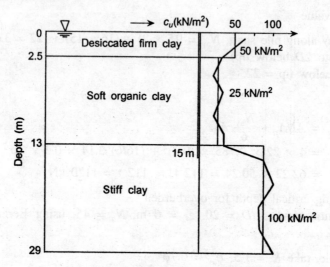

Fig. 9.47 Group of piles in a cohesive deposit.

Solution

Individual pile action

Capacity of a single pile: $Q_u = Q_f + Q_p$

Taking adhesion factors as follows:

c_u (kN/m^2)	α
50	0.9
25	1.0
100	0.45

$$Q_f = 3.14 \times 0.4(0.9 \times 50 \times 1 + 1.0 \times 25 \times 10.5 + 0.45 \times 100 \times 3.5)$$

$$= 584.2 \text{ kN}$$

$$Q_p = 0.785 \times (0.4)^2 \times 9 \times 100 = 113 \text{ kN} \quad (\text{where } N_c = 9.0)$$

∴ $\quad Q_u = 584 + 113 = 697$

∴ $\quad Q_{all} = 697/2.5 = 280$ kN

∴ $\quad$ Safe capacity of 9 piles = $9 \times 280 = 2520$ kN

Block action (Fig. 9.48)

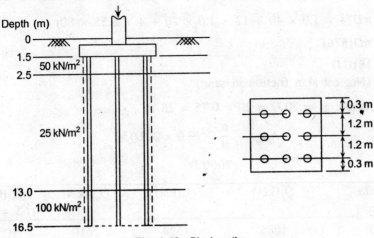

Fig. 9.48 Block action.

$$Q_{ult} = 2(2.4 + 2.4)(50 \times 1 + 25 \times 10.5 + 100 \times 3.5) + (2.4 \times 2.4 \times 9 \times 100)$$

$$= 6360 + 5184$$

$$= 11,544 \text{ kN}$$

$$Q_{all} = \frac{11,544}{2.5} = 4600 \text{ kN, which is greater than 2520 kN}$$

Hence, safe capacity of pile group = 2520 kN.

Example 9.3

A multistoreyed building is to be constructed at a site whose subsoil profile and properties are shown in Fig. 9.49. It has been decided to support the building on pile foundation, with cut-off level at 2 m below G.L. Compute the safe load of 25 m long bored piles of 650 mm, 750 mm, and 1 m diameter. Assume ground water table at ground level and average unit weight, $\gamma = 19$ kN/m³.

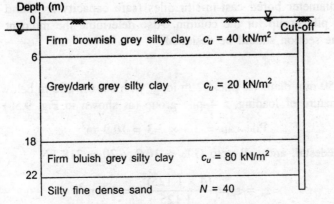

Fig. 9.49 Subsoil profile for Example 9.3.

$$Q_{ult} = \sum A_f f_s + A_p q_p$$

$$\sum A_f = \pi D(4 \times 1.0 \times 40 + 12 \times 1.0 \times 20 + 4 \times 0.55 \times 80)$$

$$= \pi D(576)$$

$$= 1810\, D$$

(Neglect skin friction in sand)

Taking critical depth, $z_c = 20\, D = 20 \times 0.75 = 15$ m

$$A_p q_p = \frac{\pi}{4} D^2 (9.0 \times 20D) 50$$

$$= 7060\, D^3$$

Pile dia. (mm)	Q_f(kN)	Q_p(kN)	Q_{ult}(kN)	Q_{all}(kN) (FS = 2.5)
600	1086	1524	2610	1050
750	1357	2978	4330	1730
1000	1810	7060	8870	3550

[*Note:* There is some uncertainty over the choice of critical depth in piles that penetrate through cohesive soil into sand. If the critical depth is not considered and full overburden is taken, end bearing in sand will increase appreciably.]

Therefore, 750 mm diameter piles with $Q_{all} = 1700$ kN are used.

Example 9.4

A typical column of the multistoreyed building in Example 9.3 is subjected to the following loads at ground level.

Vertical load V (kN)			M_{xx} (kNm)		M_{yy} (kNm)		Base shear H_x(kN) H_y(kN)	
DL	LL	SL	DL + LL	SL	DL + LL	SL	SL	SL
5500	760	657	123.5	—	115.2	985	380	—

Using 750 mm diameter bored cast-in-situ piles (safe capacity as found in Example 9.3), design a suitable pile group for the column. Also determine the moment due to horizontal load for which the section should be designed.

Solution

Safe bearing of 750 mm diameter pile, 22 m long, with cut-off at 2 m below G.L. is 1700 kN. Considering the nature of loading, a 4–pile group (as shown in Fig. 9.50) is chosen.

$$\text{Pile cap} = 3.3 \times 3.3 = 10.9 \text{ m}^2$$

Weight of cap, pedestal, and soil upto G.L. $\approx 10.9 \times 20 = 218$ kN

$$z_x = z_y = \frac{(4 \times 1.125)^2}{1.125} = 4.5 \text{ m}^3$$

(a) *DL + LL Condition*

$$\Sigma V = 5500 + 760 + 218 = 6478 \text{ kN}$$

$$M_{xx} = 123.5 \text{ kNm}$$

$$M_{yy} = 115.2 \text{ kNm}$$

∴ Maximum load in pile, $P = 647.8/4 + (123.5 + 115.2)/45 = 1672.5$ kN > 1670 kN.

(Marginal overloading on one pile may be accepted.)

(b) *DL + LL + SL Condition*

$$\Sigma V = 6478 + 657 = 7135 \text{ kN}$$

$$M_{xx} = 123.5 \text{ kNm}, \quad M_{yy} = 115.2 + 985 = 1100.2, \quad H_x = 380 \text{ kN}$$

Load per pile, $P = \dfrac{7135}{4} + \dfrac{123.5 + 45 + 110.2}{4.5} = 1845$ kN ($<1.25 \times 1700 = 2125$ kN)

(c) Horizontal load

$$H \text{ per Pile} = \frac{380}{4} = 95 \text{ kN}$$

Maximum Moment on Pile Section

(i) Using IS Code (IS 2911, Part I, Sec 4, 1984 with Amendment in 1987)

Stiffness Factor, $\quad R = \sqrt[4]{\left(\dfrac{E_p I_p}{K_2}\right)}$

$$E_p = 2 \times 10^6 \text{ t/m}^2$$

$$I_p = (\pi/64) \times (0.75)^4 = 0.015 \text{ m}^4$$

$$K_2 = 48.8 \text{ from code}$$

$$R = \sqrt[4]{\left(\frac{2 \times 10^6 \times \pi \times (0.75)^4}{64 \times 48.8}\right)} = 504 \text{ cm} = 5.04 \text{ m}$$

Embedded length, $L_e = 20$ m, $L_1 = 0$, and $L_1/R = 0$

Using the curve for fixed headed pile in preloaded clay,

$$\frac{L_f}{R} = 1.6$$

∴ Depth of fixity, $L_f = 1.6 \times 5.04 = 8.1$ m

Cantilever moment, $M_f = H(0 + L_f)/2 = (95 \times 8.1)/2 = 466$ kNm.

Corrected moment, $M = m M_f$

Using appropriate curve from code, $m = 0.7$

∴ Design moment $= 0.7 \times 466 = 326$ kNm.

Example 9.5

Compute the long term settlement of the pile group considered in Example 9.2.

The load on the pile cap is assumed to spread at a slope of 4 horizontal to 1 vertical upto the level of 2/3 pile length from the top, and then at slope of 2 vertical to 1 horizontal to a depth of 12 m below that level, which is about 1.5 times the width of equivalent imaginary footing.

Solution

Two points A and B are considered at the mid point of the relevant strata beneath the imaginary footing. Water table is assumed to be at G.L.

At A, $p_o = 12.25 \times 9 = 110$ kN/m^2

$\Delta p = 2520/(8.2)^2 = 37.4$ kN/m^2

$\delta_1 = 0.10 \times 150 \times \log_{10} [(110 + 37.4)/110] = 1.9$ cm

At B, $p_o = 19 \times 9 = 171$ kN/m^2

$\Delta p = 2520/(12.2 \times 12.2)^2 = 10.2$ kN/m^2

$\delta_2 = 0.08 \times 1200 \times \log_{10} [(171 + 10.4)/171] = 2.46$ cm

∴ Total settlement = 1.9 + 2.46 = 4.36 cm

Example 9.6

RCC bored piles, 1000 mm diameter × 20 m long are proposed to be used for a flyover passing over a canal bed with predominantly river channel deposits, as depicted in Fig. 9.50. DMC method of boring is adopted and steel liners would be used to a depth of at least 15 m below G.L. Design the safe pile capacity.

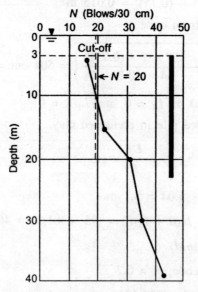

Fig. 9.50 RCC bored piles for Example 9.6.

Solution

Here, skin friction will develop between the liner and the soil.

Hence, take $K_s = 1.0$, $\delta = 0.7\phi' = 0.7 \times 34 = 23°$

End bearing: Consider reduction of ϕ' value for bored piles $= \phi' = 36 - 3 = 33°$

$$Q_{ult} = \sum A_f f_s + A_p q_p$$

Take
$z_c = 15D = 15$ m

$f_s = 1.0 \times 8 \times 15 \times \tan 23° = 51$ kN/m^2

$N_q = 43$

$Q_f = 1.0\pi \,(1/2(51 \times 15 + 5 \times 51))$

$\quad = 1600$ kN

$Q_p = \pi/4 \,(1)^2(8 \times 15 \times 43) = 3950$ kN

∴ $Q_u = 1600 + 3950 = 5950$

$Q_{all} = \dfrac{5950}{2.5} = 2380$ kN ≈ 2300 kN

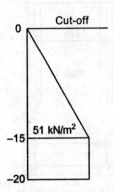

Example 9.7

If the pile in Example 9.6 is taken down to 28 m B.G.L. What will be safe capacity?

Solution

$Q_u = Q_f + Q_p$

$Q_f = \pi \times 1.0 \times (51 \times 15 + 51 \times 10 + 0.5 \times 100 \times 3)$

$\quad = 3270$ kN

$Q_p = \pi/4(1)^2 \times 9 \times 100$

$\quad = 700$ kN

$Q_u = 3270 + 700$

$\quad = 3970$ kN

∴ $Q_{all} = \dfrac{3970}{2.5} = 1590$ kN ≈ 1600 kN

[Note that increasing the pile length increases frictional resistance but there is major reduction end bearing also.]

Example 9.8

A foot bridge having the main trestle consisting of steel columns (refer Fig. 9.51(a)) is to be founded in the subsoil shown in Fig. 9.51(b). Design a suitable pile group for the trestle foundation.

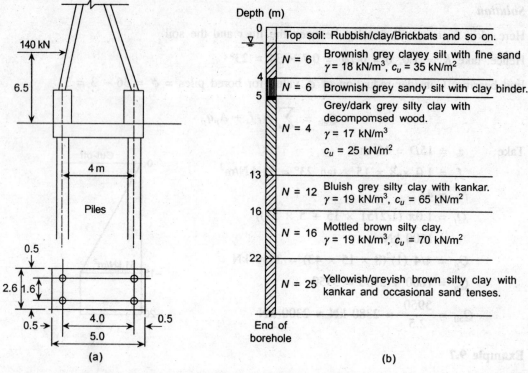

Fig. 9.51 (a) Main trestle columns and (b) subsoil type for Example 9.8.

Solution

Pile capacity

Pile diameter = 500 mm

Depth = 18 m/24 m

Cut-off level = 1 m below ground level

Pile tip at 18 m below ground level

$$Q_{ult} \text{ per pile} = \sum A_f f_s + A_p q_p$$

$$= 0.5\pi(4 \times 35 \times 1.0 + 8 \times 25 \times 1.0 + 3 \times 65 \times 0.7 + 2 \times 70 \times 0.6)$$
$$+ \pi/4(0.5)^2 (9 \times 70)$$

$$= 0.5\pi (140 + 200 + 136 + 84) + (0.2 \times 630)$$

$$= 879 + 126 = 1005 \text{ kN}$$

$$Q_{all} \text{ per pile} = \frac{1005}{3(FS)} = 333 \text{ kN} \approx 300 \text{ kN}$$

$$\text{Uplift per pile} = \frac{879}{3} = 293 \text{ kN} \approx 250 \text{ kN}$$

Main trestle foundation

Case I

Loading: vertical

$$DL = 240 \text{ kN}$$
$$LL = 520 \text{ kN}$$

Self weight of cap: $\quad\underline{\quad 100 \text{ kN}}$

$$\text{Total } 860 \text{ kN}$$

Lateral service wind = 17.5 kN

$$Q \text{ per pile} = \frac{860}{4} = 215 \text{ kN}$$

$$I_{xx} = 4 \times 2^2 = 16 \text{ m}^4$$

$$z_{xx} = \frac{16}{2} = 18 \text{ m}^3$$

∴ Maximum load on pile $= \dfrac{860}{4} + \dfrac{17.5 \times 6.5}{8}$

$$= 215 + 14 = 229 \text{ kN} < 300 \text{ kN} \quad \text{which is OK.}$$

Also, $\quad\quad\quad\quad\quad 229 - 14 = 215 \text{ kN} < 300 \text{ kN} \quad \text{which is OK.}$

Case II

Loading: vertical

$$DL = 240 \text{ kN}$$

Pile cap: $\quad\underline{\quad 100 \text{ kN}}$

$$\text{Total } 340 \text{ kN}$$

Lateral: Storm = 140 kN

$$\text{Maximum Compression in pile} = \frac{340}{4} + \frac{140 \times 6.5}{8} = 85 + 113$$

$$= 198 \text{ kN} < 300 \text{ kN, thus, accepted.}$$

$$\text{Maximum Uplift in pile} = \frac{340}{4} - \frac{140 \times 6.5}{8} = 85 - 113$$

$$= -28 \text{ kN} < 250 \text{ kN, thus, accepted.}$$

Hence, pile groups are safe for both cases.

Pile design

Consider storm load at 6.5 m above G.L. and vertical load (impact) at 0.6 m above G.L.

$$c_u = 35 \text{ kN/m}^2$$

$$K = \text{Coefficient of subgrade reaction} = 600 \text{ kN/m}^3$$

Stiffness factor, $\quad T = \sqrt[5]{\dfrac{EI}{K}}$

$$= \sqrt[5]{\dfrac{2 \times 10^6 \times \pi/64(0.5)^4}{60}}$$

$$= \sqrt[5]{102.2} = 2.52 \text{ m}$$

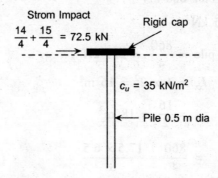

As pile depth = 18 m > $4T$. Hence, it is a long elastic pile and we use Brom's method.

$$\dfrac{H_u}{c_u B^2} = \dfrac{35}{35 \times (0.5)^2} = 4.0$$

Also, $\qquad\qquad e/B = 0$

$\therefore \qquad\qquad \dfrac{M_u}{c_u B^3} = 7$, which gives

$$M_u = 7 \times 35 \times (0.5)^3$$

$$= 30.6 \text{ kNm}$$

REFERENCES

Berezantzev, V.G., V. Khristoforov, and V. Golubkov (1961), *Load Bearing Capacity and Deformation of Piled Foundations*, Proc. 5th Int. Conf. on SMFE, Vol. 2, pp. 11–15.

Birmingham, P. and M. James (1989), *An Innovative Approach to Load Testing of High Capacity Piles*, Proceedings of the International Conference on Piling and Deep Foundations, London, Balkema, Roterdam, 1989.

Bishop, A.W. (1966), Sixth Rankine Lecture: The Strength of Soils as Engineering Materials, *Geotechnique,* Vol. 16, pp. 89–130.

Bowles, J.E. (1968), *Foundation Analysis and Design*, McGraw-Hill Inc., New York.

Bowles, J.E. (1974), *Analytical and Computer Methods in Foundation Engineering*, McGraw-Hill, New York.

Broms, B.B. and J.O. Silberman (1964), *Skin Friction Resistance for Piles in Cohesionless Soils*, Sols-Soils, pp. 10–33.

Broms, B.B. (1966), *Methods of Calculating the Ultimate Bearing Capacity of Piles—a summary*. Sols-Soils, pp. 18–19: 21–32.

Broms, B.B. and L. Hill (1973), *Pile Foundations for Kuwait Towers*, Proc. 8th Int. Conf. on SMFE, Moscow, Vol. 2–1, pp. 33–38.

Burland, J.B. (1973), *Shaft Friction of Piles in Clay—a simple fundamental approach*, Ground Engineering, Vol. 6, No. 3, pp. 30–42.

Cooke, R.W. and G. Price (1973), *Strains and Displacements Around Friction Piles*, Proc. 8th Int. Conf. on SMFE, Moscow, Vol. 2–1, pp. 53–60.

Das, B.M. (1998), *Principles of Foundation Engineering*, PWS Publishing, California, USA.

Davisson, M.T. and H.L. Gill (1963), Laterally Loaded Piles in a Layered Soil System, *Journal of the Soil Mechanics and Foundations Division,* American Society of Civil Engineers, Vol. 89, No. SM3, pp. 63–94.

De Beer, E.E. (1970), Experimental Determination of the Shape Factors and Bearing Capacity Factors of Sand, *Geotechnique,* Vol. 20, No. 4, pp. 387–411.

Flaate, K. (1972), Effects of Pile Driving in Clays, *Canadian Geotechnical Journal,* Vol. 9, No. 1.

Hanna, T.H. and R.H.S. Tan (1973), The Behavior of Long Piles Under Compressive Loads in Sand, *Canadian Geotechnical Journal,* Vol. 10, No. 3: pp. 311–340.

Janbu, N. (1953), *An Energy Analysis of Pile Driving with the Use of Dimensionless Parameters*, Norwegian Geotechnical Institute, Oslo, Publication No. 3.

IS 2911 (1979)—*Code of Practice for Design of Pile Foundations*, Bureau of Indian Standards, New Delhi.

Kerisel, J. (1961), *Foundations Profondes en Milieu Sableux*, Proc. 5th Int. Conf. on SMFE, Vol. 2: pp. 73–83.

Marsland, A. (1971), *Shear Strength of Fissured Clay, Stress-Strain Behavior of Soils*, T.G. Foulis, pp. 59–68.

McClelland, B. (1974), Design of Deep Penetration Piles for Ocean Structures, *Journal of the Geotechnical Engineering Division,* American Society of Civil Engineers, Vol. 100, No. GT7, pp. 709–747.

Meyerhof, G.G. (1959), Compaction of Sands and Bearing Capacity of Piles, ASCE, *Journal of Soil Mechanics and Foundation Division,* SM6, pp. 1–29.

Meyerhof, G.G. and L.J. Murdock (1953), An Investigation of the Bearing Capacity of Some Bored and Driven Piles in London Clay, *Geotechnique,* Vol. 3: p. 267.

Meyerhof, G.G. (1976), Bearing Capacity and Settlement of Pile Foundations, *Journal, Geotechnical Engineering Division,* ASCE, Vol. 88, SM4: pp. 32–67.

Nordlund, R.L. (1969), Bearing Capacity of Piles in Cohesionless Soil, *Journal of Soil Mechanics and Foundation,* ASCE, Vol. 89, No. SM3, pp. 1–35.

Poulos, H.G. and E.H. Davis (1980), *Pile Foundations Analysis and Design,* John Wiley and Sons, New York.

Radhakrishnan, H.S. and J.I. Adams (1973), Long-term Uplift Capacity of Augured Footings in Fissured Clays, *Canadian Geotechnical Journal,* Vol. 10, No. 4, pp. 647–652.

Rees, L.C. and H. Matlock (1956), *Non-dimensional Solutions for Laterally Loaded Piles with Soil Modulus Assumed Proportional to Depth,* 8th Texas Conf. on SMFE, Special Publication No. 29, University of Texas.

Schultze, E. and G. Sherif (1973), *Prediction of Settlement from Evaluated Settlement Observations for Sand,* Proc. 8th Int. Conf. on SMFE, Moscow, Vol. 1, p. 225.

Skempton, A.W. (1948), *The $\phi = 0$ Analysis of Stability and its Theoretical Basis,* Proc. 2nd Int. Conf. on SMFE, Rotterdam, Vol. 2.

Skempton, A.W. (1951), *The Bearing Capacity of Clays,* Building Research Congress, England.

Skempton, A.W. (1959), Cast in-situ Bored Piles in London Clay, *Geotechnique,* Vol. 9, No. 4, pp. 153–173.

Skempton, A.W. and P. La Rochelle (1955), The Bradwell Ship: A short term failure in London Clay, *Geotechnique,* Vol. 15, No. 3.

Skempton, A.W. and R.D. Northey (1952), The Sensitivity of Clays, *Geotechnique,* Vol. 3, No. 1, pp. 40–51.

Smith, E.A.L. (1960), Pile Driving Analysis by the Wave Equation, *Journal on Soil Mechanics and Foundation Division,* ASCE, Vol. 86, SM4: pp. 35–61.

Sliwinski, Z. and W.G.K. Fleming (1974), *Practical Considerations Affecting the Performance of Diaphragm Walls,* Proc. Conference on diaphragm walls and anchors, ICE, London, pp. 1–10.

Terzaghi, K. (1942), *Discussion on the Progress Report of the Committee on the Bearing Value of Pile Foundations,* Proc. ASCE, Vol. 68: pp. 311–323.

Terzaghi, K. (1943), *Theoretical Soil Mechanics,* John Wiley and Sons Ltd., p. 510.

Terzaghi, K. (1955), Evaluation of Coefficient of Subgrade Reaction, *Geotechnique,* Vol. 5: p. 297.

Terzaghi, K. and R.B. Peck (1967), *Soil Mechanics in Engineering Practice,* 2nd ed., John Wiley and Sons, New York.

Thorburn, S. and R.S.L. MacVicar (1970), Discussion, Conf. on Behaviour of Piles, ICE, London, p. 54.

Tomlinson, M.J. (1965), *Foundation Design and Construction,* 2nd ed., Pitman, London.

Tomlinson, M.J. (1977), *Pile Design and Construction Practices,* 1st ed., Viewpoint Publications, London.

Tomlinson, M.J. (1981), *Pile Design and Construction Practices*, 4th ed., E and F N Spon, London.

Touma, F.T. and L.C. Reese (1974), Behaviour of Bored Piles in Sand, *Journal on Geotechnical Engineering Division,* ASCE, Vol. 100, No. GT7: pp. 749–761.

Vargas, M. (1948), *Building Settlement Observations in Sao Paulo*, Proc. 2nd Int. Conf. on SMFE, Rotterdam, Vol. 4, p. 13.

Vesic, A.S. (1967), *A Study of Bearing Capacity of Deep Foundations*, Final Report, Project B-189, School of Civil Engineering, Georgia Institute of Technology, Atlanta.

Vesic, A.S. (1975), *Principles of Pile Foundation Design*, Duke University, School of Engineering, Soil Mechanics, Series No. 38.

Vesic, A.S. (1970), Tests in Instrumented Piles—Ogeechee River Site, *Journal of the Soil Mechanics and Foundations Division,* American Society of Civil Engineers, Vol. 96, No. SM2, pp. 561–584.

Vijayvergiya, V.N. and J.A. Focht, Jr. (1972), *A New Way to Predict Capacity of Piles in Clay*, Offshore Technology Conference Paper 1718, Fourth Offshore Technology Conference, Houston.

Whitaker, T. (1976), *The Design of Pile Foundations*, 2nd ed., Pergamon Press, London.

10 Well Foundations

10.1 INTRODUCTION

The conventional foundations discussed in the previous chapters are applicable to low and medium rise buildings. Depending on the subsoil condition at a given site, an optimum foundation design is to be achieved for the given structure. For multistorey buildings, say upto 20 storeys high, pile foundations with or without basement are generally adopted. For even heavier structural load—for bridge pier and abutments for example,—the vertical load may go upto a few thousand tonnes and major horizontal forces due to wind, current, and so on may apply. Conventional pile foundations are not suitable in such cases because of the limited capacity of piles to resist lateral forces. Large size foundations, often quite deep, are required for this type of loading. The behaviour of these foundations under large lateral forces are determined by conditions which are somewhat different from conventional foundations. Such foundations may undergo rigid body movement under the external forces and the lateral earth pressure on the embedded portion of the foundation. These foundations are commonly known as caissons and well foundations.

10.2 CLASSIFICATION

Wells or caissons are large size prismatic or cylindrical shells which are built deep into the ground to support heavy loads. Depending on the method of installation the caissons may be of three types, namely (i) open caisson or well, (ii) box caisson or floating caisson, and (iii) pneumatic caisson. The types of caissons or well foundations are depicted in Fig. 10.1. The top and bottom of *open caissons* are kept open and installed into the ground by excavation of soil within the shaft so that it may sink into the ground either under its own weight or by addition of surcharge load. The *box caisson* is a shell open at the top and closed at the bottom, built first on the solid ground and then towed to the site where it is sunk to a prepared foundation base. The *pneumatic caisson* has a working chamber at the bottom which is kept dry by maintaining a high air pressure to prevent water from entering into the chamber, thus, facilitating sinking of the caisson.

Although caissons are often used to support heavy structures including high rise buildings, they constitute by far the most common type of foundations for major bridges on

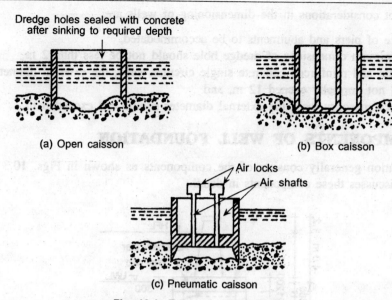

(a) Open caisson   (b) Box caisson

(c) Pneumatic caisson

Fig. 10.1 Types of well foundations.

rivers with erodable bed. Mostly open caissons or well foundations are used in India. This is because the well can be installed to sufficient depth below the maximum depth of scour, and a single well is capable of supporting large axial and lateral loads.

10.3 PHYSICAL CHARACTERISTICS—SHAPE AND SIZE

Taking into consideration the ease in sinking and construction, the most favoured shape of well, except for a very large size, is *circular*. The circular shape is equally strong in all directions, and presents a uniform surface to any direction of river current. Such wells are more suited for distribution of force uniformly over circumference during sinking, and are generally economical. Other alternatives are *double-D* and *dumb-bell* shapes. Typical cross-sections of wells of these shapes are shown in Fig. 10.2.

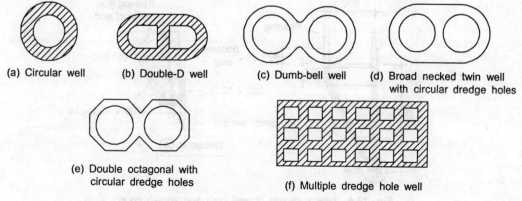

(a) Circular well   (b) Double-D well   (c) Dumb-bell well   (d) Broad necked twin well with circular dredge holes

(e) Double octagonal with circular dredge holes   (f) Multiple dredge hole well

Fig. 10.2 Typical cross-sections of well foundations.

Some important considerations in the dimensioning of wells are:

- (a) the size of piers and abutments to be accommodated,
- (b) the minimum dimensions of dredge hole should not be less than 2 m,
- (c) for plain and reinforced concrete single circular wells, the external diameter of well should not normally exceed 12 m, and
- (d) for brick masonry well, the external diameter should not exceed 6 m.

10.4 COMPONENTS OF WELL FOUNDATION

A well foundation generally consists of the components as shown in Figs. 10.3 and 10.4. This section discusses these components in detail.

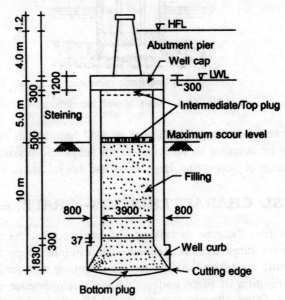

Fig. 10.3 Components of well foundation.

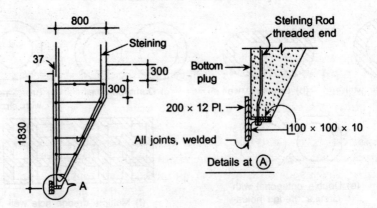

Fig. 10.4 Typical details of well curb and cutting edge.

10.4.1 Steining

The *well steining* is the main body of the well which transfers load to the subsoil. It also acts as a coffer dam during sinking, and provides the weight for sinking. In addition to the usual design forces at service condition, the steining is likely to be subjected to certain additional loading during installation. The thickness and reinforcement of well steining are generally determined as per recommendations of IRC 78–1983. The thickness of well steining should not be less than 500 mm and should satisfy the following relationship

$$h = kd\sqrt{L} \tag{10.1}$$

where h = minimum thickness of steining (in m),

d = external diameter of circular/dumb-bell shaped well or smaller dimension in plan for twin D wells (in m),

L = depth of well (in m) below LWL or ground level whichever is higher, and

k = a constant varying from 0.030 to 0.068 depending on shape of well and type of soil strata.

10.4.2 Well Curb

The wedge shaped part of the well steining at the lower end is called *well curb*. It facilitates the process of sinking. To satisfy the requirement of minimum resistance during sinking and the strength to transmit superimposed load from steining to bottom plug, the shape and outline dimensions should be as per Fig. 10.3. The concrete grade should not be lower than M20 and minimum reinforcement of 72 kg/cu m should be provided.

10.4.3 Cutting Edge

The lower most portion of the well curb is the *cutting edge*. This should be strong enough to facilitate sinking through the type of soil strata expected to be encountered during sinking and should be properly anchored to well curbs. Cutting edge should be fabricated from steel angles and plates weighing not less than 40 kg/m length of the cutting edge. Heavier section (80 kg/m) is required for hard and bouldery soil strata.

10.4.4 Bottom Plug

After the well is sunk to the required depth, the base of the well is plugged with concrete. This is called *bottom plug* which transmits the load to the subsoil. The top of the bottom plug should not be less than 300 mm above the top of well curb. A suitable sump below the level of the cutting edge should be made and cleaned thoroughly before concreting. The concrete mix used in bottom plug should not be leaner than 1:2:4, with minimum cement content of 330 kg/m^3, and should be placed by tremie or skip boxes under water.

10.4.5 Dredge Hole

The well is sunk by excavating soil from within the well. The hole so formed is called *dredge hole* which is later filled fully or partly by sand or excavated soil. The extent of filling is decided by the function it is required to do. Filling is used to increase the stability of the well against overturning. Where it is not possible to attain a positive sump for bottom plug (specially in sandy strata), complete filling of well is desirable. In case of soft/normally consolidated clay, the allowable bearing pressure can be considerably increased by keeping the well empty.

10.4.6 Intermediate/Top Plug

Intermediate plugs over the fill should be of thickness 500 mm of 1:3:6 concrete. When the well is filled upto the top, a top plug of thickness 300 mm of 1:3:6 concrete is generally provided. The top plug provides contact between the well cap and the sand filling, and helps in transferring the load through the sand filling.

10.4.7 Well Cap

Wells are provided with a properly designed RCC well cap, with its bottom surface preferably at LWL.

10.5 SINKING OF WELLS

Wells are sunk to the desired depth by excavating the soil from the dredge hole by means of manual labour, mechanical winch and grab. Divers' help and/or pneumatic sinking may be adopted in difficult situations. As far as possible, the wells shall be sunk plumb without any tilt and shift. However, a tilt of 1 in 80 and a shift of 150 mm is considered allowable and the same has to be considered in the design of well foundations.

10.6 PHYSICAL CHARACTERISTICS—DEPTH

The depth of a well foundation must be such that the following requirements are met:

- In erodable soil, there is a minimum grip length of one third the maximum anticipated depth of scour below high flood level (HFL),
- In non-erodable strata, there is adequate seating and embedment on sound rock soil, and
- The base pressure is within permissible limits.

10.6.1 Scour Depth

For natural streams in cohesionless soil, the scour depth may be determined from Lacey's formula. Accordingly, the normal depth of scour, d (in metres), below the highest flood level is given by

$$d = 1.34 \left(\frac{q^2}{f} \right)^{1/3}$$ (10.2)

where q = discharge (in m³/s) through width of the channel obtained by dividing the design discharge by linear waterway between abutments/guidebanks and

f = Lacey's silt factor which is related to the mean diameter of soil grains forming the bed, by the empirical formula, $f = 1.76 \sqrt{m}$. Here, m is the weighted mean diameter in mm. The value of f generally varies from 0.6 to 1.50. The method of determination of m is described in IRC 5–1998.

For preliminary design, the normal depth of scour below HFL may be obtained from

$$d = 0.473 \left(\frac{Q^2}{f} \right)^{1/3}$$ (10.3)

where Q = maximum flood discharge in m³/s.

The maximum depth of scour below the highest flood level (HFL) is given by,

$$D_{max} = 2d \text{ in the vicinity of piers}$$

$$\left. \begin{array}{l} = 1.27d \, (\text{scour restrained}) \\ = 2d \, (\text{scour all round}) \end{array} \right\} \text{ near abutment}$$

If the river bed is not readily susceptable to scouring effect of floods, the fomula for scour depth (Eq. (10.2)) shall not apply. In such cases, the maximum depth of scour shall be assessed from actual observation and experience.

10.7 ALLOWABLE BEARING PRESSURE

The base of a well is normally located at some depth below the maximum depth of scour. During exceptionally heavy flood, the depth of scour in sandy soil may increase, thereby, reducing the grip length. Also for some depth below the maximum scour level, there may be some gap between the well surface and the soil because of tilt under fluctuating lateral load. Hence, skin friction is normally ignored in calculating bearing capacity of wells installed in sand.

For wells embedded in sand, the depth of foundation is generally sufficient ($D/B \geq 2$) to prevent base failure. Hence, settlement is normally the guiding factor. The allowable bearing pressure for a permissible settlement may be calculated from the following (Eq. (10.4)), for footings on sand as suggested by Teng (1962).

$$q_a = 0.14 C_b (N - 3) \left(\frac{D + 0.3}{2B} \right)^2 R'_w C_d s_a$$ (10.4)

where q_a = allowable net soil pressure in t/m^2,

C_b = correction factor (Peck and Bazzaraf, 1969)

= 2.0 for gravelly soil,

= 1.5 for coarse to medium sand, and

= 1.0 for fine and silty sand,

N = standard penetration resistance corrected for overburden and dilatancy,

B = diameter or equivalent width of well in m,

R'_w = water table correction factor

= 0.5 for water level at or above the base of well,

D = depth of well from maximum scour level in m, and

C_d = depth correction factor.

Alternatively, safe bearing pressure may be obtained from the ultimate net bearing capacity of footings as per IS: 6403–1971. A factor of safety, say 2.5, may be applied and the weight of soil (removed by excavation) is added to get the gross bearing capacity.

For wells founded in overconsolidated clay and clay shale, scour is much less and effective depth of embedment is usually much more. So, skin friction may be considered over a part of embedded length with due allowance for possible gap due to tilt. However, the general practice is to ignore the skin friction and calculate allowable bearing pressure from bearing capacity and settlement analysis by treating the well as a deep footing and using relevant equations discussed in Chapters 6 and 7.

In case of rock, the allowable bearing pressure may be estimated from crushing strength of rock or from judgement based on core recovery and R & D.

10.8 FORCES ACTING ON WELL FOUNDATION

A well foundation for a bridge pier is subjected to the following forces:

(a) dead load (weight of pier/abutment, weight of the well and relevant weight of bridge structure),

(b) live load (all superimposed, vertical load including traffic load),

(c) wind force,

(d) forces due to water currents,

(e) forces due to tractive effect of vehicles, braking force and/or those caused by restraint to movement of bearings,

(f) centrifugal forces in case the well is located on a curve,

(g) buoyancy,

(h) earth pressure,

(i) temperature stress, and

(j) seismic forces

Normally, three combinations of the forces are considered for stability analysis, which are grouped as

1. N case : All forces except temperature and seismic forces.
2. (N + T) case : All forces including temperature except seismic forces.
3. (N + T + S) case : All forces including temperature and seismic forces.

The permissible stresses may be increased by 15% for (N + T) case, and by 50% for (N + T + S) case. Also when wind or earthquake forces are considered, the allowable bearing pressure may be increased by 25%.

With the knowledge of the magnitude, direction, and points of application of all the forces and for the worst combination of forces, the resultant vertical force W and resultant horizontal forces P across the pier and Q along the pier, are determined. The horizontal force acting in the direction of the transverse axis of the pier, that is, perpendicular to the flow, is more critical for stability. The forces must satisfy the following conditions of equilibrium:

$$\sum V = 0 \tag{10.5}$$

i.e., total downward force = base reaction + friction on the sides of the well

$$\sum H = 0 \tag{10.6}$$

i.e., net lateral earth pressure including friction at the base should be zero.

$$\sum M = 0 \tag{10.7}$$

i.e., the algebraic sum of moments at base due to net lateral earth pressure, friction on sides and base reaction should be zero.

10.9 LATERAL STABILITY

10.9.1 Analysis Based on Bulkhead Concept

The simplest approach to analyzing the lateral stability of a well foundation is based on Terzaghi's analysis of a free, rigid bulkhead (Terzaghi, 1943). The force system on a bulk-head with an embedment D, subjected to a horizontal force q, at the top, is shown in Fig. 10.5(a). The bulkhead tends to rotate about a point O, at a depth d from the ground surface. At failure, the soil reaches a state of plastic equilibrium and the resistance offered by the soil can be approximated by the pressure diagram shown in Fig. 10.5(b), for cohesionless soil.

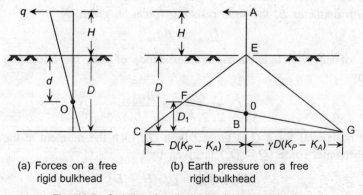

(a) Forces on a free
rigid bulkhead

(b) Earth pressure on a free
rigid bulkhead

Fig. 10.5 Stability of well: rigid bulkhead method.

Resultant pressure per unit length of bulkhead, $q_{max} = \Delta\,CEB - \Delta\,GEF$. The analysis is based on the assumption that the bulkhead is light, there is no friction at the base and the sides, and earth pressure can be calculated by Rankine's theory. This can be applied to a well foundation if the moments on account of base reaction and side friction are neglected.

A heavy well under a lateral load will rotate about its base, and the force diagram will be as shown in Fig. 10.6. For soil below scour level, submerged unit weight γ' is considered.

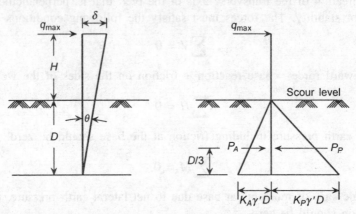

Fig. 10.6 Force and deflection diagrams of a well for rotation about base.

By calculating moment about the base,

$$q_{max}(H + D) + \frac{1}{2}DK_A\gamma'D\frac{D}{3} = K_P\gamma'D\frac{D}{2}\frac{D}{3} \tag{10.8}$$

$$\Rightarrow \qquad q_{max} = \frac{1}{6}\gamma'(K_P - K_A)\frac{D^3}{H + D} \tag{10.9}$$

If there is unscoured soil of thickness Z above the maximum scour level, then its effect on active earth pressure is considered. As a result,

$$q_{max} = \frac{1}{6}\gamma'(K_P - K_A)\frac{D^2(D + Z)}{H + D} \tag{10.10}$$

For a well with diameter B, the total resisting forces is given by

$$Q_{max} = Bq_{max} \tag{10.11}$$

With a factor of safety F against passive resistance of soil, the allowable horizontal force Q_a is given by

$$Q_a = \frac{Q_{max}}{F} = \frac{Bq_{max}}{F} \tag{10.12}$$

If the applied horizontal force Q is greater then Q_a, then the moment at the base, M_B due to unbalanced horizontal force $(Q - Q_a)$ is given by

$$M_B = (Q - Q_a)(H + D) \tag{10.13}$$

Thus, the foundation pressure (maximum f_{max} or minimum f_{min}) at the base is given by

$$f = \frac{W}{A} \pm \frac{M_B}{Z_B} \tag{10.14}$$

where W = net direct vertical load on the base taking into account buoyancy and skin friction,

A = area of the base of the well, and

Z_B = section modulus of the well base.

The above analysis is based on the assumptions that

- bulkhead is weightless,
- there is no contribution of base reactions
- there is no wall friction

10.9.2 IRC Method

Correct evaluation of passive relief is a very important factor in the design of a well foundation, because it governs the depth to which the well is to be sunk. Views differ regarding the assumptions and methods, to be adopted for working out its values. The main points to be addressed while deciding for the method are:

(a) whether the rotation of the foundation takes place only at the base or at points above or below the base,

(b) whether the skin friction on the sides can be considered in the calculation of well resistance, and

(c) what fraction of applied moment is resisted by soil resistance on the sides and base resistance.

To have a uniform practice in design, IRC Bridge Sub-committee in its Bombay meeting held in 1963, suggested the procedure for calculation of ultimate soil resistance below scour level. This is known as *Bombay method*.

Bombay method

This method is based on the following assumptions:

(a) The point of rotation of the well is at base level.

(b) The effect of skin friction on the sides of the well should be ignored.

(c) Until the question of sharing of moment between sides and base is finally resolved, only the relief from resultant passive resistance is to be considered in the design.

(d) For lateral resistance of well, only the difference between passive and active pressure is considered.

(e) Coefficients of active pressure K_A and passive pressure K_P at any depth below scour level are to be considered in the following manner:

For cohesionless soil (Coulomb's formulae)

$$K_A = \frac{\cos^2 \phi}{\cos\delta \left(1 + \sqrt{\frac{\sin(\phi + \delta)\sin\phi}{\cos\delta}}\right)^2} \tag{10.15}$$

$$K_P = \cfrac{\cos^2\phi}{\cos\delta\left(1 - \sqrt{\cfrac{\sin(\phi+\delta)\sin\phi}{\cos\delta}}\right)^2} \qquad (10.16)$$

For cohesive soil (Rankine's formulae)

$$K_A = \frac{\gamma Z}{N_\phi} - \frac{2C}{\sqrt{N_\phi}} \qquad (10.17)$$

$$K_P = \frac{\gamma Z}{N_\phi} + \frac{2C}{\sqrt{N_\phi}} \qquad (10.18)$$

where ϕ = angle of internal friction of soil,

δ = angle of wall friction (to be taken as 2/3 ϕ, limited to 22.5°),

C = cohesion,

N_ϕ = $\tan^2(45 + \phi/2)$,

γ = submerged unit weight of soil, and

Z = depth below design scour level.

The total soil resistance is obtained as algebric sum of the areas of the force diagram due to active and passive pressure.

The factors of safety for determining the net permissible soil resistance mobilized below maximum scour level are

	Load combination (excluding seismic/wind)	For load combination (including seismic/wind)
Noncohesive soil	2	1.6
Cohesive soil	3	2.4

IRC: 45-1972

The method is based on model and prototype studies. It has been observed that,

- sharing of the moment between sides and base continuously changes with increasing deformation of soil and
- elastic theory gives soil pressure at the sides and the base under design load. To determine the actual factor of safety, it would be necessary to calculate the ultimate soil resistance.

Following considerations are made in formulating the design procedure:

(a) *At elastic stage*

   (i) point of rotation is at base level,

   (ii) skin friction at base and side is taken into account, and

   (iii) sharing of moments between the base and side need to be accounted for on the basis of respective coefficients of subgrade reaction.

(b) *At ultimate stage*

    (i) the ultimate pressure distribution at front and back is calculated considering the point of rotation at $0.2D$ above base level,

   (ii) the base resisting moment is the moment of the frictional force along the surface of rupture, and

  (iii) the frictional force due to passive pressure on the front and back results in a resisting moment.

Method based on elastic theory

Assumptions

   (a) This method assumes that the soil surrounding the well and under the base is perfectly elastic, homogeneous, and obeys Hooke's law,

   (b) the lateral deflections are small and the unit soil reaction p increases with lateral deflection Z as expressed by $p = K_H Z$, where K_H is the coefficient of horizontal subgrade reaction at base,

   (c) in cohesionless soils, K_H increases linearly with depth, and

   (d) the well behaves as a rigid body and is acted upon by a unidirectional horizontal force H and moment M at scour level.

Calculation

Step 1

Having determined the minimum grip length, the applied loads (W and H) and moment M under normal load combinations without wind and seismic loads are calculated. In this case W is the total downward load acting at the base of well; H is external horizontal force acting on the well at scour level; and M is the total applied moment about the base of the well, including those due to tilts and shifts.

Step 2

Using the well dimensions, geometrical properties are calculated as:

Moment of inertia of base about an axis passing through CG and perpendicular to horizontal resultant forces,

$$I_B = \frac{\pi}{64} B^4 \tag{10.19}$$

where B = well diameter.

Moment of inertia about the horizontal axis passing through the CG of the projected area in elevation of the soil mass offering resistance,

$$I_v = \frac{LD^3}{12} \tag{10.20}$$

where L = projected width of well in contact with soil offering passive resistance, multiplied by appropriate shape factor (0.9 for circular well and 1.0 for square or rectangular well) and

D = depth of well below scour level.

$$I = I_B + mI_v \, (1 + 2\mu'\alpha) \tag{10.21}$$

where m = ratio of horizontal to vertical coefficient of subgrade reaction = $\dfrac{K_H}{K_V}$

μ' = coefficient of friction between the sides and soil

= $\tan \delta$ (δ is the angle of wall friction)

α = $B/2D$ for rectangular well and $B/\pi D$ for circular well

Step 3

Ensure that $\qquad\qquad H > M/r \, (1 + \mu\mu') - \mu w \qquad\qquad$ (10.22)

and $\qquad\qquad\qquad H > M/r \, (1 - \mu\mu') + \mu w \qquad\qquad$ (10.23)

so that the point of rotation of the well lies at the base.

Here, $r = (D/2)(I/mI_v)$ and

μ = coefficient of friction between the base and the soil

= $\tan \phi$ (ϕ is angle of internal friction of soil).

Step 4

Check that the soil on the sides remains elastic. This can be ensured by keeping

$$\frac{mM}{I} > \gamma(K_P - K_A) \tag{10.24}$$

where γ = effective unit weight of soil.

K_A and K_P = earth pressure coefficients according to Coulomb's theory. Take angle of wall friction δ as $(2/3)\phi$ but limited to 22.5°.

Step 5

Check the base pressures ensuring that the maximum compressive pressure is less than bearing capacity of soil and that the minimum pressure is not tensile.

$$\sigma_1, \sigma_2 = \frac{W - \mu'P}{A} \pm \frac{MB}{2I} \tag{10.25}$$

where σ_1 and σ_2 are maximum and minimum base pressures, respectively, as depicted in Fig. 10.7,

A = area of the base of well,

$P = M/r$, is the total horizontal reaction from the side, and

B = width of base of well in the plane of bending.

Step 6

Check minimum and maximum base pressures so that $\sigma_2 < 0$, that is, no tension and $\sigma_1 >$ allowable bearing capacity of soil.

Step 7

If any or all of the conditions in Steps 3, 4 and 6 do not satisfy, redesign the well accordingly.

Step 8

Repeat the same steps for combination with wind/seismic separately.

Derivation of the theoretical relationships can be obtained from IRC: 45–1972. The pressure distribution and deflection diagrams of a well used in the derivation are presented in Fig. 10.7.

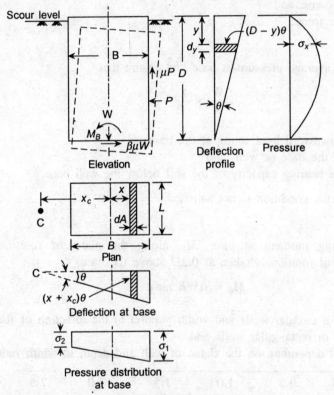

Fig. 10.7 Pressure and deflection diagrams of a well for elastic case.

Method based on ultimate soil resistance

The applied loads are increased by multiplying with suitable load factors and the ultimate resistance is reduced by appropriate strength factors. The two are then compared.

Step 1

Compute the applied load for different combinations of loading using appropriate load factors as

(a) $1.1W_D + 1.6W_L$

(b) $1.1W_D + W_B + 1.4(W_L + P_c + P_e)$

    (c) $1.1W_D + W_B + 1.4(P_c + P_e + P_w \text{ or } P_s)$

    (d) $1.1W_D + W_B + 1.25(W_L + P_c + P_e + P_w \text{ or } P_s)$

where,

        W_D = dead load,

        W_L = live load including braking force and so on,

        W_B = buoyancy,

        P_c = water current force,

        P_e = earth pressure,

        P_w = wind forces, and

        P_s = seismic force.

Step 2

Check for maximum average pressure at base and ensure that

$$\frac{W}{A} \not> \frac{q_{\text{ult}}}{2} \qquad (10.26)$$

where,

        W = total downword load acting at the base,

        A = area of the base or well, and

        q_{ult} = ultimate bearing capacity of the soil below the well base.

Increase base area if the condition is not satisfied.

Step 3

Calculate the resisting moment at base, M_b, along the plane of rotation (as already mentioned, the point of rotation is taken at $0.2D$ above the base)

$$M_b = \alpha WB \tan\phi \qquad (10.27)$$

where,

        B = diameter of circular wells and width parallel to the direction of forces in case of square or rectangular wells and

        α = a constant depending on the shape of well and depth to width ratio D/B.

D/B	0.5	1.0	1.5	2.0	2.5
α	0.41	0.45	0.50	0.56	0.64

The values are for square/rectangular well. For circular wells, the values of α are to be multiplied by 0.6.

Step 4

Compute the ultimate moment of resistance on the side of well due to passive soil resistance, M_s and due to frictional resistance on side, M_f.

For rectangular wells,

$$M_s = 0.10\gamma'D^3(K_P - K_A)L \qquad (10.28)$$

$$M_f = 0.18\gamma'(K_P - K_A)LBD^2\sin\delta \qquad (10.29)$$

For circular wells,

$$M_f = 0.11\gamma'(K_P - K_A)B^2D^2\sin\delta \qquad (10.30)$$

Step 5

Compute the total resisting moment,

$$M_r = M_b + M_s + M_f \qquad (10.31)$$

and ensure that $0.7\, M_r \not< 1\, m$

where 0.7 is a reduction factor, and m is the total applied moment.
If this condition is not satisfied, the grip length is increased and Steps 1 to 5 are repeated.
Derivation of the theoretical relationships is given in IRC: 45–1972.

10.9.3 Mode of Failure

The patterns of failure of the soil mass under the application of lateral forces to wells, with small and large depths of embedment, are shown in Fig. 10.8.

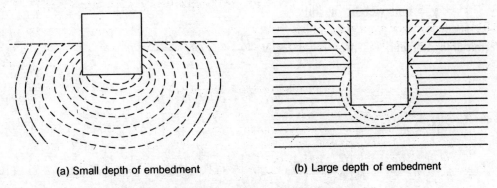

(a) Small depth of embedment   (b) Large depth of embedment

Fig. 10.8 Patterns of failure of soil supporting well.

In either case, the soil fails over a circular/cylindrical path with centre of rotation somewhere above the base. For wells with large depth of embedment, the plastic flow at the side follows the same concept as that of a rigid bulkhead.

10.9.4 Recommendations of IRC: 78–1983

Guidelines for calculation of passive pressure on abutment wells in any type of soil, and on wells in cohesive soils are given in IRC: 78–1983. The basic assumptions and procedure of design are in agreement with "Bombay method" except that the effect of skin friction is allowed to be considered in calculating resistance against moment and direct load.

The expressions of permissible resisting moments for pier and abutment wells in cohesive soil are presented under this subsection.

Case 1: Pier well (Fig. 10.9)

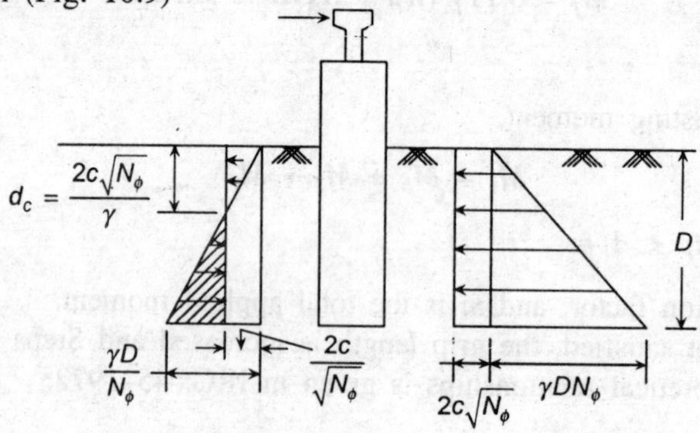

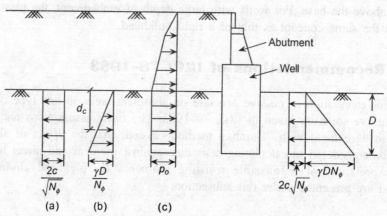

$$d_c = \frac{2c\sqrt{N_\phi}}{\gamma}$$

Fig. 10.9 Active and passive pressure diagram for pier well in cohesive soil.

Depth of tension crack, $d_c = \dfrac{2c\sqrt{N_\phi}}{\gamma} \geq D$

where D = depth below scour line,
        c = cohesion
        γ = unit weight of soil

Permissible moment, $M_p = \dfrac{1}{F}\left(2c\sqrt{N_\phi}\,\dfrac{D^2}{2} + \gamma DN_\phi\,\dfrac{D}{2}\,\dfrac{D}{3}\right)B$

$$= \frac{1}{F}\left(\frac{1}{6}\gamma N_\phi D^3 + c\sqrt{N_\phi}\,D^2\right)B \tag{10.32}$$

For $d_c < H$

$$M_p = \frac{1}{F}\left[\frac{1}{6}\gamma D^3\left(N_\phi - \frac{1}{N_\phi}\right) + cD^2\left(\sqrt{N_\phi} + \frac{1}{\sqrt{N_\phi}}\right)\right]B \tag{10.33}$$

Case 2: Abutment well (Fig. 10.10)
Tension crack extends down to a depth d_c where net $P_a = 0$

Fig. 10.10 Active and passive pressures on abutment well in cohesive soil.

That is, (b) + (c) = (a)

or,

$$p_o + \frac{\gamma d_c}{N_\phi} = \frac{2c}{\sqrt{N_\phi}}$$

or,

$$d_c = \frac{2c\sqrt{N_\phi}}{\gamma} - p_o \frac{N_\phi}{\gamma} \qquad (10.34)$$

For $d_c \geq D$

then $P_a = 0$, that is, same as in case 1 of pier well.

For $d_c < D$, that is, (a) > (b) but < (b) + (c)

P_a exists below d_c.

At base,

$$M_p = \frac{1}{F}\left[\frac{1}{6}\gamma D^3\left(N_\phi - \frac{1}{N_\phi}\right) + cH^2\left(\sqrt{N_\phi} + \frac{1}{\sqrt{N_\phi}}\right)\right]B - p_o\frac{D^2}{2}B \qquad (10.35)$$

For (a) < (b)

Then net P_a is trapezoidal.

At base,

$$M_p = \frac{B}{F}\left[\frac{1}{6}\gamma D^3\left(N_\phi - \frac{1}{N_\phi}\right) + cD^2\left(\sqrt{N_\phi} + \frac{1}{\sqrt{N_\phi}}\right)\right] - p_o\frac{D^2}{2}B \qquad (10.36)$$

10.10 WELL SINKING

The sinking of a well foundation is done by excavating the soil within the dredged hole. This can be done manually or by mechanical dredger. A mechanical dredger consists of prongs with hard steel teeth which are pushed into the soil. When the dredger is pulled up, the prongs close to form a bucket full of excavated soil. The mechanical dredger should be adapted to suit the soil condition.

With increasing depth of well during well sinking, the friction on the sides of the well increases. If the weight of the steining is not adequate to overcome the friction, suitable kentledge may be placed on a platform to increase the load on the steining. Air and water jets are also used to minimize friction. Compressed air is used when well sinking is done below water level to counter the water pressure inside the well.

Example 10.1

A bridge 120 m long, is to be constructed over a river having Q_{max} = 2418 m³/s, HFL = 81.17 m; LWL = 73.00 m and existing bed level = 72.00 m. The subsoil consists of loose silty sand layer (N_{corr} = 10), 3.5 m thick, underlain by a thick stratum of medium to coarse sand (N_{corr} = 24). Determine the founding level and allowable bearing capacity of a 4.5 m diameter abutment well. The weighted mean diameter of the bed material upto relevant depth is 0.275 mm, and permissible settlement is 45 mm.

Solution

$m = 0.275$ mm

Silt factor, $f = 1.76\sqrt{m} = 0.923$

Discharge intensity, $q = \dfrac{2418}{120} = 20.15$ m³/s/m

Normal scour depth, $d = 1.34\left(\dfrac{q^2}{f}\right)^{1/3} = 1.34\left(\dfrac{(20.15)^2}{0.923}\right)^{1/3} = 10.17$ m

Maximum scour depth (near abutment) = $1.27 \times 10.17 = 12.92$ m

∴ Scour level below *HFL* = $81.17 - 12.92 = 68.25$ m

Grip length = $\dfrac{1}{3} \times 12.92 = 4.30$ m

∴ Required bottom level of well = $68.25 - 4.30 = 64.95$ m

Depth below existing bed level = $72.00 - 64.95 = 7.05$ m

The embedment in dense sand layer = $7.05 - 3.5 = 3.55$ m

For better embedment in dense sand, provide grip length which is 1.5 times well diameter = 6.75 m

Thus, founding level provided = $68.25 - 6.75 = 61.50$ m

Allowable bearing pressure, q_a can be calculated using Eq. (10.4) as

$$q_a = 0.14 \times 1.5(24 - 3)\left(\dfrac{4.5 + 0.3}{2 \times 4.5}\right)^2 0.5 \times 45$$

$$= 555 \text{ kN/m}^2$$

Example 10.2

The subsoil at the typical pier location of a major bridge consists of medium to coarse sand (N_{corr} = 11) upto a depth of 6 m from bed level (RL + 9.20 m). This is underlain by 9 m thick layer of very stiff to hard sandy silty clay (N_{corr} > 30), overlying highly weathered rock ($RQD \approx 0$).

Using Lacey's formula calculate the maximum scour depth and determine the founding level of the well. Also, estimate the allowable net bearing pressure if the diameter of the well is 6 m.

Given: Maximum flood discharge = 10,465 m³/s

Length of bridge = 382.5 m; *HFL* = 13.00 m; silt factor, f = 1.053;

submerged unit weight of soil and rock = 10 kN/m³; c' = 0; and ϕ' = 35° for weathered rock.

Solution

Discharge intensity, $q = \dfrac{10,465}{382.5} = 27.36$ m³/s/m.

Normal scour depth, $d = 1.34 \left(\dfrac{q^2}{f}\right)^{1/3} = 1.34 \left(\dfrac{(27.36)^2}{1.053}\right)^{1/3}$

$$= 11.70 \text{ m}$$

Maximum scour depth = 2 × 11.70 = 23.40 m

∴ Computed maximum scour level = 13.00 − 23.40 = − 10.40 m (RL)

But as stiff clay overlying weathered rock exists at 6 m below bed level (+9.20 m), the above calculation of scour depth is not applicable.

Now take maximum scour level = 9.2 − 6.0 = 3.20 m

Below this level there is 9 m thick clay followed by weathered rock. For restricting settlement, it is preferable to take bottom of well 3 m into the rock.

∴ Provide founding Level = 3.2 − (9 + 3) = − 8.8 m

For bearing capacity:

B = 6 m, D_f = 12 m, γ' = 10 kN/m³, c' = 0, and ϕ' = 35° for weathered rock

Now $q_{ult} = \gamma' D_f (N_q - 1)s_q d_q + 0.5\gamma' BN_\gamma s_\gamma d_\gamma w' + \gamma D_f$

For ϕ = 35°

N_q = 32, N_γ = 33,

s_q = 1.2, s_γ = 0.6

$d_q = d_\gamma = 1 + 0.1 \left(\dfrac{3}{6}\right) \tan\left(45 + \dfrac{35}{2}\right)^{\circ}$

≈ 1

or q_{ult} = (10 × 12 (32 − 1) × 1.2 × 1) + (0.5 × 10 × 6 × 33 × 0.6 × 1 × 0.5) + (10 × 12)

$= 4460 + 297 + 120$

≈ 4870 kN/m²

∴ $q_{safe} = \dfrac{4870}{3} \approx 1620$ kN/m²

As the well is founded on weathered rock, settlement will be negligible.
Hence, allowable bearing pressure = 1620 kN/m^2

Example 10.3

Using IRC methods, check the adequacy of design of the pier well of a bridge (Fig. 10.11(a)). Use the following data on loads and soil properties:

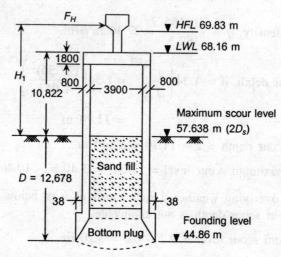

Fig. 10.11(a) Pier well design for Example 10.3.

Loading:

Vertical Load:

DL = 12,400 kN

LL = 1390 kN

∴   Total load = $DL + LL$ = 13,790 kN

Above well cap—8100 kN

Below well cap—5690 kN

Total 13,790 kN

Total horizontal force, F_H = 1755 kN

Moment at base = 43,180 kNm

Consider Tilt: 1 in 60

Shift   : 0.15 m

Soil Properties:

N_{corr} = 20; ϕ = 33°; γ = 19.2 kN/m^3; δ = 20°

Soil type: Coarse to medium sand

Consider permissible settlement of 1.5% of well diameter.

Solution

$$e = \frac{4318}{175.5} = 24.603 \text{ m}$$

$$D = 12.678 \text{ m}$$

Also, $e = H_1 + D$

∴ $H_1 = 11.925$ m

Moment due to tilt and shift

For load above well cap: $8100 \left(0.15 + \dfrac{23.5}{60}\right) = 4390$ kNm

For load below well cap: $5690 \left(\dfrac{23.5}{120}\right) = 1110$ kNm

$$\text{Total} = 5500 \text{ kNm}$$

Total moment at base = 43,180 + 5500 = 48,680 kNm

Total vertical load = 8100 + 5690 = 13,790 kNm

Bombay method (Fig. 10.11(b))

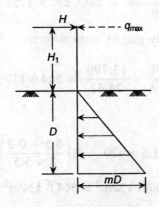

Fig. 10.11(b) Bombay method.

$$K_A = \frac{\cos^2 \phi}{\cos\delta \left(1 + \sqrt{\dfrac{\sin(\phi + \delta)\sin\phi}{\cos\delta}}\right)^2}$$

$$= \frac{0.703}{0.94 \left(1 + \sqrt{\dfrac{0.799 \times 0.545}{0.94}}\right)^2}$$

$$= 0.265$$

$$K_P = \frac{\cos^2 \phi}{\cos\delta\left(1 - \sqrt{\frac{\sin(\phi + \delta)\sin\phi}{\cos\delta}}\right)^2}$$

$$= \frac{0.703}{0.94(1 - 0.68)^2} = 7.30$$

$$K_P - K_A = 7.30 - 0.265 = 7.035$$

$$m = \gamma'(K_P - K_A)\cos\delta = 9.2 \times 7.035 \times 0.94 = 61.8$$

$$q_{max} = \frac{1}{6}m\left(\frac{D^3}{D + H}\right) = \frac{1}{6} \times 61.8 \times \frac{(12.678)^3}{12.678 + 11.925}$$

$$= 839.3 \text{ kN/m}^2$$

Allowable horizontal force,

$$Q_{max} = \frac{q_{max}B}{FS} = \frac{839.3 \times 5.5}{1.6} = 2885 \text{ kN} > 1755 \text{ kN (applied force)}$$

Hence, base pressure (f_{max}) is only due to vertical force.

$$f_{max} = \frac{W}{A} = \frac{13,790}{24.42} = 564.6 \text{ kN/m}^2$$

Allowable bearing capacity, q_a

Using Eq. (10.4), $q_a = 0.14 \times 1.5 \times (20 - 3)\left(\frac{5.5 + 0.3}{2 \times 5.5}\right)^2 0.5 \times 2 \times 82.5$

$$= 81.9 \approx 800 \text{ kN/m}^2 > 564.6 \text{ kN/m}^2$$

Hence, this is acceptable.

$$N = 20,$$
$$C_{b1} = 1.5$$
$$R_W = 0.5$$
$$B = 5.5$$
$$C_d = 2$$
$$S_a = 1.5 \% \text{ of } 5500 = 82.5 \text{ mm}$$

Method Based on IRC: 45 (elastic)

Step 1 $W = 13,790$ kN, $M = 48,680$ kNm, $H = 1755$ kN

Step 2 $D_B = 5500 + 2 \times 38 = 5576$ mm

$$I_B = \frac{\pi}{64} (5.576)^4 = 47.43 \text{ m}^4$$

$$I_v = \frac{LD^3}{12} = \frac{0.9 \times 5.5 \times (12.678)^3}{12} = 840.574 \text{ m}^4$$

$$\mu' = \tan \delta = \tan 20° = 0.364$$

$$\mu = \tan \phi = \tan 33° = 0.649$$

$$\alpha = \frac{B}{\pi D} = \frac{5.5}{3.14 \times 12.678} = 0.138$$

$$m = \frac{K_h}{K_v} = 1$$

$$I = I_B + mI_v (1 + 2 \ \mu'\alpha)$$

$$= 47.43 + 1 \times 840.574 (1 + 2 \times 0.364 \times 0.138) = 972.45 \text{ m}^4$$

Step 3

$$r = \frac{D}{2}\left(\frac{I}{mI_v}\right) = \left(\frac{12.678}{2}\right)\left(\frac{972.45}{1 \times 840.574}\right) = 7.33$$

$$\frac{M}{r}(1 + \mu\mu') - \mu W = \frac{48,680}{7.33}(1 + 0.649 \times 0.364) - 0.649 \times 13,790$$

$$= 8210 - 8950 = -740 \text{ kN} < H$$

$$\frac{M}{r}(1 - \mu\mu') + \mu W = \frac{48,680}{7.33}(1 - 0.649 \times 0.364) + 0.649 \times 13,790$$

$$= 5072.3 + 8950.7 = 14,023 \text{ kN} > H$$

Step 4

$$\frac{mM}{I} = \frac{1 \times 48,680}{972.45} = 50; \ \gamma'(K_P - K_A) = 9.2 \times 7.035 = 64.7$$

with seismic, $1.25\gamma'(K_P - K_A) = 1.25 \times 64.7 = 80.88 < 50$

Step 5

$$\sigma_1, \ \sigma_2 = \frac{W - \mu'p}{A} \pm \frac{MB}{2i}$$

$$= \frac{13,790 - 0.364 \times 6641}{24.42} \pm \frac{48,680 \times 5.5}{2 \times 972.45}$$

$$= 465.8 \pm 137.7$$

$$= 603.5, \ 328.1 \text{ kN/m}^2 < 800 \text{ kN/m}^2. \text{ Hence, it is acceptable.}$$

Method based on IRC: 45 (Ultimate)

Step 1

$$W = 1.1 \ DL + 1.6 \ LL$$
$$= 1.1 \times 12{,}400 + 1.6 \times 1390 = 15{,}865 \text{ kN}$$

Step 2

For $\phi = 33°$; $N_q = 20$, $N_\gamma = 15$

For circular section, $s_q = 1.2$, $s_\gamma = 0.6$

$$d_q = d_\gamma = 1 + 0.1\left(\frac{12.678}{5.5}\right)\tan\left(45 + \frac{33}{2}\right)^{°} = 1.4$$

$W' = 0.5$; $I_q = I_\gamma = 1$

$$q_{ult} = q(N_q - 1) \ s_q d_q I_q + 0.5\gamma B N_\gamma s_\gamma d_\gamma I_\gamma W' + \gamma D$$

$$= 9.2 \times 12.678 \ (20 - 1) \times 1.2 \times 1.4 \times 1 + (0.5 \times 19.2 \times 5.5 \times 15 \times 0.6 \times 1.4 \times 0.5) + (19.2 \times 12.678)$$

$$= 3723 + 332.6 + 243.4 \approx 4300 \text{ kN}$$

$$\frac{W}{A} = \frac{15{,}865}{24.42} = 649.7 \text{ kN} < \frac{430}{2}$$

Hence, it is justified.

Step 3

$$M_B = \alpha W B \tan\phi \quad \text{(where } \alpha = 0.6 \times 0.61 = 0.366\text{)}$$
$$= 0.366 \times 15{,}865 \times 5.5\tan 33°$$
$$= 20{,}739 \text{ kNm}$$

Step 4

$$M_s = 0.1\gamma'(K_P - K_A)LD^3 = 0.1 \times 9.2 \times 7.035 \times 0.9 \times 5.5 \times (12.678)^3$$
$$= 65{,}280 \text{ kNm}$$

For circular well, $M_f = 0.11\gamma'(K_P - K_A)B^2D^2 \sin\delta$
$$= 0.11 \times 9.2 \times 7.035 \times 5.5^2 \times 12.678^2 \times \sin 20°$$
$$= 11{,}839 \text{ kNm}$$

Total resisting moment, $M_r = 20{,}739 + 65{,}280 + 11{,}839$
$$= 97{,}850 \text{ kNm}$$
$$M_t = 0.7 \times 97{,}850 = 68{,}500 \text{ kNm}$$

Moment due to tilt and shift = 5500 kNm

Equivalent $H = 1.1\left(\dfrac{5500}{24.603}\right) + 1.25 \times 1755 = 245.9 + 2193.7$
$$= 2439.6 \text{ kN}$$

Applied moment = 2439.6(24.603 − 0.2 × 12.678)
(Point of rotation 0.2*D* above base)

$$= 53,830 \text{ kNm} < 68,500 \text{ kNm. Therefore, accepted.}$$

Example 10.4

For the bridge considered in Example 10.3, check the adequacy of the abutment well whose elevation and nature of earth pressure diagrams have been shown in Fig. 10.12

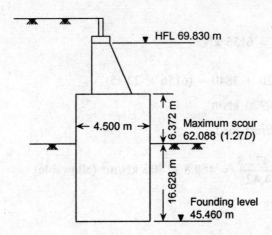

Fig. 10.12 Elevation and earth pressure for bridge in Example 10.3.

Given:

Total Vertical Load *V* = 11,472.8 kN

Horizontal Force F_H = 3446.5 kN

Moment *M* = 77,020 kNm

Moment due to tilt and shift = 3840 kNm

Normal scour depth d_s = 6.096 m

p_o = 47.66 kN/m^2
(refer Fig. 10.10)

Solution

Bombay method

$$e = H_1 + D = \frac{M}{F_H} = \frac{77,020}{3446.5} = 22.35 \text{ m}$$

$$D = 16.628 \text{ m}$$

$$q_{max} = \frac{D^2}{6(H_1 + D)}[\gamma'(K_P - K_A)D\cos\delta - 3p_o]$$

$$= \frac{D^2}{6(H_1 + D)}[mD - 3p_o]$$

$$= \frac{(16.628)^2}{6 \times 22.35}[(9.2 \times 7.035 \times 16.628 \times 0.94) - (3 \times 47.66)]$$

$$= 1791 \text{ kN/m}$$

$$Q_{max} = \frac{1791 \times 5.5}{(FS =)1.6} = 6156 \text{ kN}$$

Net moment $= 77,020 + 3840 - (6156 \times 22.35)$

$$= -56,720 \text{ kNm}$$

$\therefore$ Consider only direct load.

Base pressure $= \dfrac{11,472.8}{24.42} = 469.8 < 800 \text{ kN/m}^2$ (allowable)

Method based on IRC : 78

$$q_{max} = \frac{D^2}{6(H_1 + D)}(mD - 3p_o)$$

$$= \frac{mD^3}{6(H_1 + D)} - \frac{p_oD^2}{2(H_1 + D)}$$

$$= \frac{60.8 \times 16.628^3}{6 \times 22.35} - \frac{47.66 \times 16.628^2}{2 \times 22.35} \qquad \text{where } m = 9.2 \times 7.035 \times 0.94 = 60.8$$

$$= 2084.4 - 294.8$$

With FS (= 1.6) on 1st term only,

$$q'_{max} = \frac{2084.4}{1.6} - 294.8 = 1008 \text{ kN/m}^2$$

$\therefore$ $Q_{max} = 1008 \times 5.5 = 5540 \text{ kN}$

Balance moment $= 80,860 - 5540 \times 22.35 = -429.5 \text{ kNm}$

$\therefore$ Base pressure $= \dfrac{11,472.8}{24.42} = 469.8 < 800 \text{ kN/m}^2$ (allowable)

REFERENCES

IRC: 5–1998, *Standard Specifications and Code at Practice for Road Bridge*, Section 1, General Features of Design, The Indian Roads Congress, New Delhi, 1998.

IRC: 45–1972, *Recommendations for Estimating the Resistance of Soil below the Maximum Scour Level in the Design of Well Foundation of Bridges,* The Indian Roads Congress, New Delhi, 1972.

IRC: 78–1983, *Standard Specifications and Code of Practice for Road Bridges*, Section: VII, Foundation and Substructure, 1st Revision, The Indian Roads Congress, New Delhi, 1994.

IS: 6403–1971, *Indian Standard Code of Practice for Determination of Allowable Bearing Pressure on Shallow Foundations*, Indian Standard Institution, New Delhi, 1972.

Peck, R.B. and A.R.S.S. Bazzaraf (1969), Discussion on Paper by D'Appolonia, et al., *Journal Soil Mech. and Found. Div.* ASCE, Vol. 95, SM. No. 3.

Teng, W.C. (1962), *Foundation Design*, Prentice Hall, Englewood Cliffs, New Jersey.

Terzaghi, K. (1943), *Theoretical Soil Mechanics*, John Wiley and Sons, New York.

11 Foundations on Expansive Soils

11.1 INTRODUCTION

Some cohesive soils undergo swelling when they come into contact with water and shrink when water is squeezed out. Foundations built on these soils undergo movement with swelling and shrinkage of the soil. As a result, there is considerable cracking and other forms of distress in the buildings.

Expansive soils are mostly found in the arid and semi-arid regions of the world. They cover large areas of Africa, Australia, India, United States, South America, Myanmar and some countries of Europe. In India, expansive soils cover nearly 2 percent of the land area and are called *black cotton soil* because of their colour and cotton growing potential. Large areas of Deccan plateau, Andhra Pradesh, Karnataka, Madhya Pradesh, and Maharastra have deposits of expansive soil.

Expansive soils derive their swelling potential mainly from montmorillonite mineral which is present in these soils. They are residual soils formed by the weathering of basaltic rocks under alkaline environment. Extended periods of dry climate cause desiccation of the soil but rains cause swelling near the ground surface. The depth of the swelling zone is not much—being less than 5 m in most cases. However, the active zone over which there are seasonal changes in moisture content, varies from 1.5 m to 4 m (O'Neill and Poormoayed, 1980).

11.2 NATURE OF EXPANSIVE SOIL

The swelling characteristics of a soil depend largely on the type of clay mineral present in the soil. Differential Thermal Analysis (DTA), X-ray diffraction, and electron microscopy are common methods of determining the proportion of different minerals present in a soil. Some simple laboratory tests are often used to determine the swelling potential of natural soil.

11.2.1 Free-swell Test

This test is performed by pouring 10 g of dry soil, passing 425 micron sieve, in a 100 cc graduated cylinder filled with water. The soil collects at the bottom of the cylinder and gradually increases in volume. After 24 hours, the volume of the soil is read from the graduations in the cylinder. The percent free swell of the soil is given by

$$\text{Free swell (\%)} = \frac{\text{final volume} - \text{initial volume}}{\text{initial volume}} \times 100$$

The percent free swell of predominant clay minerals is given in Table 11.1. Kaolinite and illite have percent free swell less than 100 and are generally regarded as non-swelling minerals. Montmorillonite, on the other hand, has high swelling potential.

Table 11.1 Free swell of clay minerals

Mineral	*Percent free swell*
Montmorillonite (Bentonite)	1200–2000
Kaolinite	80
Illite	30–80

Gibbs and Holtz (1956) suggested that soils having free swell above 50% may be expected to cause problem to light structures.

11.2.2 Differential Free Swell

Two samples of oven-dried soil (10 g) passing 425 micron sieve are taken and poured into 100 cc graduated glass cylinders—one filled with water and the other with kerosene. Kerosene, being a non-polar liquid does not cause any volume change in the soil. After 24 hours, the volumes of soil in the two cylinders are measured and the differential free swell, *DFS* is obtained

$$DFS = \frac{\text{soil volume in water} - \text{soil volume in kerosene}}{\text{soil volume in kerosene}} \times 100$$

IS 2720 (Part III–1980) gives the degree expansiveness of a soil in terms of the differential free swell, Table 11.2.

Table 11.2 Degree of expansiveness and *DFS*

Degree of expansiveness	*DFS (%)*
Low	Less than 20
Moderate	20–35
High	35–50
Very high	Greater than 50

11.2.3 Unrestrained Swell Test

This test is done in the standard odometer. The soil specimen is given a small surcharge load (say, 5 kN/m^2) and submerged in water. The volume expansion of the specimen is measured in terms of the increase in thickness of the specimen—the cross-sectional area remaining constant. The percent swell is expressed as

$$s_w(\%) = \frac{\Delta H}{H} \times 100$$

where, $s_w(\%)$ = free swell,

ΔH = increase in height of soil sample, and

H = original height.

Vijayvergiya and Ghazzali (1973) gave a correlation between free swell, natural moisture content, and liquid limit of some clays. This correlation is presented in Fig. 11.1.

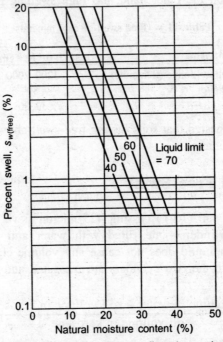

Fig. 11.1 Relationship between free swell, water content, and liquid limit (after Vijayvergiya and Ghazzali, 1973).

11.2.4 Swelling Pressure

The swelling pressure indicates the external pressure required to prevent swelling of a soil when the latter comes into contact with water. The test is done in the consolidation apparatus. Water is added to the specimen and as it starts to swell, pressure is applied in small increments to prevent swelling. The test is continued till the sample just begins to settle. This gives the swelling pressure (p_s) of the sample (Sridharan et al. 1986). As a general rule, a swelling pressure of 20–30 kN/m² is considered low. A highly swelling soil, for example, bentonite may have swelling pressure as high as 1500–2000 kN/m².

Some empirical formulae have been suggested for estimating the swelling pressure of a soil from the void ratio and plasticity index of the soil (Pidgeon, 1987).

$$p_s = 2.7 - 24\left(\frac{e_i}{PI}\right)$$

where, p_s = swelling pressure (kg/cm^2),
e_i = initial void ratio, and
PI = plasticity index (%).

11.2.5 Classification of Swelling Potential

Potential swell is defined as the vertical swell under a pressure equal to overburden pressure. A number of classification systems for the swelling potential have been proposed (Seed et al. 1962; Sowers and Sowers, 1970; Chen, 1988; Vijayvergiya and Ghazzali 1973). These are generally based on the Atterberg Limits of the soil. O'Neill and Poormoayed (1980) summarized the U.S. Army Waterways Experiment Station criterion based on plasticity index and the potential swell. This classification of expansive soil is expressed in Table 11.3.

Table 11.3 Classification of expansive soil

Liquid limit	Plasticity index	Potential swell	Swelling potential
< 50	< 25	< 0.5	Low
50–60	25–35	0.5–1.5	Marginal
> 60	> 35	> 1.5	High

USBR (1960) gives similar criteria for clays based on colloid percent (less than 0.001 mm) and shrinkage limit of the soil, which are shown in Table 11.4.

Table 11.4 Criteria for expansive clays

Colloid content (%)	Plasticity index	Shrinkage limit	Probable expansion (%)	Degree of expansiveness
> 28	> 35	< 11	> 30	Very high
20–30	25–41	7–12	20–30	High
13–23	15–28	10–16	10–20	Medium
< 15	> 10	> 15	< 10	Low

11.3 EFFECT OF SWELLING ON BUILDING FOUNDATIONS

In tropical countries, the soil near the ground surface dries up during the summer as a consequence of intense heat and recession of ground water table. The soil becomes stiff and cracks and fissures open up. When rains come, the soil gets wet by the precipitation and with time, the ground water table also rises. This causes increase in water content of the soil and hence, swelling of the soil. The structure built on the soil protects the soil from the heat but nonetheless water tends to accumulate from the surrounding areas and contributes to swelling. The depth over which such variation in water content occurs depends on the nature of soil and climatic conditions but an active zone of 3.5–4 m has generally been observed.

The movement of the foundation with swelling and shrinkage of the soil causes the floor slabs of buildings to lift up and develop a dome shaped deformation pattern. This leads to cracks in the floor and external walls. The differential movement also causes diagonal cracks in the walls and at the corner of doors and windows, as shown in Fig. 11.2 Utilities buried in the soil get damaged and the leakage of water into the soil results in further swelling. The effect is more pronounced in the one or two storey buildings where the foundation pressure is often less than the swelling pressure. For tall structures, foundation pressure generally exceeds the swelling pressure and the swelling potential reduces.

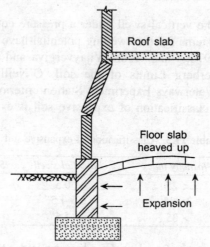

Fig. 11.2 Distress in buildings due to swelling.

11.4 FOUNDATION DESIGN IN EXPANSIVE SOIL

Foundation design in expansive soil needs a different approach as compared to that in non-swelling soil. It must be realized that limiting the safe bearing pressure to a low value does not help to counteract the swelling pressure. Rather, design should be done with a high enough bearing pressure and following the criteria of bearing capacity failure and permissible of settlement, so that chances of swelling are minimized.

Foundation design in expansive soil can be done in the following ways:

1. Isolating the foundation from the swelling soil,
2. Taking measures to prevent the swelling, and
3. Employing measures to make the structure withstand the movement.

11.4.1 Isolating the Foundation from the Swelling Zone: Under-reamed Piles

A common method of building foundation in expansive soil is to provide under-reamed piles below the foundation. Here, the structural load is transferred to the soil beneath the zone of fluctuation of water content. The piles are taken to depths of 5–6 m, that is, well beyond the expansive zone. These piles are bored cast-in-situ piles with the lower end enlarged to form under-reamed bulbs with the help of special tools, Fig. 11.3. The piles generally have shaft diameter of 300 mm and bulb diameter of 750 mm, as depicted in Fig. 11.4. The piles are fixed at the top to RCC plinth beams. A gap of 75–100 mm is kept between the plinth beam and the soil which is filled with granular material to permit swelling of the underlying soil without straining the plinth beams. The piles are adequately reinforced to take care of the uplift forces caused by the swelling action of the soil.

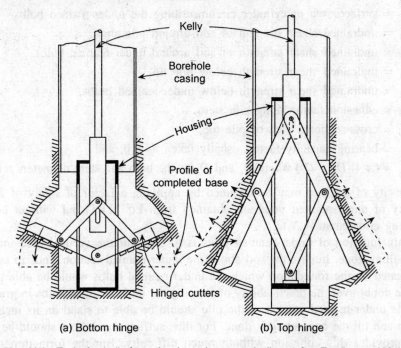

Fig. 11.3 Tools for bulb formation in under-reamed piles. (after Tomlinson, 1994)

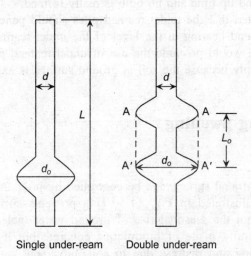

Fig. 11.4 Typical under-reamed pile foundation.

The ultimate being capacity of an under-reamed pile is determined from static analysis as already discussed in Chapter 9.

$$Q_u = \alpha C_{a1} A_{f1} + C_{a2} A_{f2} + A_p N_c C_{p1} + A_u N_c C_{p2} \qquad (11.1)$$

where, A_{f1} = surface area of pile stem,

A_{f2} = surface area of cylinder circumscribing the under-reamed bulbs,

C_{a1} = undrained shear strength of soil around pile shaft,

C_{a2} = undrained shear strength of soil around under-reamed bulbs,

C_{p1} = undrained shear strength below pile tip,

C_{p2} = undrained shear strength below under-reamed bulbs,

α = adhesion factor along pile stem,

A_p = cross-sectional area of pile toe,

N_c = bearing capacity factor, usually taken as 9.0, and

$A_u = (\pi/4)(D_u^2 - D^2)$ where D_u and D are the bulb and stem diameters respectively.

The safe capacity of the pile may be obtained by applying a factor of safety of 2.5–3. The uplift capacity of under-reamed piles are obtained from Eq. (11.1) but without considering the end bearing component $A_p N_c C_{p1}$.

The main function of under-reamed piles is to transmit the vertical load into the soil below the swelling zone. In case the soil above the under-reamed section tends to swell, uplift forces are created on the foundation which the under-reamed bulbs would be able to counter. There is some doubt over the practicability of forming the under-reamed bulbs in granular soil. Obviously, the under-reamed section of the pile should be able to stand on its inclined faces during boring and till the concreting is done. For this, sufficient cohesion should be available. Clayey soil provides this cohesion without much difficulty. But the formation of bulb in granular soil is not free from uncertainties. Field experiments have shown that the sand tends to disintegrate during the stand up time and no bulb is really formed.

It should also be noted that the under-reamed piles should penetrate well into the firm ground to give sufficient end bearing at the level of the under-reams. Presence of soft clay beneath the expansive soil would preclude the use of under-reamed piles. Indiscriminate use of under-reamed piles simply because the soil at ground surface is expansive in nature, often serves no useful purpose.

11.4.2 Controlling Swelling

Impervious apron

Swelling of soil near the ground surface can be controlled by providing an impervious apron around the structure as illustrated in Fig. 11.5. This prevents surface precipitation from penetrating into the soil but the seasonal rise of ground water table is not controlled. The impervious apron is generally made of bituminous concrete but it should be sufficiently flexible to prevent cracking and distress due to soil movement caused by swelling. It is necessary for the apron to penetrate sufficiently into the foundation to prevent ingress of water into the soil during inundation.

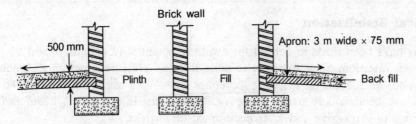

Fig. 11.5 Impervious apron for controlling swelling.

Surcharge loading

Swelling can also be controlled by applying a pressure on the ground at least equal to the swelling pressure of the soil. Attempts have been made to apply a surcharge on the footing area but this does not prevent the swelling between foundations. If surcharge loading is to be applied, this should be done to cover the entire building area by a suitable non-swelling soil. The depth of surcharge should, of course, be such that the surcharge pressure is at least equal to the swelling pressure of the soil. Surcharge loading is depicted in Fig. 11.6.

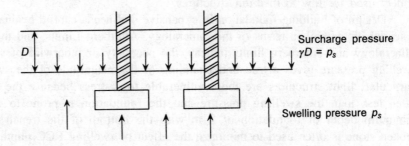

Fig. 11.6 Surcharge loading to control swelling.

CNS layer

Katti (1979) proposed the use of a cohesive non-swelling (CNS) layer to reduce the effect of swelling. A clay soil of adequate thickness having non-swelling clay minerals, is placed on the subgrade and the foundations are placed on this layer. The optimum thickness of the CNS layer is to be determined from large scale tests. The method has been used in canal lining works, as shown in Fig. 11.7 but its use in foundations is still limited.

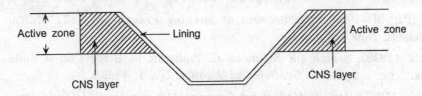

Fig. 11.7 Use of CNS layer in canal lining.

Chemical Stabilization

Attempts have been made to control the swelling potential of expansive soil by lime-slurry or lime-flyash injection. Such injection is done through pressure grouting technique on a close grid around the foundation to cover the entire foundation area. But the method is expensive and has not been widely used. However, lime slurry injection has been used to stabilize foundations on expansive soil after occurrence of distress.

11.4.3 Measures to Withstand Settlement

The foundation may be made sufficiently rigid by providing interconnected beams and band lintel to withstand the effects of differential movement on the structure. Stiffened mat foundations (Lytton, 1972) have also been adopted to counter the effect of differential ground movement. Premlatha (2002) carried out a study on the use of stiffened mat for low rise structures in Chennai, based on evaluation of heave of the soil. A three dimensional soil structure interactive analysis was done to suit the loading, climate, and environmental conditions of Chennai. However, these methods are rather expensive and have not been widely used for low to medium structures.

Design of building foundation in expansive soil needs careful evaluation of the swelling potential of the soil in terms of the mineralogy, Atterberg Limits, and the swelling pressure. Mineralogy and Atterberg limits indicate the necessity or otherwise of special design. Only swelling pressure gives a true understanding of the magnitude of the swelling potential. In particular, light structures are more vulnerable to distress because the bearing pressure is often less than the swelling pressure and the foundation is prone to uplift forces. For a minimum depth of foundation of 2 m with the bottom of the trench filled with sand, a broken stone is often used to minimize the effect of swelling RCC plinth beams. Band lintel also helps to withstand the effect of swelling.

Under-reamed piles and surcharge loading seem to be the most suitable methods of countering the effect of swelling. In addition, good surface drainage and impervious paving around the site help to prevent water percolation in the soil.

REFERENCES

Chen, F.H. (1988), *Foundations on Expansive Soils,* Elsevier, Amsterdam.

Gibbs, H.J. and W.G. Holtz (1956), *Engineering Properties of Expansive Clays,* Transactions ASCE, New York. Vol. 121, pp. 641–663, Paper No. 2814, Discussion, pp. 664–777.

IS 2720 (Part III–1980), *Measurement of Swelling Pressure of Soils,* Bureau of Indian Standards, New Delhi.

Katti, R.K. (1979), Search for Solutions to Problems in Black Cotton Soils, First IGS Annual Lecture, *Indian Geotechnical Journal,* No. 1, Vol. 9.

Lytton, R.L. (1972), *Design Method for Concrete Mats on Unstable Soils,* Proceedings 3rd American Conference on Materials Tech., Rio-de-Janeiro, Brazil, pp. 171–177.

O'Neill, M.W. and N. Poormoayed, (1980), Methodology for Foundations on Expansive Clays, *Journal of the Geotechnical Engineering Division,* American Society of Civil Engineers, Vol. 106, No. GT12, pp. 1345–1367.

Pidgeon, J.T. (1987), *A Comparison of Existing Methods for the Design of Stiffened Raft Foundation on Expansion Soil,* Proceeding 7th Regional Conference of Africa on SMFE, pp. 277–289.

Premlatha, K. (2002), *Prediction of Heave Using Soil–Water Characteristics and Analysis of Stiffened Raft on Expansive Clays of Chennai,* Ph. D. Thesis, Anna University, Chennai.

Seed, H.B., R.J. Woodward, Jr., and R. Lundgren (1962), Prediction of Swelling Potential for Compacted Clays, *Journal of the Soil Mechanics and Foundations Division,* American Society of Civil Engineers, Vol. 88, No. SM3, pp. 53–87.

Sowers, G.B. and G.F. Sowers (1970), *Introductory Soil Mechanics and Foundations,* 3rd ed., New York, Macmillan.

USBR (1960), *Earth Manual,* U.S. Bureau of Reclamation, Denver, Colorado, July, 1960.

Vijayvergiya V.N. and O.I. Ghazzali (1973), *Prediction of Swelling Potential of Natural Clays,* Proceedings 3rd International Research and Engineering Conference on Expansive Clays, pp. 227–234.

Ground Improvement Techniques

12.1 INTRODUCTION

Soils are deposited or formed by nature under different environmental conditions. Man does not have any control on the process of soil formation. As such the soil strata at a site are to be accepted as they are and any construction has to be adapted to suit the subsoil condition. The existing soil at a given site may not be suitable for supporting the desired facilities such as buildings, bridges, dams, and so on because safe bearing capacity of a soil may not be adequate to support the given load. Although pile foundations may be adopted in some situations, they often become too expensive for low to medium-rise buildings. In such cases, the properties of the soil within the zone of influence have to be improved in order to make them suitable to support the given load.

Ground improvement for the purpose of foundation construction essentially means increasing the shear strength of the soil and reducing the compressibility to a desired extent. A number of ground improvement techniques have been developed in the last fifty years. Some of these techniques need specialized equipment to achieve the desired result. In this chapter, only the common ground improvement techniques which use simple mechanical means to improve soil properties for low to medium-rise structures are considered. For tall structures, pile foundations with or without basement would generally give the most economic foundations.

12.2 PRINCIPLES OF GROUND IMPROVEMENT

The mechanics of ground improvement depends largely on the type of soil—its grain-size distribution, water content, structural arrangement of particles and so forth. In general, ground improvement is called for in soft cohesive soil with low undrained shear strength ($c_u < 2.5$ t/m^2) and loose sand ($N < 10$ blows per 30 cm). The mechanics of ground improvement can be understood in terms of the structural arrangement of particles constituting the soil deposit.

(a) Cohesive soil

Sedimentary (alluvial or marine) clays during deposition under flowing water have the flexible flake shaped particles arranged at random flocculated structure with large void spaces filled with water. Figure 12.1(a) depicts the flocculated structure of cohesive soil. Such structural arrangement with high water content is unstable and gives high compressibility. Under the influence of increasing overburden pressure or external load, the soil consolidates and the particles tend to re-orient themselves along horizontal planes (that is, perpendicular to the line of action of the applied load). Such a dispersed structure is more stable and the reduction of water content brings the particles closer together to reduce compressibility, as shown in Fig. 12.1(b). Thus, reduction of water content through application of external load would cause improvement of engineering properties of cohesive soil.

(a) Flocculated structure.

(b) Dispersed structure.

Fig. 12.1 Structure of cohesive soil.

Cohesive soil can be improved using

(a) Preloading with vertical drains and
(b) Soil reinforcement with stone columns.

(b) Granular soil

Particles of granular soil such as sand and gravel, have three-dimensional structure. For the purpose of understanding, they can be represented by spheres which in loose condition are arranged one on top of the other as shown in Fig. 12.2(a). Granular soils in this condition have low relative density. The shear strength is also low because of the tendency of the particles to roll over one another under the influence of shearing stresses. If the same particles are rearranged as shown in Fig. 12.2(b), the void space decreases and the relative

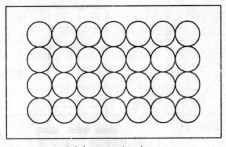

(a) Loose structure.

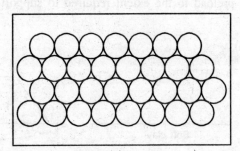

(b) Dense structure.

Fig. 12.2 Structure of granular soil.

density increases with corresponding increase of shear strength. Thus, properties of granular soil can be improved by increasing its relative density by external means. Following ground improvement techniques are used for granular soil:

(a) Compaction with drop hammer and
(b) Deep compaction by compaction piles or vibrofloatation.

Apart from these, soil reinforcement by inserting stiffer materials within the soil fabric such as metallic strips, compacted granular piles, geotextiles, and so on would also improve the properties of the soil which then behave as reinforced mass. Figure 12.3 depicts soil reinforcement technique.

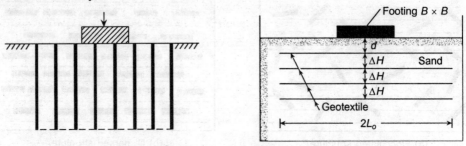

Fig. 12.3 Soil reinforcement.

Chemical injection and grouting are also adopted as methods of ground improvement. In these cases, the chemicals penetrate into the voids and get set to strengthen the soil fabric. Normally, these techniques require specialized equipment to achieve success in the field. These methods are not discussed in this chapter.

12.3 GROUND TREATMENT IN COHESIVE SOIL

12.3.1 Preloading with Vertical Drains

The most common method of ground treatment in cohesive soil is to reduce the void ratio or water content of the soil by preconsolidation. This increases the shear strength of the soil and reduces the compressibility even before construction of the building is commenced. Figure 12.4 displays preloading for building foundations. Soil properties are improved under the preload to the extent required to support the building.

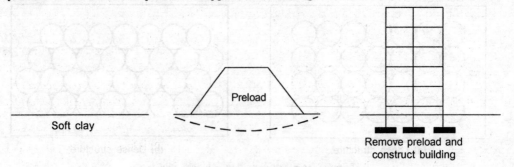

Fig. 12.4 Preloading for building foundations.

The mechanics of ground improvement by preloading may be discussed with reference to the void ratio versus effective stress relationship of a normally consolidated clay, as shown in Fig. 12.5.

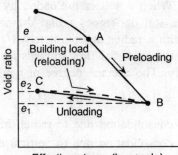

Fig. 12.5 Pressure versus void ratio relationship for normally consolidated clay.

The pressure void ratio relationship of a normally consolidated clay follows the virgin consolidation curve from A to B during preloading. There is large settlement consequent upon the change of void ratio from e to e_1. When the preload is removed, the soil undergoes swelling from B to C and the void ratio increases from e_1 to e_2. Thereafter, when the building is erected, the same intensity of pressure Δp is applied but now the settlement is a function of the change of void ratio corresponding to reloading, that is, e_2 to e_1. So the settlement of the building is reduced considerably. In effect, the potential settlement of the ground under the building load is made to occur under the preload prior to construction of the building.

In direct preloading, the time for consolidation may run into years because of the low permeability of the clay and long drainage path. To reduce the time for consolidation, vertical drains with a drainage blanket on top are used. This is illustrated in Fig. 12.6.

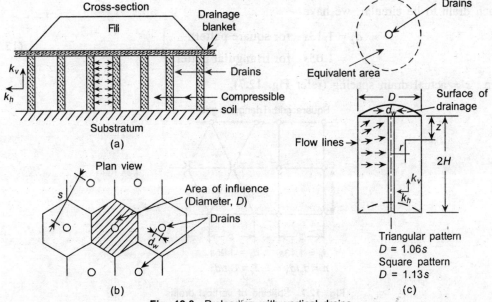

Fig. 12.6 Preloading with vertical drains.

In the earlier days, sand drains, 300–500 mm diameter, were installed by filling with sand the vertical holes made into the soil at predetermined spacing. Nowadays sand wicks and prefabricrted vertical drains (PVD), e.g., band drains are mostly used to have more efficient consolidation under the preload. When a soil is preloaded by dead weight, the horizontal drainage path is reduced and the soil undergoes radial drainage. Each drain well has an axisymmetric zone of influence with a radius approximately 1/2 times the well spacing. The flow within the zone is a combination of radial flow towards the sand drain and vertical flow towards the free-draining boundary. The average degree of consolidation is then given by,

$$\bar{U} = 1 - (1 - U_R)(1 - U_Z) \tag{12.1}$$

where U_R = average degree of consolidation due to radial drainage and

U_Z = average degree of consolidation due to vertical drainage.

Assuming uniform vertical strain at the surface, Barron (1948) gave the expression for degree of consolidation due to radial drainage as,

$$U_R = 1 - \exp\left(-\frac{8T_r}{F_a}\right) \tag{12.2}$$

where $F_a = \left(\dfrac{n^2}{n^2 - 1}\right) \log_e n - \left(\dfrac{3n^2 - 1}{4n^2}\right)$,

$n = \dfrac{d_e}{d_w} = \dfrac{\text{equivalent drain spacing}}{\text{drain diameter}}$, and

$T_r = \dfrac{C_r t}{d_e^2}$ = radial time factor.

The drains may be installed in either square or triangular grid. Considering the influence area of each drain to be circular, we have

$$\left. \begin{aligned} d_e &= 1.13s \quad \text{for square pattern} \\ &= 1.05s \quad \text{for triangular pattern} \end{aligned} \right\} \tag{12.3}$$

where s = actual drain spacing (refer Fig. 12.7).

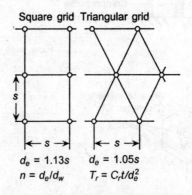

Square grid Triangular grid

$d_e = 1.13s$ $d_e = 1.05s$
$n = d_e/d_w$ $T_r = C_r t/d_e^2$

Fig. 12.7 Spacing of vertical drains.

Coefficient of radial consolidation,

$$C_r = \frac{k_h}{\gamma_w m_v} \tag{12.4}$$

In field problems, U_Z is small compared to U_r and is often neglected. Therefore, for a time t and knowing (C_r), the time factor (T_r) can be calculated and degree of consolidation obtained from Eq. (12.5) as,

$$U = 1 - U_R \tag{12.5}$$

The relationship between U and T_r can be obtained by solving Eq. (12.2) for different values of n, as shown in Fig. 12.8. In practical design, for a given drain diameter d_w, a spacing d_e is chosen, and the time required to achieve 90% degree of consolidation is calculated. If the time is too little or too much, the spacing is changed and calculations are repeated.

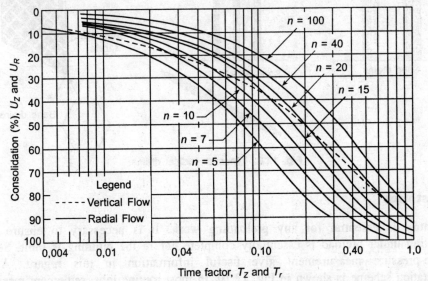

Fig. 12.8 Degree of consolidation versus time factor for radial average (Barron, 1948).

Types of vertical drains

(a) *Sand drains:* Sand drains have been used extensively for preloading since the early thirties. Sand drains (diameter 300–500 mm) were mostly installed by driving a steel casing with an expendable shoe and filling the hole with sand. Smaller diameter sand drains would not ensure integrity of the sand drain due to arching action of the sand.

(b) *Sand wicks:* Sand wicks were introduced by Dastidar et al. (1969). These consist of cylindrical bags made of jute or any permeable fabric, for example, HDPE, which is filled with sand and stitched along the sides and ends. The wicks are prefabricated on the ground by manual sand filling and kept in a water vat to saturate. A 65–75 mm diameter pipe casing with a shoe at the bottom is driven into the soft clay and the wick introduced into the hole at the top as the casing is withdrawn. Sand wicks are kept projected above the ground and covered with a drainage blanket. The sand wicks may be of 55–75 mm diameter and can be installed at spacing of 1–2 m.

(c) *Band drains:* Band drains or fabric drains are usually 75–100 mm wide and 3–5 mm thick made of synthetic fabric with high permeability. They are installed in the ground by special mandrel and cranes.

In addition to the above cardboard drains (Kjellman, 1948) and rope drains have also been used. Different types of vertical drains are shown in Fig. 12.9.

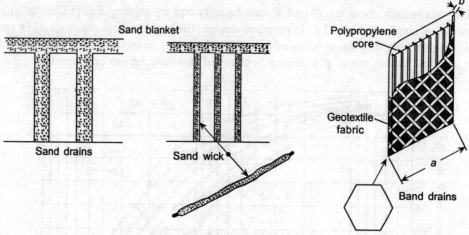

Fig. 12.9 Types of vertical drains.

Case histories

Field control is essential for any preloading work. It is necessary to ensure that the consolidation under preload is essentially complete before the building is built. Settlement and pore-pressure measurement give useful information in this regard. A simple instrumentation scheme is shown in Fig. 12.10. In most routine jobs, settlement measurement on vertical rods and pore-pressure measurement with a few standpipe piezometers, around the periphery of the preload should be sufficient.

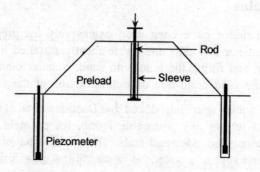

Fig. 12.10 Preloading: field measurement scheme.

Dastidar et al. (1969) described preloading with sand wicks for four storey residential buildings at Salt Lake, Kolkata. Salt Lake is a vast stretch of low lying marshy land in the eastern side of Kolkata which originally served as a natural drainage outfall of the city. The area was reclaimed by filling with dredged fine sand and silt from the river Hooghly in the early sixties.

The reclaimed fill varies in thickness from 1.5–3.0 m. The subsoil, at a depth of about 12 m from the present GL is soft and organic in nature with shear strength seldom exceeding 1.5 t/m^2, as depicted in Fig. 12.11. Below this layer, there is stiff clay with shear strength of 5–10 t/m^2. This is underlain by medium/dense brown silty sand below a depth of 16 m.

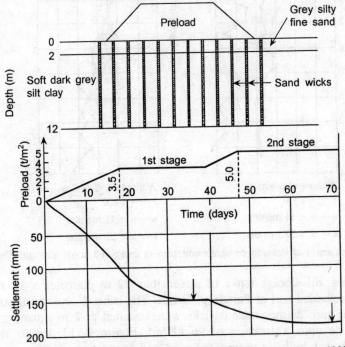

Fig. 12.11 Preloading at Salt Lake, Kolkata (Dastidar et al. 1969).

It is obvious that the soft clay would be liable to undergo excessive settlement even under low to medium-rise buildings on shallow foundations. Preloading is primarily aimed at consolidating the soft clay and making it strong enough to support the building. As an experiment, 75 mm diameter sand wicks were provided at 1.5 m square grid to accelerate the consolidation. Loading was done upto 50 kN/m^2 in two stages. It was found that consolidation under each stage of loading was completed in 5–6 weeks. The buildings founded on spread footings on the preloaded soil.

Pilot (1977) reported the case history of preloading the Palavas embankment. Vane shear test was done before and after preloading to determine the gain in shear strength of the soil. The measurement showed increased undrained shear strength of the soil throughout the depth of sand drains while in the area without sand drains, the strength gain was limited to the top 4 m of the soil where consolidation was only effective in the first 26 months. This is presented in Fig. 12.12.

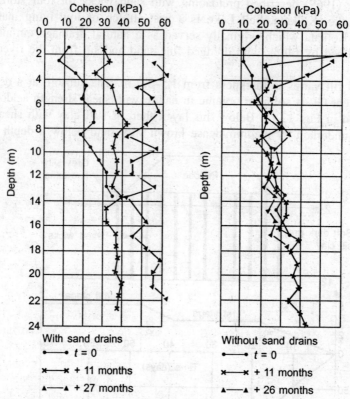

Fig. 12.12 Preloading of Palavas embankment: gain in undrained shear strength (Pilot, 1977).

In recent times, oil storage tanks of dimensions 22 m diameter × 15 m height have been founded on preloaded soil at Paradeep, Orissa. The subsoil consists of soft marine clay in the top 10 m. 65 mm diameter sand wicks were installed at 2 m square grid. Two stage preloading was done upto a pressure of 96 kN/m^2. Figure 12.13 depicts the preloading. Thereafter, the tank was built on compacted sand pad foundation. The gain in strength was sufficient to give adequate bearing capacity of the soil to support the tank load of 160 kN/m^2. Also, settlement of the untreated soil under the design load was estimated as 1000 mm. With a preload of 96 kN/m^2, 600 mm settlement was made to occur during preloading. The tank settlement during hydro test was, thus, restricted to 350 mm at the centre and 200 mm at the periphery. Further, the settlement of the tank during hydro test indicated fairly uniform settlement within permissible limits.

Dastidar (1985) reported extensive preloading at Salt Lake, Calcutta and Haldia. He suggests a consolidation time of 20–120 days for spacing of 55 mm sand wicks between 1.2 m to 2.5 m in the predominantly alluvial deposits of Haldia and Calcutta. Table 12.1 presents the relevant data.

Table 12.1 Spacing of sand wicks and time for settlement

Location	Salt Lake, Calcutta		Haldia	
Spacing of 55 mm diameter sand wicks	1.2 m	1.5 m	2.2 m	2.5 m
Time for 95% consolidation (days)	20	24	80	120

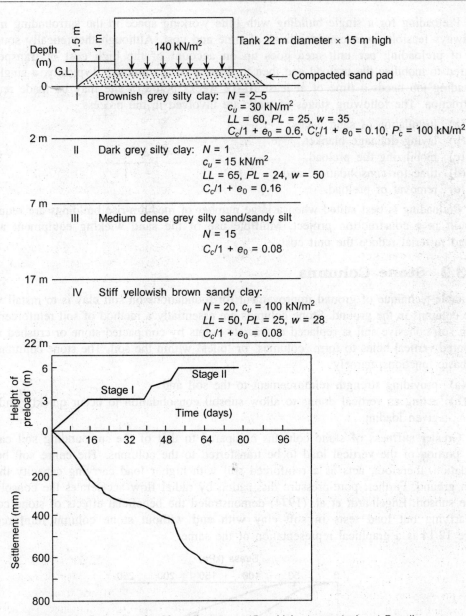

Fig. 12.13 Preloading for 22 m diameter × 15 m high storage tanks at Paradip.

Field control

Preloading is done in the field by assembling dead weight in the form of earth fill, sand bags, brick, stone blocks and other construction materials. Sand bags are the most common method of applying the preload. However, it is difficult to assemble very high preload, requiring more that 5 to 6 m of fill. This would restrict the preload intensity 80–100 kN/m² which should be sufficient for low to medium-rise buildings. Tall buildings would require much higher preload which is difficult to mobilize.

Preloading for a single building with little working space in the surrounding may not be always feasible from considerations of time and cost. Although theoretically sound, the cost of preloading per unit area goes up on account of the high cost of transportation required to mobilize the preload and then to dispose of the preload. Typically, a single stage preloading job needs a time of at least 10–12 weeks before a site may be made ready for construction. The following stages of work are involved in the process:

(a) installation of sand wicks,
(b) laying drainage blanket,
(c) mobilizing the preload,
(d) time for consolidation, and
(e) removal of preload.

Preloading is best suited when a large number of medium-rise buildings are required to be built in a construction project. Multiple use of the sand wicking equipment and the preload material reduce the unit cost.

12.3.2 Stone Columns

A suitable technique of ground improvement for foundations on soft clay is to install vertical stone columns in the ground. Stone columns are essentially a method of soil reinforcement in which soft cohesive soil is replaced at discrete points by compacted stone or crushed rock in pre-bored vertical holes to form 'columns' or 'piles' within the soil. The stone columns serve two basic functions, namely

(a) providing strength reinforcement to the soil and
(b) acting as vertical drains to allow subsoil consolidation to occur quickly under any given loading.

Greater stiffness of stone columns compared to that of the surrounding soil causes a large portion of the vertical load to be transferred to the columns. The entire soil below a foundation, therefore, acts as a reinforced soil with higher load carrying capacity than the virgin ground. Further, pore-pressure dissipation by radial flow accelerates the consolidation of the subsoil. Engelhardt et al. (1974) demonstrated the beneficial effects of stone columns by carrying out load tests in soft clay with and without stone column reinforcement. Figure 12.14 is a graphical representation of the same.

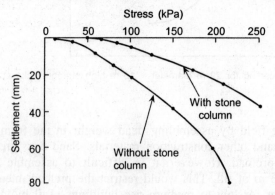

Fig. 12.14 Effectiveness of stone columns (Engelhardt et al., 1974).

Construction technique

Installation of stone columns in soft clay may be done in two ways:

  (a) Vibratory technique using vibroflot and
  (b) Rammed stone column technique.

Vibroflotation

The basic tool used in these techniques is a poker vibrator or vibroflot, as shown in Fig. 12.15, which is 2.0–3.0 m long with a diameter varying between 300 mm to 500 mm. Extension tubes are attached to the vibroflot whenever greater depth of treatment is needed. The vibroflot is a hollow steel tube containing an eccentric weight mounted at the bottom of a vertical shaft; the energy is imparted by rotational motion through the shaft while the eccentric weight imparts vibration in a horizontal plane. Vibration frequencies are fixed at 30 Hz or

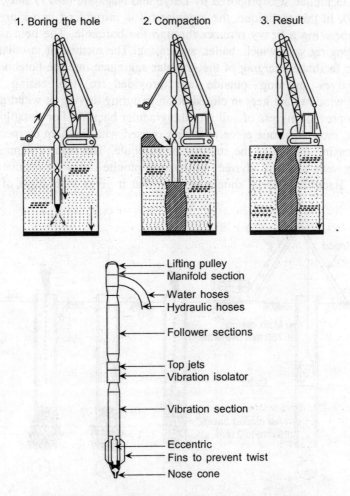

1. Boring the hole   2. Compaction   3. Result

- Lifting pulley
- Manifold section
- Water hoses
- Hydraulic hoses
- Follower sections
- Top jets
- Vibration isolator
- Vibration section
- Eccentric
- Fins to prevent twist
- Nose cone

Fig. 12.15 Stone column installation by vibroflotation.

50 Hz to suit electric power cycles. The free fall amplitude varies between 5–10 mm. The machine is suspended from a vibration damping connector by follower tubes through which power lines and water pipes pass. These allow simultaneous release of water jets to remove the soil around the vibroflot as the latter makes its way into the hole under vertical pressure from the top. When the vibroflot reaches the desired depth, the water jet at the lower end is cut off and granular backfill is poured through the annular space between the hole and the vertical pipe by head load or conveyor as the vibratory poker is withdrawn. Well graded stone backfill of size 75 mm to 2 mm is used and compaction is achieved by vibration of the poker as it is lifted up. Due to compaction, the stones are pushed sideways into the soft soil to produce a stone column of diameter larger than the diameter of the borehole. Normally, 600–900 mm diameter stone column can be obtained for 300–500 mm diameter vibroflot.

Rammed stone column

This installation technique was proposed by Datye and Nagaraju (1977) and developed further by Nayak (1983). In this technique, the granular fill is introduced into a pre-bored hole and compacted by operating a heavy rammer through the borehole. The hole is made by using normal bored piling rig with winch, bailer, and casing. The method of installation is illustrated in Fig. 12.16. To facilitate charging of the granular aggregate into the borehole, windows with hinged flap valves opening outside are provided to the casing at interval of 2 m or so. These windows are kept in closed position during driving or withdrawal of casing by screwing nuts to prevent ingress of soil into the granular backfill. For installing stone columns to greater depths, more than one piece of casing is used with the help of special quick release couplings. The casing maintains the stability of borehole. The stone columns are required to function as drain wells and it is advised not to use bentonite slurry for maintaining the stability of the borehole. Backfill material should be such that it gives high angle of internal friction

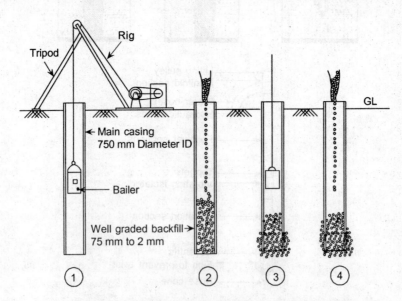

Fig. 12.16 Stone column installation by ramming method.

under given energy of compaction. Sometimes the mixtures of stone aggregate and sand, generally in proportion of 2:1, are used as backfill material. It is observed that sand is utilized mainly in filling the voids in gravel skeleton. Gravel backfill of aggregate size 75 mm to 2 mm is generally recommended. The gravel should be well graded and preferably angular shaped for good interlock. The main purpose of compaction is to rearrange the stone particles so that very good interlocking between particles is obtained to give high angle of internal friction. Too much ramming, however, crushes the aggregate. For a given compaction energy, greater weight and smaller drop of the rammer give better results.

Comparison of construction techniques

All the installation techniques for stone columns in soft clay are self-adjusting in the sense that enlargement of the column during ramming or vibration occurs depending on the soil consistency. Figure 12.17 shows the range of soil suitable for such a treatment by stone columns.

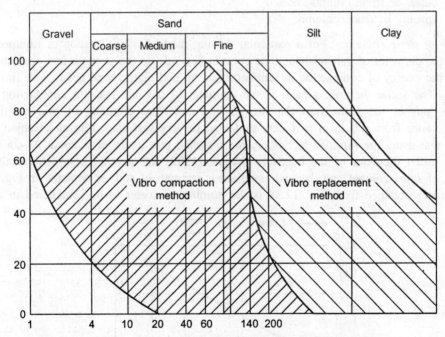

Fig. 12.17 Range of soil suitable for treatment by stone columns.

Rammed stone columns have been used extensively in India. They are found to be quite suitable for all kinds of soil (Datye 1982). Nayak (1982) has suggested that the angle of internal friction, ϕ' may be as high as 45° for compacted granular fill in rammed stone column, whereas for vibrofloted stone column ϕ' ranges between 38°–42°. It is also to be noted that in vibrofloted stone column, the top about 1 m deep does not get properly compacted for lack of confinement near the surface, whereas in practice this portion of the stone column is required to take greater load intensity. In case of rammed stone column, proper compaction can be achieved even for this length because of lateral confinement of the casing pipe. With vibroflot, there is no harm in using high energy of compaction. Rather, it results in larger diameter of the stone column and better compaction of aggregate to give higher value of angle of internal friction. Net effect of this is to increase the load carrying capacity of the stone column. But for rammed stone column, such high compaction energy may crush the aggregates, resulting in lower value of ϕ' and lower capacity of the stone columns. In general, however, vibroflotation needs skilled labour and better quality control while the installation of rammed stone columns needs greater manpower. Overall, rammed stone column appears to be more economical although it is a very slow process compared to vibroflotation.

Design principles

The design of stone column foundation involves the assessment of

  (i)   diameter of stone column,
 (ii)   depth of stone column, and
(iii)   spacing of stone column.

Diameter of stone column: For a particular driving method, vibroflotation or rammed stone column, the diameter of the finished stone column depends on the strength and consistency of the soil, the energy of compaction in rammed column, and diameter of poker with fins of the vibroflot. The softer the soil, greater is the diameter of the pile because compaction of the aggregate pushes the stone into the surrounding soil. The diameter of pile installed by vibroflot varies from 0.6 m in stiff clay to 1.1 m in very soft clay. Rao and Ranjan (1985) reported that using the rammed technique, the installed pile diameter is about 20–25% more than the initial diameter of the borehole. Nayak (1982) has given a chart to correlate the diameter of stone column and the undrained shear strength of soil, as depicted in Fig. 12.18. The diameter obtained from Fig. 12.18 is the nominal diameter to be considered in design.

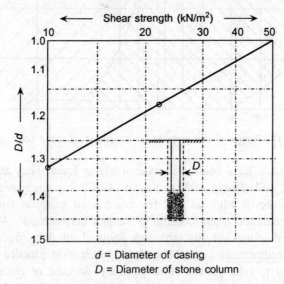

d = Diameter of casing
D = Diameter of stone column

Fig. 12.18 Shear strength of soil versus diameter of stone column (Nayak, 1982).

Depth of stone column: The stone column is installed below a foundation upto the depth of soft compressible strata within the zone of influence in the subsoil. In addition to carrying vertical load, stone columns function as drainage path to dissipate excess pore water pressure and hence, accelerate the rate of consolidation. This requires the stone columns to be taken down to the depth of major compressible strata which make significant contribution to the settlement of the foundation.

This point can best be understood by determining the contribution of each layer of soil towards the settlement of the foundation. For example, in the subsoil condition at Haldia (refer Fig. 12.19), the depth of soil contributing to 90% of the total settlement works out to 17 m for 50 m diameter storage tanks. Accordingly, stone column depths of 17 m were adopted for these foundations. However, in some stratified deposits, the nature of stratification more or less determines the depth of stone column. For example, a 10 m thick clay deposit followed by hard clay or dense sand would require stone column depth of 10 m irrespective of the size of foundation.

Depth (m)		Description	Soil properties
0		Top soil	—
1 – 4	I	Brownish grey silty/clayey silt	N = 5 blows/30 cm LL = 50% PL = 22% w = 29% c_u = 50 kN/m^2 M_v = 0.0003 m^2/kN
4 – 10	II	Grey silty clay with decomposed wood and silt laminations	N = 3 blows/30 cm LL = 50% PL = 24% w = 40% c_u = 30 kN/m^2 M_v = 0.0007 m^2/kN
10 – 14	III	Grey clayey silt/sandy silt with laminations of silty clay	N = 5 blows/30 cm LL = 50% PL = 22% w = 29% c_u = 50 kN/m^2 M_v = 0.0004 m^2/kN
14 – 17	IV	Grey/Brownish grey silty clay with calcareous nodules	N = 9 blows/30 cm M_v = 0.0002 m^2/kN
17 – 25	V	Mottled brown/yellowish brown silty clay with nodules and rusty brown patches	N = 20 blows/30 cm LL = 46% PL = 18% w = 24% c_u = 85 kN/m^2 M_v = 0.0001 m^2/kN
25 –	VI	Brown silty fine sand with lenses of clay	N = 40 blows/30 cm
	VII	Brown silty fine sand	N = 50 blows/30 cm
29		Very stiff to hard silty clay	—

Fig. 12.19(a) Settlement calculation for 50 m diameter × 11.4 m height tank at Haldia.

Stratum	Thickness H_z (m)	$\Delta\sigma_z = I_q \times q_{net}$ (kN/m^2)	M_v (m^2/kN) $\times 10^{-4}$	δ (m)	Cumulative settlement (%)
I	1.0	96.6	3.0	0.030	4.0
I	3.0	91.4	3.0	0.083	14.2
III	6.0	87.7	7.0	0.368	65.0
IV	4.0	83.0	4.0	0.133	83.0
V	3.0	78.5	2.0	0.047	89.3
VI	8.0	69.2	1.0	0.055	96.7
VII	4.0	54.1	0.5	0.011	98.2
VIII	11.0	41.5	0.3	0.014	100.0
				0.740	

Fig. 12.19(b) Table for settlement calculation for 50 m diameter × 11.4 m height tank at Haldia.

Spacing of stone columns: The design of stone column foundation primarily involves determination of a suitable spacing of stone column for a chosen diameter and length of the latter. It depends on the required load bearing capacity of the foundation and the allowable time for consolidation by radial drainage through stone columns. It can be worked out in terms of the degree of improvement required for providing a satisfactory foundation for the design load. The settlement improvement ratio of the reinforced ground to untreated ground is a function of pile spacing as shown in Fig. 12.20 (Greenwood, 1970). Mitchell and Katti (1981) have suggested typical pile spacing for rectangular and square grid depicted in Fig. 12.21.

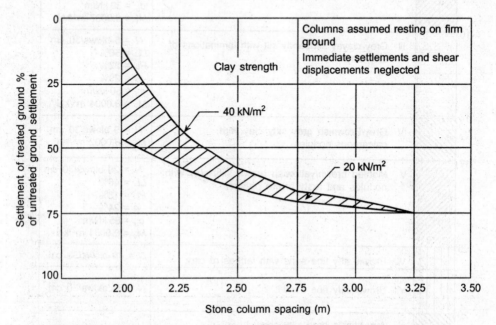

Fig. 12.20 Effect of stone column on anticipated settlement (Greenwood, 1970).

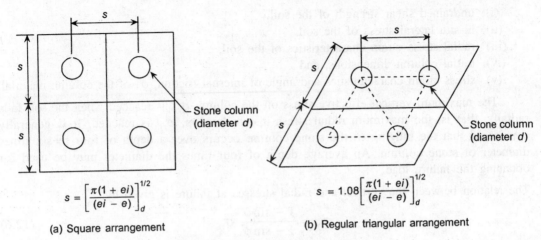

(a) Square arrangement

$$s = \left[\frac{\pi(1 + ei)}{(ei - e)}\right]^{1/2}_d$$

(b) Regular triangular arrangement

$$s = 1.08\left[\frac{\pi(1 + ei)}{(ei - e)}\right]^{1/2}_d$$

Fig. 12.21 Spacing of stone columns.

Analysis of granular pile foundations for triangular grid of piles show that significant reduction in settlement occurs only if the spacing of stone column is close ($s/d \leq 4$) and the piles are installed to full depth of consolidating layer. However, too close a spacing ($s/d \leq 2$) is not feasible from construction point of view. Thus, a stone column spacing (s/d) of 2.5–4 is adopted for most practical problems. Also it has been recognized that closer spacing is preferred under isolated footing than beneath large rafts (Greenwood, 1970).

Load carrying capacity of individual stone columns

A stone column is subjected to a stress condition much alike that imposed in the standard triaxial test as shown in Fig. 12.22. A vertical stress, σ_v is applied by surface loading, and a radial effective stress, σ_r results from the horizontal reaction of the ground. Therefore, the factors which govern the soil–column behaviour are (Hughes et al., 1975):

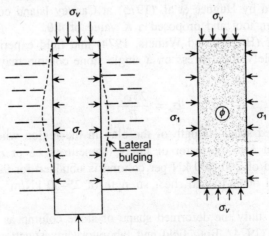

Fig. 12.22 Stresses acting on stone column.

(i)   undrained shear strength of the soil,
(ii) in-situ lateral stress of the soil,
(iii) radial stress–strain characteristics of the soil,
(iv) initial column dimensions, and
(v)   stress–strain characteristics and angle of internal friction, ϕ', of the column material.

The maximum vertical effective stress on the column σ_{vf}, is reached when the soil fails radially, that is, the maximum radial stress it can develop, σ_{rf}, is reached. It is generally considered that the bulging in the stone column occurs over a depth of four to six times diameter of stone column. An average depth of four times the diameter may be used for obtaining the failure load.

The relation between the vertical and radial stresses at failure is given by,

$$\sigma_{vf} = \frac{1 + \sin\phi'}{1 - \sin\phi'}\,\sigma_{rf} \tag{12.6}$$

The value of σ_{rf} can be expressed in terms of the initial radial stress, σ_{ro} as

$$\sigma_{rf} = \sigma_{ro} - u + Kc_u \tag{12.7}$$

From Eqs. (12.6) and (12.7),

$$\sigma_{vf} = \frac{1 + \sin\phi'}{1 - \sin\phi'}\,(\sigma_{ro} - u + Kc_u) \tag{12.8}$$

where c_u is the undrained shear strength of the clay,
σ_{ro} is the initial total radial stress,
u is the pore-pressure,
ϕ' is the angle of internal friction of the material of the column, and
K is an earth pressure coefficient.

Measurements carried out by Hughes and Withers (1974) in soft clay using a selfboring pressuremeter (Camkometer) yielded K values of about 4.0, whereas Menard had earlier used a conventional pressuremeter to obtain values of about 5.5. The full-scale loading tests on stone columns performed by Hughes et al. (1975) at Canvey Island confirmed the reliability of Eq. (12.8) as a design tool and proposed a K value of 4.0.

Both limit analysis (Hughes and Withers, 1974) and field experience (Thorburn 1975) suggest that the allowable vertical stress on a single stone column may be obtained from the empirical expression

$$\sigma_v = \frac{25\,c_u}{F} \tag{12.9}$$

where c_u is the undrained shear strength of the clay in the region where the bulging of the stone column occurs and F is the factor of safety (sometimes FS or Fs).

Typical design load of 200–400 kN per column is obtained for 900 mm diameter stone columns in cohesive soil of undrained shear strength of 25–50 kN/m^2 and a factor of safety of 2.0.

It is interesting to study the deformed shape of stone column as observed after failure by Hughes and Withers (1974). Both field and laboratory investigations show geometrically similar deformed shapes with bulging in the upper region, as visible in Fig. 12.23. The data appear to confirm the concept by which stone columns are believed to improve the bearing capacity of soft clays.

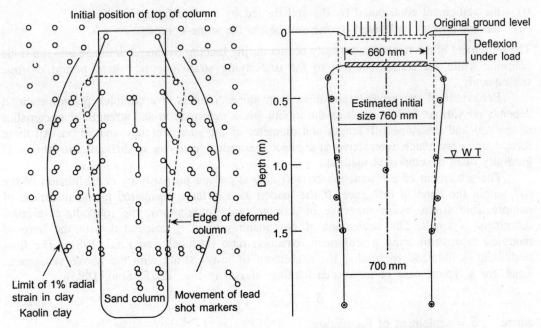

Fig. 12.23 Deformation of stone column (Hughes and Withers, 1974).

Figure 12.24 shows a typical load versus settlement curve of a test column loaded to failure at a site in Haldia. The failure load compares well with the predictions based on Eq. (12.8). A factor of safety of 2 gives the safe load on the test column.

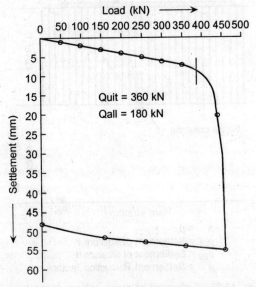

Fig. 12.24 Load settlement relationship of single stone column.

Settlement of stone column foundations

The settlement of stone column foundations consists of two components:

(i) the settlement contributed by the soil treated by stone columns and
(ii) the settlement contributed by the soil below the stone columns.

The settlement of the treated soil mostly occurs during loading by lateral drainage into the stone columns while the settlement due to the underlying strata occurs over a period of time subsequently.

Behaviour of the ground reinforced with stone columns is a complex phenomenon. It depends on various factors, such as the in-situ stress deformation and strength characteristics of the soil and the stone fill, length and diameter of the stone column, and so on. All these factors are very much interactive and exact theoretical analysis is difficult. So design is generally done by empirical methods.

The settlement of a foundation on soft clay is caused primarily by consolidation of the soil within the zone of influence. If the loaded area is large compared to the thickness of compressible strata, as in the case of storage tank foundations, the immediate (elastic) settlement is small. The settlement of composite ground is calculated with the help of modified expression using a settlement reduction ratio (Mitchell and Katti, 1981). The final settlement is obtained by adding the settlement of lower strata with the above settlement. Thus, for a typical case of large area loading, shown in Fig. 12.25 (Som, 1995),

$$\delta = \beta\rho_{C1} + \rho_{C2} \tag{12.10}$$

where δ = settlement of foundation

ρ_{C1} = settlement of untreated soil within the depth of stone column treatment
ρ_{C2} = settlement of untreated soil below the stone columns, and
β = settlement reduction ratio for stone column treatment.

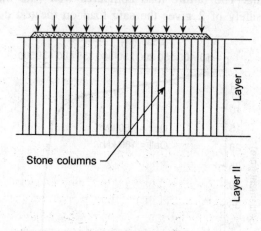

δ = $\beta\rho_{C1} + \rho_{C2}$
ρ_{C1} = Settlement of stratum I
ρ_{C2} = Settlement of stratum II
β = Settlement Reduction factor

Fig. 12.25 Settlement of stone column foundation.

The settlement reduction ratio is a function of the area ratio of stone column installation A_c, and the stress concentration ratio η, Eq. (12.11).

$$\beta = \frac{1}{1 + (\eta - 1)A_c} \tag{12.11}$$

The stress concentration factor gives the ratio of the stresses in the stone column and the surrounding soil for equal deformation. Mitchell and Katti (1981) gave values of $n = 3$–5 for practical design. Som (1995) obtained values of n varying between 4–7 from field observations of settlement of stone column foundations at Haldia and Kandla.

The rate of settlement is usually determined by the theory of radial consolidation (Barron, 1948) for the chosen diameter and spacing of stone column.

Greenwood (1970) suggested a relationship between the settlement of treated ground as a function of soil strength and column spacing, as shown in Fig. 12.19. To these settlements should be added any anticipated settlement contributed by the underlying strata. Figure 12.19 should, however, be taken as indicative only because the diameter of the stone column and the area ratio is not taken into consideration. It will be apparent from Eqs. (12.10) and (12.11) that even for a spacing as close as twice the diameter of the columns, the area ratio A_c comes to only 20–25% and for a stress concentration ratio of 4–6, the settlement reduction ratio will vary in the range of 40–50%. In fact, this is the order of settlement reduction that can be expected for large area loading on stone column foundation. Further, reduction does not appear possible. Therefore, one cannot expect a drastic reduction of settlement by providing stone column foundation beneath a foundation.

Practical applications

Flexible large area loading provides ideal situation for stone column foundations on soft cohesive soil. Thick deposits of clay are abundant along the coastal and alluvial plains of India, as shown in Fig. 12.19. In most cases, the clay layer is 10–15 m thick, somewhat desiccated near the ground surface and underlain by firm clay/dense sand/weathered rock, and so on. Such clay deposits present perennial foundation problems on account of low undrained shear strength and high settlement potential. They are characterized by water content close to the liquid limit often with high organic content. Any large area loading by way of rigid or flexible raft foundation has typical safe bearing capacity of 50 kN/m^2 from shear failure consideration. Even under this kind of pressure, the long term consolidation settlement is likely to be of the order of 200–400 mm depending on the size of foundation. If the foundation pressure increases to, say 100 kN/m^2, not only does the safe bearing capacity falls appreciably short but the estimated settlement increases to 500–1000 mm which will not generally be acceptable even for a highly flexible foundation. In order to have optimum improvement of the soil, for example, to raise the safe bearing capacity of the foundation to say 100–150 kN/m^2 and to bring down the settlement to 250–500 mm various alternative methods of ground improvement, including preloading and stone columns have been tried. In recent years, stone columns have been extensively used to provide foundations for large area loading in the coastal and alluvial plains of India (Som, 1999). Some typical settlement data for a 50 m diameter × 10 m high tank at Kandla are shown in Fig. 12.26.

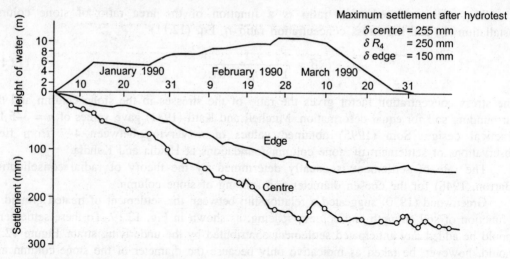

Fig. 12.26 Stone column foundation for large diameter storage tanks at Kandla: settlement during hydrotest.

12.4 GROUND IMPROVEMENT IN GRANULAR SOIL

Numerous methods of ground improvement in predominantly cohesionless soil using the principle of vibration are in practice today. Loose granular soil with 'N' less than 10, is characterized by low shear strength and low bearing capacity. In order to improve its strength, the soil is compacted. This increases its relative density with consequent increase of angle of shearing resistance.

The compaction of soil is achieved either by repeated hammer blows on the ground or by insertion of probes within the soil which are then vibrated. The depth and extent of vibration to be imparted depend on the nature of foundation problem encountered in a given situation.

12.4.1 Heavy Tamping or Drop Hammer

The heavy tamping or drop hammer method of ground compaction employs repeated blows of a heavy weight on the ground surface. In small works, a hammer (20–80 kN) is dropped from a height of 4–8 m with the help of a driving rig or a tripod stand. The hammer is made of RCC in the shape of a truncated cone with a low centre of gravity, as shown in Fig. 12.27. The repeated blows of the hammer transmit vibration into ground which improves the relative density of the soil in the upper layers and to a lesser degree in the lower layers. Ramming is associated with gradual depression of the ground surface. With each successive impact, the depression reduces in magnitude. After 5–10 blows, it remains essentially constant. The ramming is then continued in the adjacent areas. The ground is considered to be compacted if the dry density of the soil achieves the following values:

$$\text{Sandy silt} \quad : \quad 15.5\text{–}16 \text{ kN/m}^3$$
$$\text{Sand} \quad : \quad 16\text{–}16.5 \text{ kN/m}^3$$

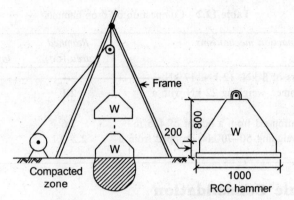

Fig. 12.27 Compaction by drop hammer.

Ground improvement by the method of drop hammer is usually achieved in the top 2.5–3 m of the soil. Deep compaction is not possible by this method. This is clear from the pre and post compaction density tests reported by Tsytovich et al. (1974) at a site in Russia, which are presented in Fig. 12.28. The method is, therefore, suitable for small foundations only.

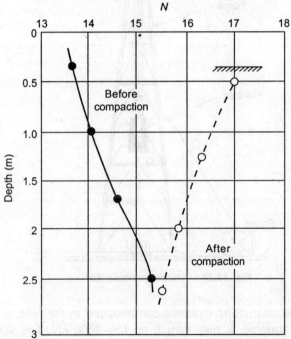

Fig. 12.28 Density versus depth in compaction by drop hammer (Tsytovich, 1974).

Based on experiences in Russia, Tsytovich (1974) suggested the depth to which significant improvement can be achieved by drop hammer technique. Table 12.2 presents compaction by drop hammer in a tabulated manner.

Table 12.2 Compaction by drop hammer

Ground compaction mechanisms	Rammed area (m²)	Depth of layer compaction (m)
Drop-weight rollers of 8 kN, 12 kN, 17 kN	—	1.0–1.5
Double-acting hammer, weighing 22 kN with a metal bottom plate	2.1	1.2–1.4
Hammers 24 kN, dropped from a height of 4–5 m	1.6	1.6–2.2
Heavy hammers weighing 50–70 kN, dropped from a height of 6–8 m	2.2–3.1	2.7–3.5

12.4.2 Dynamic Consolidation

Dynamic consolidation is the name given to the procedure of deep compaction of soils by repeated blows of a hammer falling from a height of 30–40 m. The technique, pioneered by Louis Menard, is used for compaction of soils 30 m below GL. Figure 12.29 is a diagrammatic representation of dynamic consolidation set up.

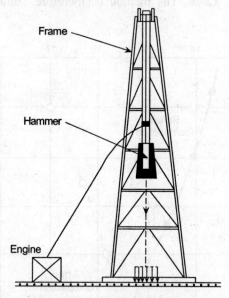

Frame

Hammer

Engine

Fig. 12.29 Dynamic consolidation.

Hammer weight used to perform dynamic consolidation in the field is many times more than that used for heavy tamping. It may vary from 150–2000 kN. The shape of the contact surface and the cross-sectional area of the hammer are chosen to suit the soil type and reaction of the ground to the impact energy. Dynamic consolidation can also be done in saturated cohesive soil provided the formation is varved and contains continuous sand partings to facilitate pore-pressure dissipation. In homogeneous cohesive soil, dynamic consolidation may be done with pre-installed vertical sand drains to effect drainage, as depicted in Fig. 12.30.

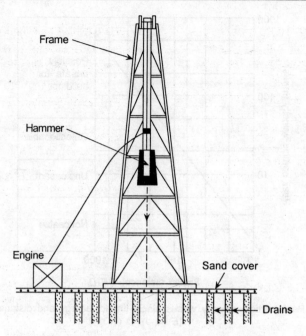

Fig. 12.30 Dynamic consolidation with vertical drains.

The depth of influence for compaction by dynamic consolidation is given by the expression:

$$D = \alpha \sqrt{\frac{E}{10}} \qquad (12.12)$$

where D = depth of compaction (in metres),
        E = energy developed by each impact (in KJ), and
        α = a factor that varies between 0.5–0.8.

The extent of ground improvement by dynamic consolidation as given by the improvement of bearing capacity is generally of the order of 2 in clays, 3 in silts, and 4 in sand. Heavy tamping and the associated vibration often leads to damages to adjoining structures. Figure 12.31 shows variation of the peak particle velocity v, as a function of the scaled energy factor ($\sqrt{E/D}$). A safe upper limit of 'v' is generally taken as:

$$v = 70 \left(\frac{E}{10D^2} \right)^{0.56} \qquad (12.13)$$

where E = energy per blows (in KJ) and
        D = distance between impact centre and observation point.

If the permissible maximum peak particle velocity is taken as 50 mm/s, the following relationship may be used to limit the effect of applied energy on nearby buildings.

$$E \leq 5.5 \, D^2 \qquad (12.14)$$

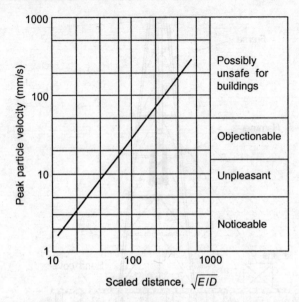

Fig. 12.31 Peak particle velocity versus single impact energy and distance from impact point
[E (in J), D (in m)].

12.4.3 Vibrocompaction

This is method of deep compaction of granular soils where a probe is inserted into the soil and then vibrated. Compaction is brought about by vibration, as displayed in Fig. 12.32. The process is essentially similar to vibroflotation as adopted for installation of stone columns in cohesive soil described earlier.

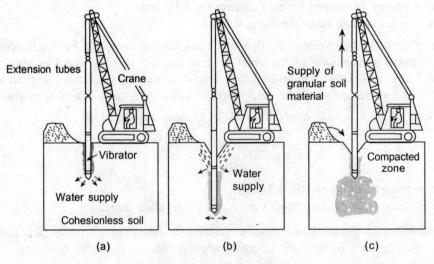

Fig. 12.32 Vibrocompaction.

The range of soil particle size that can be compacted by vibrocompaction is shown in Fig. 12.33 (Brown, 1978). During penetration of the probe, the sand below WT liquefies due to increase of pore water pressure. As the probe is withdrawn, the soil particles rearrange themselves to give a dense packing. The presence of fines, however, may inhibit the compaction. Figure 12.34 illustrates the particle size distribution suitable for the process.

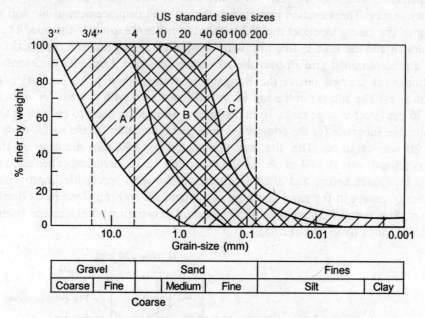

Fig. 12.33 Particle size distribution suitable for vibrocompaction.

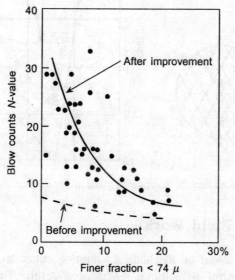

Fig. 12.34 Effect of fines content on the extent of improvement by vibrocompaction.

12.4.4 Compaction Piles

Densification of loose sand by compaction piles has been done for many years. Loose sand strata are liable to liquefaction during earthquake and they often need densification by artificial means to improve their relative density. The process essentially involves driving a casing that has an expendable shoe into the soil—as in the case of cast-in-situ driven piles—by repeated blows of a hammer. The vibration imparted to the soil and displacement of the soil caused by the driving of the casing compact the soil within the influence zone of the pile. The casing is then withdrawn and the hole is filled up with compacted aggregate. The compaction piles are driven on a predetermined grid all over the area to be densified. Dhar (1976) describes the soil densification work carried out at the Bongaigaon refinery and petrochemical complex by compaction piles. The subsoil in the top 10 m had loose sand with average 'N' value less than 10 blows/30 cm. Steel storage tanks 16 m diameter × 10 m high, were to be built at the site. To obtain a suitable subgrade for the compacted sand pad foundation for the tanks, an average 'N' value of 20 was required. The site was compacted by 450 mm diameter × 10 m deep compaction piles driven at 2.67 m c/c, as shown in Fig. 12.35. Standard penetration tests done in adjacent boreholes before and after densification showed appreciable improvement in the relative density except in the top 1 m of the soil where the soil remained loose due to lack of confinement. Substantial improvement of the standard penetrating resistance has been achieved by compaction piles (refer Fig. 12.35).

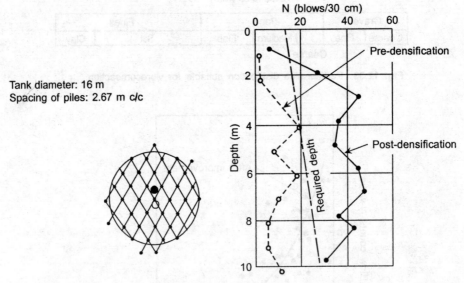

Fig. 12.35 Ground improvement by compacting piles at Bongaigaon Refinery (Dhar, 1976).

12.4.5 Control of Field Work

Field compaction of granular soil by mechanical vibration either at the surface or inside the ground needs adequate control to achieve the desired result. There is no mathematical approach to determine the nature and extent of ground improvement required. The problem has to be approached on a trial and error basis with past experiences serving as useful guide. A suitable approach may be to

(a) carry out detailed soil exploration by conducting standard penetration tests or cone penetration tests within the influence zone below the proposed foundations.

(b) determine the extent of improvement required in terms of increased N or N_c value from appropriate foundation analysis.

(c) choose the method of ground improvement, say, dynamic consolidation, vibrocompaction, or compaction piles depending on availability of facilities.

(d) select a representative area close to the construction site and carry out trial ground improvement work varying the construction control parameters, such as weight of hammer, height of fall, number of blows, spacing of compaction piles, and so on. In general, the weight of hammer and height of fall are predetermined as per equipment capacity. Before starting the trial ground improvement scheme, make sure that predensification properties of the soil, namely SPT, CPT data, and so forth have to be determined.

(e) carry out field tests to determine the post densification parameters after completion of trial ground improvement work and determine the extent of improvement as a function of spacing of compaction piles.

(f) choose the spacing of compaction piles on the basis of pre and post densification data.

(g) carry out field tests to check the efficacy of compaction at different locations.

A trial ground improvement work has recently been done for the proposed construction of a 120 m high temple about 75 km from Kolkata. The subsoil consists of a thin layer of firm silty clay/clayey silt followed by a deep deposit of loose to medium silty fine sand to 25 m below GL. Thereafter, a thin layer of stiff clay is found and the same is followed by dense to very dense sand. The soil condition at the temple site is shown in Fig. 12.36. In view of the high foundation loading anticipated from the temple structure, the sandy soil immediately beneath the top silty clay 12–14 m below GL required improvement.

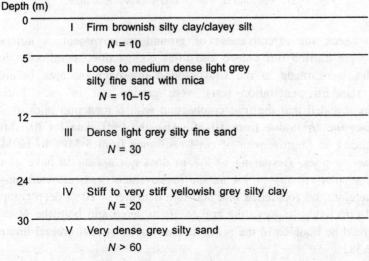

Fig. 12.36 Soil condition at temple site.

Trial ground improvement was done at a suitable location near the proposed temple structure. The area was divided into three sections shown in Fig. 12.37 and compaction was done with pile spacing of 2.8 m × 2.8 m, 2 m × 2 m, and 1.4 m × 1.4 m. The method essentially consisted of driving a steel casing, 500 mm diameter × 8–12 m long with expendable shoe at the bottom and then filling the hole with compacted stone as the casing was withdrawn. The loose stone inside the casing was compacted in layers by 30 blows of a 10 kN hammer falling through a height of 2 m.

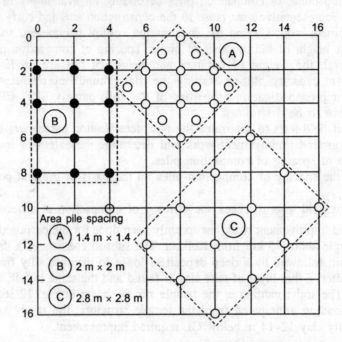

Fig. 12.37 Planning of trial ground improvement work.

In order to check the effectiveness of ground improvement, a number of cone penetration tests were carried out before and after compaction, at adjacent locations. In addition, boreholes were made in the trial ground improvement area before and after compaction and standard penetration tests were carried out in each borehole (refer Fig. 12.37). It was revealed that the trial compaction with compaction piles at 2 m spacing effectively increases the 'N' value from 10–30. Also the CPT data for the three different spacings clearly shows an improvement of cone resistance from 5 MPa to 10 MPa for 2 m spacing of compaction piles. A spacing of 2.8 m does not appear to have any significant effect while for a smaller spacing of 1.4 m, the improvement was restricted only in the top 4–5 m. It may, therefore, be concluded that 500 mm diameter × 10 m deep compaction piles at 2 m c/c would effectively improve the soil to 10 m depth and both the N value and the cone resistance would be doubled in the process. The results of trial ground improvement are shown in Fig. 12.38.

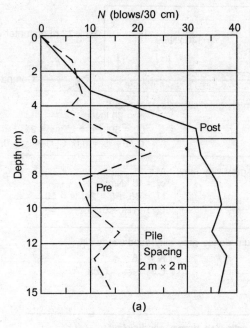

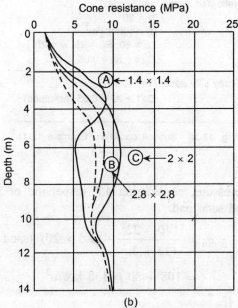

Fig. 12.38 Results of trial ground improvement.

Example 12.1

A 22 m diameter × 15 m high steel storage tank is proposed to be built in a port area. The subsoil condition is shown in Fig. 12.39. Preloading with sand wicks is proposed to be done. Design a suitable preloading scheme and estimate the settlement of the tank during hydrotest.

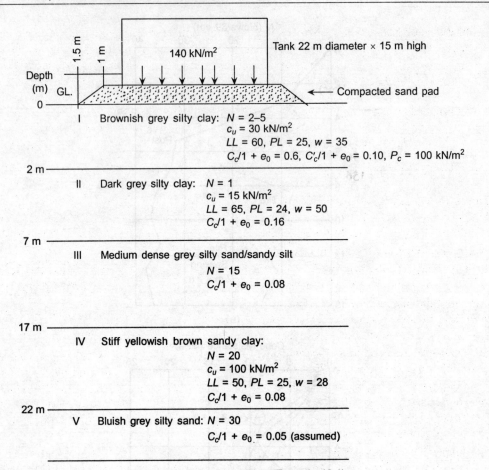

Fig. 12.39 Subsoil condition (Example 12.1).

Solution

(a) Net foundation pressure, considering 45° dispersion of water load (140 kN/m²) through compacted sand pad.

$$q_n = \frac{140 \times 22^2}{(22 + 3)^2} + (1.5 \times 20) \text{ (sand pad)}$$

$$= 108 + 30 = 138 \text{ kN/m}^2$$

Average c_u of strata I and II

$$= \frac{(30 \times 2) + (15 \times 5)}{7}$$

$$= 19.3 \text{ kN/m}^2$$

$$q_{\text{ult}}(n) = c_u N_c$$

$$= 19.3 \times 6 = 116 \text{ kN/m}^2 < 138 \text{ kN/m}^2$$

Settlement of untreated soil, $\delta = \sum \dfrac{C_c}{1 + e_0} H \log \dfrac{p_o + \Delta p}{p_o}$

Layer	H(m)	p_o (kN/m²)	Δp (kN/m²)	$\dfrac{C_c}{1 + e_0}$	δ (m)	
I	2	8	138	0.06	0.151	
II	5	36	132	0.16	0.535	0.686
III	10	96	89	0.08	0.227	
IV	5	156	44	0.08	0.043	
V	15	236	40	0.05	0.051	0.321
					Σ 1.007 m	
					($\approx$ 1000 mm)	

The subsoil of strata I and II are too weak to support the given load. Also strata I and II would contribute 70% of the total settlement.

(b) Consider preloading the soil by installing 65 mm diameter sand wicks, 8 m long. Use two stage preloading (3 m + 3 m) with sand bags or sand fill.

The average consolidation pressure in strata I and II (refer Fig. 5.12, Chapter 5)

Therefore, $P_c = 50 + 50 = 100$ kN/m²

Stratum I

$\Delta p_1 = 100$ kN/m²

Increased shear strength after consolidation by two stage preloading,

$\Delta c_u = 0.3 \times 100 = 30$ kN/m²

Stratum II

$\Delta p_1 = 0.96 \times 100 = 96$ kN/m²

Increased shear strength, $\Delta c_u = 0.3 \times 96 = 28.8$ kN/m²
Additional load during hydrotest = 138 − 100 = 38 kN/m²

Consider 90% consolidation during hydrotest. Sand wicks are to be designed accordingly. After consolidation,

Stratum I

$$\Delta p = \left(\dfrac{38 \times 25^2}{26^2} \right) \times 0.9 = 31.6 \text{ kN/m}^2$$

$$\Delta c_u = 0.3 \times 31.6 = 9.5 \text{ kN/m}^2$$

Stratum II

$$\Delta p = \frac{38 \times 25^2}{29.5^2} \times 0.9 = 24.6 \text{ kN/m}^2$$

$$\Delta c_u = 0.3 \times 24.6 = 7.4 \text{ kN/m}^2$$

Hence, strength after hydrotest for

Stratum I : $30 + 30 + 9.5 = 69.5$ kN/m^2

Stratum II : $15 + 28.8 + 7.4 = 51.2$ kN/m^2

Weighted average of $c_u = \dfrac{69.5 \times 2 + 51.2 \times 5}{7} = 56.4$ kN/m^2

$$q_{\text{ult}}(n) = 56.4 \times 6 = 338 \text{ kN/m}^2$$

$$FS = 338/138 = 2.5$$

(c) Settlement during preload

Stage I: $P_c = 50$ kN/m^2

$$\delta = \sum \frac{C_c}{1 + e_0} H \log \frac{p_o + \Delta p}{p_o}$$

Layer	H(m)	p_o (kN/m^2)	Δp (kN/m^2)	$\dfrac{C_c}{1 + e_0}$	δ (m)	
I	2	8	48	0.06	0.102	
II	5	36	46	0.16	0.288	0.390
III	10	96	31	0.08	0.096	
IV	5	156	15	0.08	0.016	
V	15	236	14	0.05	0.018	0.130
					Σ 0.520 m	
					(≈ 520 mm)	

Consider 90% consolidation of strata I and II, 30% in stratum III, and 10% in stratum IV and V.

Settlement at tank centre = $(0.390 \times 0.9) + (0.3 \times 0.096) + (0.1 \times 0.034)$

$$= 0.383 \text{ m} = 383 \text{ mm}$$

(d) Settlement during preload

Stage II: $P_c = 50$ kN/m^2

$$\delta = \sum \frac{C_c}{1 + e_0} H \log \frac{p_o + \Delta p}{p_o}$$

Layer	H(m)	p_o (kN/m^2)	Δp (kN/m^2)	$\dfrac{C_c}{1 + e_0}$	δ (m)	
I	2	56	48	0.6	0.031	
II	5	82	46	0.16	0.152	0.183
III	10	127	31	0.08	0.076	
IV	5	171	15	0.08	0.015	
V	15	250	14	0.05	0.018	0.109
					Σ 0.292 m	
					($\approx$ 290 mm)	

[*Note:* Effective overburden pressure p_o after stage I preload has been determined for full consolidation under the preload. In reality, for strata III, IV and V, p_o values would be less due to lower degree of consolidation.]

Consider residual settelement of stage I preload occurring in stage II and settlement for stage II preload as

> *Stage I* : 10% of stratum I and II
> 30% of stratum III
> 10% of stratum III and IV
>
> *Stage II* : 90% of stratum I and II
> 30% of stratum II
> 10% of stratum III and IV

$$\text{Settlement at centre} = (0.1 \times 0.390) + (0.3 \times 0.096) + (0.1 \times 0.034)$$
$$+ (0.9 \times 0.183) + (0.3 \times 0.76) + (0.1 \times 0.033)$$
$$= 0.262 \text{ m} \approx 260 \text{ mm}$$

Total settlement during stage I and stage II preload at centre is

$$383 + 260 = 643 \text{ mm}$$

(e) Settlement of tank centre during hydrotest: $\Delta p = 38$ kN/m^2

$$\delta = \sum \frac{C_c}{1 + e_0} H \log \frac{p_o + \Delta p}{p_o}$$

Layer	H(m)	p_o (kN/m^2)	Δp (kN/m^2)	$\dfrac{C_c}{1 + e_0}$	δ (m)	
I	2	104	37	0.10	0.026	
II	5	128	35	0.16	0.084	0.110
III	10	158	20	0.08	0.04	
IV	5	186	17	0.08	0.03	
V	15	266	11	0.05	0.025	0.095
					Σ 0.205 m	
					($\approx$ 205 mm)	

[Note given in (d) above is valid for this case also.]

Total settlement during hydrotest at centre of tank

30% of stage I preload (strata II), $0.3 \times 0.096 = 0.029$ m

10% of stage I preload (strata III and IV), $0.1 \times 0.034 = 0.0034$ m

10% of stage II preload (strata I and II), $0.1 \times 0.183 = 0.018$ m

20% of stage II preload (strata III, IV, V), $0.2 \times 0.110 = 0.022$ m

90% of hydrotest load (strata I and II), $0.9 \times 0.110 = 0.099$ m

30% of hydrotest load (strata III), $0.3 \times 0.04 = 0.012$ m

20% of hydrotest load (strata IV and V), $0.2 \times 0.055 = \underline{0.011 \text{ m}}$

$$= 0.195 \text{ m}$$
$$(\approx 200 \text{ mm})$$

(f) **Design of sand wicks**

Diameter of sand wicks = 65 mm, length = 8 m

Take spacing at 1.2 cm c/c square grid

$$n = \frac{d_e}{d_w} = \frac{1.13 \times 1.2}{0.065} = 20.9$$

For 90% consolidation, $U = 0.9$

Therefore, time factor, $T_{90} = 0.7$

Radial coefficient of consolidation, $C_r = 0.075$ m²/day

Therefore, time for 90% consolidation (for $n = 20.9$ and $T_{90} = 0.7$) is

$$t_{90} = \frac{T_{90} \times d_e^2}{C_r}$$

$$= \frac{0.7 \times 1.26^2}{0.075} = 15 \text{ days}$$

Example 12.2

A steel storage tank, 24 m diameter × 18 m high, is to be founded on the subsoil shown in Fig. 12.40. The formation level of the ground is to be raised by a 2 m sand fill prior to construction. The soil is proposed to be treated by 0.5 m diameter stone column. Design suitable foundation for the storage tank.

Solution

(a) The subsoil will be preconsolidated under the load of 2 m sand fill with preinstalled stone columns accelerating the consolidation by sand drain effect. This will increase the shear strength of the subsoil and also help in reducing consolidation under tank load.

(b) The tank load (that is hydrotest load) will be dispersed through the sand pad/sand fill and the subsoil will be consolidated under the dispersed load during hydrotest. The reinforcing action of the stone columns will help to increase the overall bearing capacity of the soil and reduce settlement.

(c) Ground settlement under 2 m sand fill for $\Delta p = 40$ kN/m² is

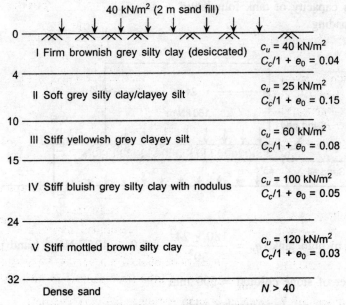

Fig. 12.40 Subsoil condition (Example 12.2).

$$\delta = \sum \frac{C_c}{1 + e_0} H \log \frac{p_o + \Delta p}{p_o}$$

Layer	H(m)	p_o (kN/m^2)	Δp (kN/m^2)	$\frac{C_c}{1 + e_0}$	δ (m)	
I	4	16	40	0.04	0.087	
II	6	56	40	0.15	0.210	0.297
III	5	100	40	0.08	0.058	
IV	9	156	40	0.05	0.045	
V	8	224	40	0.03	0.017	0.120

$$\Sigma\ 0.417\ m$$
$$(\approx 420\ mm)$$

(d) Shear strength after consolidation under sand fill

Layer	Original c_u (kN/m^2)	Δc_u (kN/m^2)	Increased c_u (kN/m^2)
I	40	$0.3 \times 40 = 12$	52
II	25	12	37
III	60	12	72
IV	100	12	112
V	120	12	132

$$[c_u]_{AVG} = \frac{(4 \times 52) + (6 \times 32) + (5 \times 72)}{15} = 52\ kN/m^2$$

(15 m depth)

(e) Bearing capacity of tank foundation
Tank loading

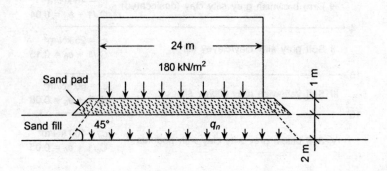

For 18 m high tank, $q_n = \dfrac{180 \times 24^2}{30^2} + 2.0 = 135$ kN/m^2 (sand pad)

(f) Diameter of stone column = 500 mm

Spacing: 1.0 m c/c triangular grid

Area = 0.2 m^2

Influence area = 0.866 $(1)^2$ = 0.866 m^2

Area ratio = $\dfrac{0.2}{0.866} \times 100$ = 23%

Capacity of stone column

$$q_{ult} = 25c_u = 25 \times 40 = 1000 \text{ kN/m}^2$$

Using Eq. (12.8)

$$\sigma_{vf} = \frac{1 + \sin\phi'}{1 - \sin\phi'}(\sigma_{ro} - u + Kc_u)$$

$$= 4.26[(18 \times 2 - 10) \times 1.5 - 10 + 4 \times 40] = 805 \text{ kN/m}^2$$

(GWT 1 m BGL, $K = 4$; $\phi' = 38°$, and depth of bulging 2 m]

Q_{ult} per stone column = $0.2 \times 805 = 161$ kN

Q_{all} per stone column = $\dfrac{Q_{ult}}{FS} = \dfrac{161}{2} = 80.5$ kN

Therefore, take it to be 100 kN per stone column.

(g) Bearing capacity of treated ground,

$$q_{all} = [100 + (0.866 - 0.2)(6 \times 50)/2.5]/0.866$$

$$= 208 \text{ kN/m}^2 > 135 \text{ kN/m}^2$$

(h) Settlement of untreated soil

$$q_n = 135 \text{ kN/m}^2$$

Effective loaded area = 30 m diameter

$$\delta = \sum \frac{C_c}{1 + e_0} H \log \frac{p_o + \Delta p}{p_o}$$

(where p_o = effective overburden pressure after consolidation under sand fill.)

Layer	H(m)	p_o (kN/m²)	Δp (kN/m²) [Boussinessq]	$\dfrac{C_c}{1 + e_0}$	δ (m)	
I	4	56	135	0.04	0.085	
II	6	106	128	0.15	0.309	0.394
III	5	140	115	0.08	0.104	
IV	9	196	95	0.05	0.077	
V	8	264	74	0.03	0.026	0.207

$$\Sigma\ 0.601 \text{ m}$$
$$(\approx 600 \text{ mm})$$

(i) Settlement of treated soil
Settlement reduction ratio

$$\beta = \frac{1}{1 + (n - 1)A_R}$$

where
n = stress concerntration ratio = 5
A_R = area ratio = 23%

(j) Settlement during hydrotest

Consider 100% for 10 m depth (by radial consolidation through stone columns) and 20% below 10 m depth

$$\delta_{\text{centre}} = (0.52 \times 394) + (0.2 \times 207)$$
$$= 208 + 40 = 248 \text{ mm} \approx 250 \text{ mm}$$

(k) Long-term settlement

Additional settlement will occur due to remaining 80% consolidation of soil below 10 m depth but under reduced operation load of oil having specific gravity 0.8).

$$\delta_{\text{centre}} = 248 + (0.8 \times 200 \times 0.8)$$
$$= 376 \approx 375 \text{ mm.}$$

REFERENCES

Barron, R.A. (1948), Consolidation of Fine-grained Soils by Drain Wells, *Trans. ASCE*, Vol. 113, pp. 718–734.

Brown, R.E. (1978), Vibroflotation Compaction of Cohesionless Soils, *Journal of Geotechnical Division, ASCE*, Vol. 103, No. GT 12, pp. 1437–51.

Dastidar, A.G., S. Gupta, and T.K. Ghosh (1969), *Application of Sandwick in a Housing Project,* Proceedings 7th International Conference on SMFE, Mexico, 2: pp. 59–64.

Dastidar, A.G. (1985), Treatment of Weak Soils—An Indian Perspective, *Geotechnical Engineering,* Vol. 1, pp. 179–229.

Datye, K.R. (1982), Simpler Techniques for Ground Improvement, 4th IGS Annual Lecture, *Indian Geotechnical Journal* Jan., 12(1): pp. 1–82.

Datye and Nagaraju (1977), Design Approach and Field Control of Stone Columns, Proceedings 10th ICSMFE, Stockholm, Vol. 3.

Dhar, P.R. (1976), *Soil Densification by Compaction Piles at a Refinery Site,* Symposium on Foundations and Excavations in Weak Soil, Calcutta.

Engelhardt, K., W.A. Flynn, and A.A. Bayak (1974), Vibro-replacement Method to Strengthen Cohesive Soils in-situ, *ASCE National Structural Engineering Meeting*, Cincinnati, p. 30.

Greenwood, D.A. (1970), *Mechanical Improvement of Soil below Ground Surface,* Proceedings Ground Engineering Conference, I.C.E., London, pp. 11–22.

Hughes, J.M.O. and N.J. Withers (1974), Reinforcing of Soft Cohesive Soil with Stone Column, *Ground Engineering*, Vol. 7, No. 3, pp. 42–49.

Hughes, J.M.O., N.J. Withers, and D.A. Greenwood (1975), A Field Trial of the Reinforced Effect of a Stone Column in Soil, *Geotechnique*, Vol. 25, pp. 34–44.

Kjellman, W. (1948), *Accelerating Consolidation of Fine-grained Soils by means of Cardboard Wicks,* Proceedings 2nd International Conference on SMFE, Rotterdam, 2: pp. 302–305.

Mitchell, J.K. and R.K. Katti (1981), *Soil improvement: State-of-the-art Report,* Proceedings 10th International Conference on SMFE, Stockholm, p. 163.

Nayak, N.V. (1982), *Stone Columns and Monitoring Instruments,* Proceedings Symposium on soil and rock improvement: geotextiles, reinforced earth and modern piling techniques, Asian Institute of Technology, Bangkok, December 1982.

Nayak, N.V. (1983), *Structures on Ground Improved by Stone Columns,* Proceedings International Symposium on Soil Structure Interaction, Roorkee, India.

Pilot, G. (1977), *Methods of Improving the Engineering Properties of Soft Clay,* State-of-the-art Report, Symposium on soft clay, Bangkok.

Som, N.N. (1995), *Consolidation Settlement of Large Diameter Storage Tank Foundations on Stone Columns in Soft Clay,* Proceedings International Symposium on compression and consolidation of clayey soils, Hiroshima, p. 653, Balkema Publishers.

Som, N.N. (1999), *Stone Column Foundation for Large Area Loading—Some Case Studies,* Symposium on Thick Deltaic Deposits, Asian Regional Conference on Soil Mechanics and Geotechnical Engineering, Seoul, Korea.

Thorburn, S. (1975), *Building Supported by Stabilised Ground,* Symposium on ground treatment by deep compaction, Institute of Civil Engineers, London.

Tsytovich, N., V. Berezantzev, B. Dalmatov and M. Abelev (1974), *Foundation Soils and Substructures,* MIR Publishers, Moscow.

Earthquake Response of Soils and Foundations

13.1 INTRODUCTION

Earthquakes can cause extensive damage to foundations and structures built on them. Earthquake motions are initiated in the soil and they are instantaneously transmitted to the foundation causing adverse effect on the behaviour of the superstructure. Damage may occur due to instability of the soil which results in extensive ground movement including differential movement. While structural damage is ultimately manifested in tilt or damage or even collapse of the superstructure, the initiating cause can often be identified as the adverse response of the soil–foundation system under seismic forces.

Geotechnical considerations are, therefore, important in the development of an earthquake resistance design. It is not only the type of soil deposit that determines the kind of response to be expected during a strong earthquake. The type of structure also influences the seismic response. The field of geotechnical earthquake engineering is quite complex. Much of its applications are based on empirical studies made on the basis of case histories which illustrate the effect of earthquake on engineering structures.

13.2 EARTHQUAKE CHARACTERISTICS

13.2.1 Magnitude

The most widely used magnitude scale to define the severity of an earthquake was developed by Richter (1958). Accordingly, the magnitude of an earthquake is given by the logarithm of the amplitude on a Wood–Anderson torsion seismogram located at a distance of 100 km from the earthquake source. Thus,

$$M = \log(A/T) + f(\Delta, h) + C_s + C_r \tag{13.1}$$

where

A = amplitude in (0.001) mm,
T = period of seismic wave in seconds,
$f(\Delta, h)$ = correction factor for epicentral distance (Δ) and focal depth (h),
C_s = correction factor for seismological station, and
C_r = regional correction factor.

Natural seismic events may have magnitude as high as 8.5 or 9. Magnitudes below 2.5 are not generally felt by humans. The frequency of occurrence of earthquake in a global scale (based on observations since 1990) is shown in Table 13.1.

Table 13.1 Frequency of occurrence of earthquakes (since 1990)

Description	Magnitude	Average (Annual)
Great	8 and higher	1
Major	7–7.9	18
Strong	6–6.9	120
Moderate	5–5.9	820
Light	4–4.9	6200 (estimated)
Minor	3–3.9	49,000 (estimated)
Very Minor	<3.0	Magnitude 2–3 about 1000 per day
		Magnitude 1–2 about 8000 per day

13.2.2 Energy Release

The energy released by an earthquake has been related to the magnitude M by the equation

$$E = 10^{4.8 + 1.5M} \text{ Joules} \tag{13.2}$$

This energy is comparable to that of nuclear explosions. For example, a nuclear explosion of one mega ton releases energy of 5×10^{15} J. An earthquake of magnitude 7.3 would also release the energy equivalent of one mega ton nuclear explosion.

13.2.3 Intensity

The magnitude of an earthquake as obtained by the Richter's scale gives measure of the amount of energy released by the earthquake, not its damage potential. The intensity of an earthquake is a measure of the effect of an earthquake at a given location. Several intensity scales have been proposed, the most widely used being the Modified Mercalli Intensity (MMI) scale, as given in Table 13.2 (Newmann, 1954). The value assigned to the MMI scale gives qualitative description of the damages based on physical verification at site. The equivalent value of the magnitude by Richter's scale is given alongside the MMI in Table 13.2.

Table 13.2 Modified Mercalli intensity (MMI) scale (abbreviated version)

Intensity	Evaluation	Description	Magnitude (Richter scale)
I	Insignificant	Only detected by instruments	1–1.9
II	Very light	Only felt by sensitive persons; oscillation of hanging objects	2–2.9
III	Light	Small vibratory motion	3–3.9
IV	Moderate	Felt inside buildings; noise produced by moving objects	4–4.9
V	Slightly strong	Felt by most persons; some panic; minor damages	
VI	Strong	Damages to nonseismic resistant structures	
VII	Very strong	People running; some damages in seismic resistant structures and serious damages to nonreinforced masonry structures	5–5.9

(Cont.)

Table 13.2 Modified Mercalli intensity (MMI) scale (abbreviated version), *Cont.*

Intensity	Evaluation	Description	Magnitude (Richter scale)
VIII	Destructive	Serious damage to structures in general	
IX	Ruinous	Serious damage to structures; almost total destruction of nonseismic resistant structures	6–6.9
X	Disastrous	Only seismic resistant structures remain standing	
XI	Disastrous in extreme	General panic; almost total destruction; the ground cracks and opens	7–7.9
XII	Catastrophic	Total destruction	8–8.6

13.2.4 Ground Acceleration

The intensity of ground motion during an earthquake is represented by the horizontal ground acceleration produced. The predominant effect of an earthquake is the horizontal forces that are produced in a structure. The horizontal ground acceleration α gives a measure of this force, which can be expressed by αW, (where W = weight of the structure) and acts at the centroid of the structure. This horizontal force is depicted in Fig. 13.1. I.S. 1893 (1983) gives the earthquake zones of India based on the horizontal ground acceleration and the vulnerability of an area to earthquakes.

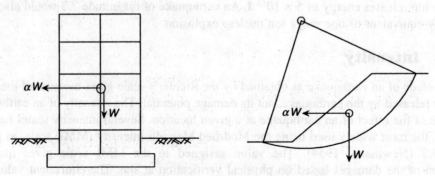

Fig. 13.1 Horizontal force due to earthquake.

13.2.5 Response Spectrum

Response spectra are typically used to portray the characteristics of the earthquake shaking at a site. Response spectrum shows the maximum response induced by the ground motions in damped single degree-of-freedom structures of different fundamental periods. Each structure has a unique fundamental period at which the structure tends to vibrate when it is allowed to vibrate freely without any external excitation. The response spectrum indicates how a particular structure with its inherent fundamental period would respond to an earthquake ground motion. For example, measurement of ground motion in the 1985 Mexico City earthquake, response spectra shown in Fig. 13.2, shows that a low-period structure (say, $T = 0.1$ s) experienced a maximum acceleration of $0.14g$, whereas a higher-period structure (say, $T = 2.0$ s) experienced a maximum acceleration of $0.74g$ for the same ground motions.

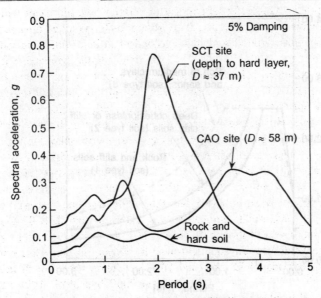

Fig. 13.2 Acceleration response spectrum as recorded in Mexico City earthquake, 1985 (Seed et al., 1987).

The response spectra shown in Fig. 13.2 illustrate the pronounced influence of local soil conditions on the characteristics of the observed earthquake ground motions. Since Mexico City was located approximately 400 km away from the epicentre, the observed response at rock and hard soil sites was fairly low (that is, the spectral accelerations were less than $0.1g$ for all periods). Damage was correspondingly negligible at these sites. At the Central Market site (CAO), spectral accelerations were significantly amplified to 0.3–$0.35g$ at periods of around 1.3 s and within the range 3.5–4.5 s. Since buildings at the CAO site did not generally have fundamental periods within these ranges, damage was minor. The motion recorded at the SCT building site, however, indicated significant amplification of the underlying bedrock motions with a maximum horizontal ground acceleration (the spectral acceleration at a period of zero) over four times that of the rock and hard soil sites and with a spectral acceleration at $T = 2.0$ s over seven times that of the rock and hard soil sites. Major damage including collapse, occurred to structures with fundamental periods ranging from about 1 s to 2 s near the SCT building site and in areas with similar subsurface conditions. At these locations, the fundamental period of the soil almost matched with that of the overlying structures, creating a near resonance condition that amplified the shaking and caused heavy damage.

The Uniform Building Code of USA (UBC, 1991) recommends the response spectrum, shown in Fig. 13.3, to determine the peak ground acceleration for a given fundamental period T for different soil conditions. For $T > 0.5$ s, ground acceleration for deep soil strata is considerably higher than that for rock and hard soils. It is to be noted that the period of ground motion at a particular site is important in determining the effect of earthquake motion on the structure. If the fundamental period of a building is close to that of the site, a resonant condition is created. This amplifies the shaking and increases the potential to damage.

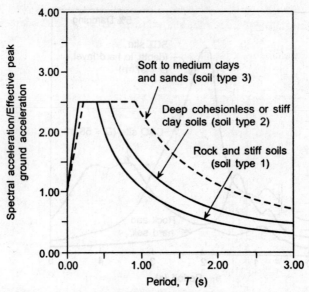

Fig. 13.3 UBC normalized acceleration response spectra (UBC, 1991).

13.3 EFFECTS OF EARTHQUAKE

Earthquakes affect soils in many ways. The major effects of earthquake on soils are:

1. Loose granular soils are compacted by ground vibration which cause large subsidence of the ground surface,
2. Compaction of loose granular soil may result in development of excess pore water pressure to cause liquefaction of the soil and lead to settlement and tilting of structures,
3. Combination of dynamic stress and induced pore water pressure may result in reduction of soil strength and cause bearing capacity failure and landslides in the earthquake area, and
4. Ground vibrations and shaking may cause structural damage even though the soils underlying the structure may remain stable during the earthquake.

13.4 GROUND SETTLEMENT

Vibration has long been recognized as an effective means of compacting granular soils. However, such compaction is associated with volume change of the soil and associated settlement of the ground surface. A measure of the ground subsidence caused by earthquake was obtained in the Alaska earthquake in 1964 (Grantz et al., 1964). A steel casing installed in firm rock in Homer, Alaska projected 30 cm above the ground surface. After the earthquake, the casing was seen to project 1.1 m above the ground indicating a subsidence of 0.8 m of the ground surface, as depicted in Fig. 13.4. Geological studies revealed that the rock surface had subsided by 0.6 m due to tectonic movements indicating a ground subsidence of 1.4 m. A combination of 1.3 m settlement of rock due to tectonic movement and 1.3 m due to compaction of overlying soil led to a ground settlement of more than

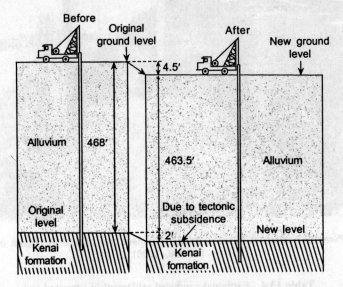

Fig. 13.4 Ground settlement around well casing at Homer in Alaska earthquake 1964 (Grantz et al., 1964).

2.6 m at Portage in the same earthquake in Alaska. This caused widespread flooding in the area during high tide periods and the town had to be relocated at a new location. Similar subsidence was noticed in other earthquakes also and the data is given in Table 13.3.

Table 13.3 Ground subsidence due to earthquake

Earthquake	*Year*	*Rock subsidence (Tectonic)* (m)	*Ground subsidence* (m)
Homer, Alaska	1964	0.6	1.4
Portage, Alaska	1964	1.3	2.6
Validina, Chile	1960	1.8	2.8
Niigata, Japan	1964	—	3.0

Ground settlement due to compaction of granular soil often leads to differential settlement of structures. A differential movement of more than a metre was noticed between a railroad bridge abutment, founded on deep piles, and the backfill placed directly on the ground surface during the Niigata earthquake of 1964. This is shown in Fig. 13.5. The bridge abutment, being founded on piles did not undergo much settlement but the granular backfill experienced major subsidence due to compaction by seismic vibrations. Field measurements have shown that vibrations induced by earthquakes are often responsible for causing significant structural damage resulting from differential settlement in a building frame. Field observations of earthquake induced settlement in saturated sandy soil are summarized in Table 13.4.

Fig. 13.5 Differential settlement between bridge abutment and backfill at Niigata earthquake 1964 (Ohsaki, 1966).

Table 13.4 Earthquake induced settlement in saturated soil

Location and year	Magnitude of earthquake	Site	Thickness of sand layer (m)	Observed settlement (cm)
Mexico, 1957	7.5	Mexico City	—	40
Tokachioki, 1968	7.9	Hachinohe	5.0	35
Niigata, 1964	7.5	Niigata C	9.0	20
Miagiken-oki, 1968	7.4	Arahama	9.0	20

Much of this settlement occurs due to horizontal motions induced by earthquakes. The stresses obtained by ground response analysis and the settlement obtained by cyclic shear stress give a basis for evaluation of the settlement potential of the ground during earthquakes (Seed and Silver, 1972).

13.5 LIQUEFACTION

A major damage to structures during earthquakes is caused by liquefaction in saturated fine sand and silt. This is seen as *sand boils* or *mud spouts* with associated ground cracks and development of quick sand-like condition over wide areas (Seed, 1968; Ambreseys and Sarma, 1969). When liquefaction occurs buildings may sink into the ground. Lightweight structures may even 'float' up to the ground surface. A most vivid illustration of liquefaction was found in Niigata earthquake, Japan in 1964, shown in Fig. 13.6. The epicentre of the earthquake was located about 60 km from Niigata. Extensive liquefaction of the soil caused water to flow out of cracks and boils. Structures settled more than 1 m accompanied by severe tilting (refer Fig. 13.6). The recent Bhuj earthquake in India also presented vivid illustrations of liquefaction (Rao, 2001). Figure 13.7 depicts liquefaction in Bhuj earthquake, India, 2001.

Fig. 13.6 Liquefaction in Niigata earthquake, 1964 (Ohsaki, 1966).

Liquefaction is caused in sand by ground vibration which tends to compact the sand and decrease its volume. If drainage does not occur, the tendency to decrease in volume results in increased pore water pressure. If the pore pressure builds up to an extent which is equal to the overburden pressure, the effective stress becomes zero and the soil loses its strength completely and gets into a liquefied state. Liquefaction may be initiated at the surface or at some depth below the ground surface. Once liquefaction occurs at some depth, the excess pore-pressure tends to dissipate by upward flow of water which, in turn, induces liquefaction in the upper layers of the soil.

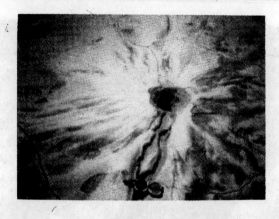

Fig. 13.7 Liquefaction in Bhuj earthquake, India, 2001 (Rao, 2001).

13.5.1 Liquefaction Potential

The liquefaction potential of a soil depends on the relative density of the soil, percentage of fines present in the soil, effective confining pressure, depth of water table, and the ground acceleration produced by the earthquake.

For practical use, the liquefaction potential has been studied from a comparison of the shear stress developed in the soil at a given depth by an earthquake and the shear stress required to cause liquefaction, the latter being related to the relative density of the soil as measured by the SPT blow count N (blows/30 cm), as presented in Fig. 13.8 (Byrne, 1976). Kishida (1969) proposed a relationship between N value and depth, and indicated a boundary line between liquefaction and nonliquefaction, which is shown in Fig. 13.9. The data clearly show that liquefaction occurs mostly in soil with low relative density ($D_r < 70\%$). Based on such field observations, a relationship between the cyclic stress ratio causing liquefaction and N value of sands containing different percentages of fines has been proposed, which is depicted in Fig. 13.10 (Seed et al., 1984).

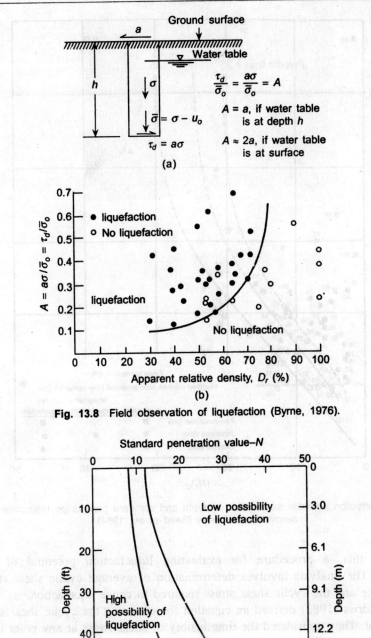

Fig. 13.8 Field observation of liquefaction (Byrne, 1976).

Fig. 13.9 Relationship between liquefaction potential and *N* value as a function of depth (Kishida, 1969).

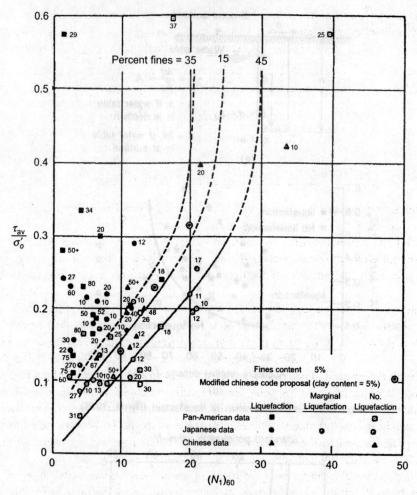

Fig. 13.10 Liquefaction potential as function of depth and standard penetration resistance for different percentages of fines (Seed et al., 1984).

Based on this, a procedure for evaluating liquefaction potential of a soil has been proposed. The analysis involves determination of average cyclic shear stress caused by the earthquake and the cyclic shear stress required to cause liquefaction.

Seed and Idriss (1982) derived an equation for obtaining the cyclic shear stress caused by an earthquake. They considered the time history of shear stress at any point in a soil due to an earthquake. A typical time history curve has an irregular distribution as shown in Fig. 13.11. A weighted average of the individual stress cycles gives the average shear stress τ_{av} which is about 65 percent of the maximum shear stress. Hence,

$$\left(\frac{\tau}{\sigma_o'}\right)_d = 0.65\left(\frac{\alpha}{g}\right)\left(\frac{\sigma_o}{\sigma_o'}\right)r_d \tag{13.3}$$

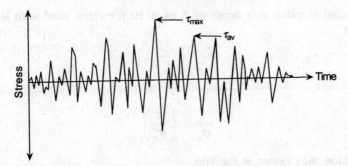

Fig. 13.11 Time history of shear stress during earthquake (Seed and Idriss, 1982).

where,

$$\left(\frac{\tau}{\sigma_o'}\right)_d = \text{average cyclic shear stress developed during earthquake,}$$

α = maximum ground acceleration,

g = acceleration due to gravity,

σ_o = total overburden pressure in the sand layer under consideration,

σ_o' = initial effective overburden pressure at depth under consideration, and

r_d = stress reduction factor (1.0 at ground surface to 0.8 at 5 m depth) (refer Fig. 13.12).

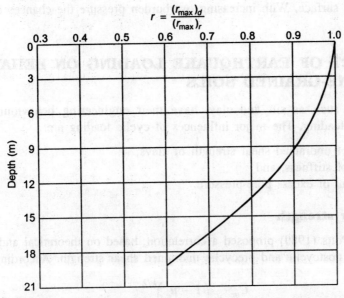

Fig. 13.12 Stress reduction factor for liquefaction analysis (Prakash, 1980).

For example, let us consider a site where the ground water table is at the ground surface and $\alpha = 0.2\ g$. At a depth of 5 m, the average cyclic stress ratio during earthquake works out as 0.29 [refer Eq. (13.3)].

The corrected N value at a depth of 5 m is 10 for clean sand with less than 5% fines. From Fig. 13.9,

$$\left(\frac{\tau}{\sigma_o}\right)_1 = 0.11$$

This gives,

$$\frac{\tau}{\sigma_o'} > \left(\frac{\tau}{\sigma_o'}\right)_1$$

Hence, liquefaction may occur at the site.

Table 13.5 gives the data on grain-size of the soil and depth of liquefaction in some well known earthquakes.

Table 13.5 Liquefaction data

Location and year	Magnitude of earthquake	Grain-size D_{10} (mm)	Depth of liquefaction (m)
Niigata, 1964	7.5	0.07–0.25	5
Mino-Owan, Japan, 1969	7.4	0.05–0.25	9
Jaltipan, Mexico, 1959	6.9	0.01–0.10	7
Alaska, 1964		0.01–0.1	8

It appears that liquefaction generally occurs in fine to medium sand within a depth of 10 m from ground surface. With increasing overburden pressure the chances of liquefaction usually decrease.

13.6 EFFECT OF EARTHQUAKE LOADING ON BEHAVIOUR OF FINE-GRAINED SOILS

Fine-grained soils such as silt and clay, have their engineering behaviour significantly affected by cyclic loading. The major influences of cyclic loading are:

- reduction of undrained shear strength of clays,
- reduction of stiffness, and
- development of excess pore-pressure.

Undrained shear strength

Van Eekelen and Potts (1989) proposed a correlation, based on theoretical and experimental work, between the postcyclic and precyclic undrained shear strength. Accordingly,

$$\frac{C_{uc}}{c_u} = \left(\frac{1 - u_e}{\sigma_c'}\right)^{k/\lambda} \qquad (13.4)$$

where,

C_{uc} = postcyclic undrained shear strength,

c_u = precyclic undrained shear strength,

u_e = excess pore-pressure due to cyclic loading,

σ'_c = initial effective confining pressure,

k = rebound or recompression index expressed on the natural logarithmic scale ($= C_r/2.3$), and

λ = compression index expressed on the natural logarithm scale ($= C/2.3$).

The effect of cyclic stress on the strength reduction of a clay with initial zero shear stress has been related to plasticity index by Ishihara and Yasuda (1980). Almost 50% reduction in strength is indicated for low plasticity soil ($PI < 15$) for 30 cycles of loading and this variation is expressed through Fig. 13.13.

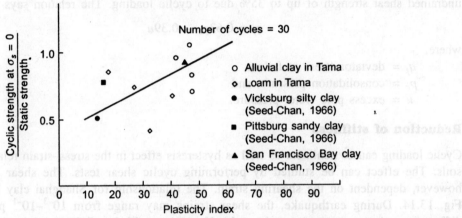

Fig. 13.13 Variation of cyclic strength ratio with plasticity index (Ishihara and Yasuda, 1980).

In general, the effect of cyclic stress on undrained shear strength has been found to be small when the cyclic strain is kept below half the precyclic failure strain (Thiers and Seed, 1969).

Pore-pressure

Excess pore-pressure developed under cyclic loading may cause marked reduction in undrained shear strength and the stiffness of clay. A number of empirical correlations have been proposed to determine the excess pore-pressure developed in cyclic loading. One such relationship proposed by Matsui et al. (1980) relate the residual pore-pressure to the over consolidation ratio,

$$\frac{u_r}{\sigma_c} = \beta \left[\log_{10} \left(\frac{\gamma_{c\max}}{A_1(OCR - 1)} + B_1 \right) \right] \tag{13.5}$$

where,

u_r = residual pore-pressure,

σ_c = effective confining pressure,

$\gamma_{c\max}$ = single amplitude maximum cyclic shear strain,

OCR = overconsolidation ratio, and

β = 0.45 (found experimentally).

Also, A_1, $B_1 = f(PI)$ such that

PI	A_1	B_1
20	0.4×10^{-3}	0.6×10^{-3}
40	1.1×10^3	1.2×10^{-3}
55	2.5×10^{-3}	1.2×10^{-3}

Togrol and Guler (1984) proposed an empirical relation for normally consolidated clays to relate the deviator stress at failure to the excess pore-pressure and found reduction of undrained shear strength of up to 35% due to cyclic loading. The relation says

$$q_f = 0.63p_c - 0.39u \qquad (13.6)$$

where,

q_f = deviator stress at failure,
p_c = consolidation pressure, and
u = excess pore water pressure.

Reduction of stiffness

Cyclic loading causes nonlinear as well as hysteresis effect in the stress–strain relationships of soils. The effect can be studied by performing cyclic shear tests. The shear modulus is, however, dependent on the shearing strain. The relationship for Shanghai clay is shown in Fig. 13.14. During earthquake, the shear strains may range from 10^{-3}–10^{-1} percent with different maximum strain in each cycle. Hence, two-thirds of the modulus measured at maximum strain is normally taken for design.

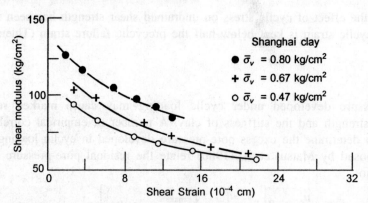

Fig. 13.14 Variation of shear modulus with strain (Seed and Idriss, 1969).

13.7 BUILDING DAMAGE DUE TO LIQUEFACTION

Seed et al. (1989) reported extensive study of the building damages that occurred in the Niigata earthquake, Japan, 1964. The subsoil consisted of sand having D_{10} of 0.07–0.25 mm and uniformity coefficient of 10. The buildings suffered on spread footings and piles. While 64% of buildings on spread footings suffered medium to heavy damage, 55% of the buildings on piles also reported similar damage. Apparently, the piles had little effect in reducing damage by the earthquake. Figure 13.15 shows a correlation of damage to the standard penetration resistance of the soil at the base of the foundation or the pile tip. When the sand underlying the footings had $N < 15$, the buildings suffered heavy damage but for N values between 20–25 the damage was less. Similar data were obtained for buildings supported on piles. Heavy damage occurred when the N value at the pile tip was less than 15.

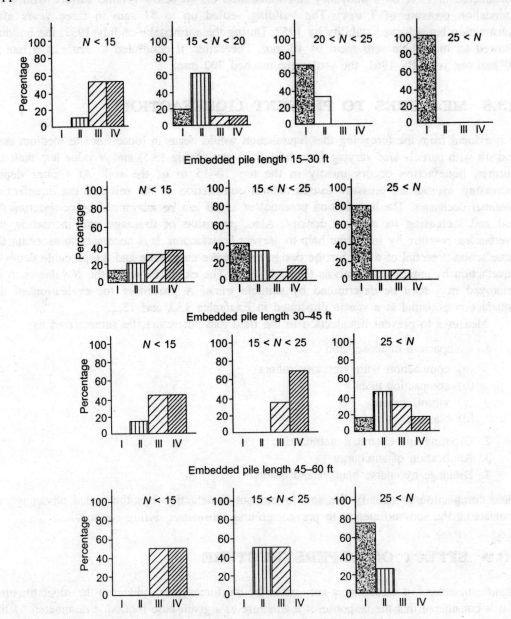

Fig. 13.15 Building damages due to liquefaction (Kishida, 1965).

For $N > 25$, the damage to the buildings was less. Similar data have been reported for building damages in Alaska earthquake (Ross et al., 1969) and Jaltipan earthquake (Marsal, 1961). The field data clearly indicate that liquefaction has generally occurred in cohesionless soil with 10% size varying 0.01 to 0.25 mm with uniformly coefficient of 2–10. Also, such sand having $N < 15$ appears most vulnerable to liquefaction. Zeevaert (1989) describes a case history of foundation settlement of a raft foundation in Mexico City. The building was constructed in 1952 on a buoyancy raft foundation 6.5 m below ground surface with a net foundation pressure of 7 t/m^2. The building settled up to 51 mm in three years after construction but achieved stability by 1957. During the earthquake of July 1957, the building showed an immediate settlement of 40 mm. Thereafter, it continued to settle at a rate of 30 mm per year. By 1961, the settlement reached 700 mm.

13.8 MEASURES TO PREVENT LIQUEFACTION

It is evident from the foregoing that liquefaction would occur in loose fines to medium sand and silt with particle size varying from 0.01–0.25 mm (Table 13.5) and N value less than 15. Further, liquefaction occurs mostly in the top 10–15 m of the soil. At greater depth, increasing overburden pressure causes natural compaction of the soil and the liquefaction potential decreases. The liquefaction potential of a soil can be minimized by compacting the soil and increasing its relative density. Also, provision of drainage and increasing the overburden pressure by surcharge help to prevent liquefaction. It is necessary to ascertain the liquefaction potential of a site for the design earthquake magnitude and the probable depth of liquefaction by appropriate analysis (Section 13.4). The extent to which the N value is to be improved may then be determined by suitable trials. A procedure for evaluation of the liquefaction potential at a site is illustrated in Examples 13.1 and 13.2.

Measures to prevent liquefaction in the field may, therefore, be summarized as

1. Compaction of loose sand
   (a) compaction with vibratory rollers
   (b) compaction piles
   (c) vibroflotation
   (d) blasting
2. Grouting and chemical stabilization
3. Application of surcharge
4. Drainage by coarse blanket and drains

Field compaction to densify the soil to prevent liquefaction has the added advantage of compacting the soil sufficiently to prevent ground subsidence during earthquake.

13.9 EFFECT ON SUPERSTRUCTURE

Significant effects of earthquakes are caused by the forces they induce on the superstructure. If it is considered that the response of a structure to a given base motion is dominated by the influence of the first mode then the maximum lateral forces on a structure would have the approximate distribution, as shown in Fig. 13.16, decreasing from a maximum at the top and zero at the base. The potential damaging effect of a base motion may be considered to be proportional to the product of the force developed and the period for which it sets, that is,

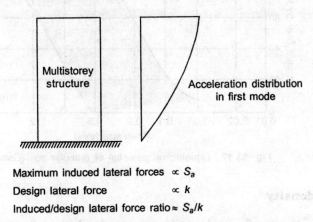

Maximum induced lateral forces $\propto S_a$
Design lateral force $\qquad \propto k$
Induced/design lateral force ratio $\approx S_a/k$

Fig. 13.16 Schematic representation of first mode forces on building.

Potential damaging factor $\propto WS_aT$
$\qquad\qquad\qquad\qquad\quad \propto WS_v$

where,

S_a = spectral acceleration and
S_v = spectral velocity.

Most building codes used for earthquake-resistant design require that buildings of a given type be designed to withstand a static lateral force having a magnitude given by a lateral force coefficient k times the weight of the building. The magnitude of the coefficient usually varies with the fundamental period of the building, or the number of storeys in the building. In general, the lateral force coefficient decreases with increasing values of the fundamental period or increasing numbers of storeys. Thus, buildings are not generally designed to be equally resistant to the same induced forces.

13.10 DYNAMIC PROPERTIES OF SOIL

In order to carry out seismic design of foundation in an earthquake region, soil parameters required for dynamic analysis should be determined. The important soil parameters relevant to earthquake design form a subject matter of this section.

Particle-size distribution

Figure 13.17 gives the liquefaction potential of granular soil based on grain-size distribution as obtained from experiences of past earthquakes (Oshaki, 1970). The grain-size distribution of the soil may be determined by standard tests.

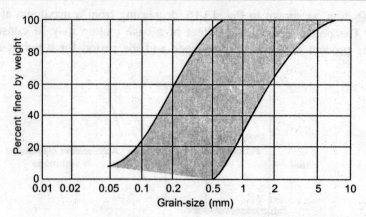

Fig. 13.17 Liquefaction potential of granular soil (Oshaki, 1970).

Relative density

The relative density of granular soil expresses the degree of compaction and is given by the expression

$$D_r = \frac{e_{max} - e}{e_{max} - e_{min}} \tag{13.7}$$

$$= \frac{\gamma_{max}(\gamma - \gamma_{min})}{\gamma(\gamma_{max} - \gamma_{min})} \tag{13.8}$$

where,

e_{max}, e_{min} = maximum and minimum void ratios,
γ_{max}, γ_{min} = maximum and minimum unit weights of soil
e = in-situ void ratio, and
γ = in-situ unit weight.

The maximum and minimum void ratio of the soil may be determined in the laboratory with the relative density apparatus and the in-situ density from field density tests.

Shear modulus

The shear modulus G of the soil is determined from the modulus of elasticity obtained from triaxial tests as

$$G = \frac{E}{1 + \nu} \tag{13.9}$$

where, ν = Poisson's ratio.

The shear modulus can also be determined from cyclic triaxial tests. The test gives the cyclic stress–strain relationship of the soil and the shear modulus is determined for the appropriate stress level. Normally two-thirds of the modulus measured at maximum strain is used for earthquake design. Typical relationship between shear modulus and shear strain is given in Fig. 13.18. During earthquake, the developed shear stress may range from 10^{-3}–10^{-1}%.

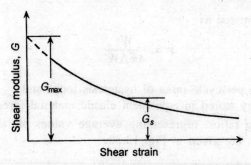

Fig. 13.18 Shear modulus versus shear strain (Seed and Idriss, 1971).

In the field, shear modulus can also be determined from shear wave velocity test by charging an explosive and measuring the wave velocity. The mathematical relationship is then given by

$$G = \rho V_s^2 \qquad (13.10)$$

where,

ρ = density of soil and

V_s = shear wave velocity.

Damping factor

Two different damping phenomena are related to soils—material damping and radiation damping. Material damping takes place when any vibration wave travels through the soil. It is related to the loss of vibration energy resulting from hysteresis in the soil. Damping is generally expressed as a fraction of critical damping and thus, referred to as damping ratio. Figure 13.19 is a graphical representation of calculation of the damping ratio.

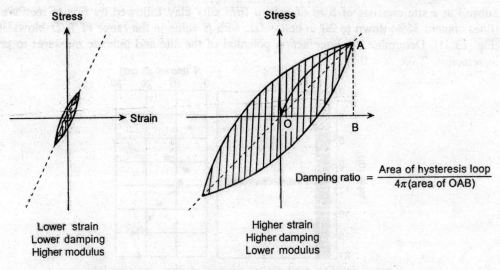

$$\text{Damping ratio} = \frac{\text{Area of hysteresis loop}}{4\pi\,(\text{area of OAB})}$$

Fig. 13.19 Calculating damping ratio (Seed and Idriss, 1970).

The damping ratio is expressed as

$$\varepsilon = \frac{W}{4\pi \Delta W}$$

where,

W = energy loss per cycle (area of hysteresis loop) and

ΔW = strain energy stored in equivalent elastic material (area OAB in Fig. 13.19).

Typical material damping ratios, representing average values of laboratory test results on sands and saturated clays, are given in Fig. 13.20.

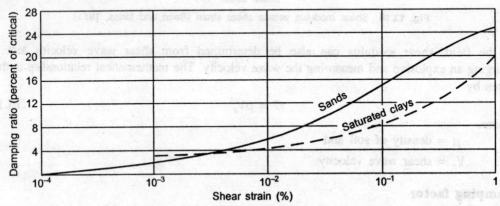

Fig. 13.20 Material damping ratio (Seed and Idriss, 1970).

Radiation damping is a measure of the loss of energy through radiation of vibration waves from the source. It is related to the geometrical properties of the foundation. This is generally used for design of machine foundations.

Example 13.1

Subsoil at a site consists of 5 m of soft to firm silty clay followed by fine to medium sand (fines content 35%) down to 20 m below GL with N value in the range of 7–22 blows/30 cm, (Fig. 13.21). Determine the liquefaction potential of the site and indicate measures to prevent liquefaction. Take $\alpha = 0.2g$.

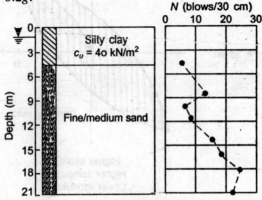

Fig. 13.21 Subsoil condition.

Liquefaction analysis may be done using the procedure of Seed et al. (1989).

Solution

The cyclic stress ratio at a given depth during an earthquake is given by Eq. (13.3). For a horizontal ground acceleration of 0.2g, the depthwise variation of cyclic stress ratio $(\tau/\sigma'_o)_d$ is calculated, as in Table 13.6. To calculate the effective stress, the ground water table has been taken at 1 m below GL. Now, from Fig. 13.10, for silt content 35%, the cyclic stress ratio required to cause liquefaction for different N values is obtained as

N (blows/30 cm) :	10	15	20
τ/σ'_o :	0.18	0.25	0.4

With factor of safety of 1.2, the allowable cyclic stress ratio would be 0.15, 0.208, and 0.33 for N values of 10, 15, and 20 respectively.

Table 13.6 Calculation of cyclic stress ratio

Depth below GL (m)	N (blows/30 cm)	σ_o (kN/m²)	σ'_o (kN/m²)	$(\sigma_o/\sigma'_o)_d$	r_d	$(\tau/\sigma'_o)_d$
6	6	111	61	1.82	0.97	0.230
8	12	148	78	1.90	0.94	0.232
10	7	185	95	1.95	0.9	0.23
12	8	222	112	1.98	0.86	0.221
14	16	259	129	2.01	0.80	0.209
16	18	296	146	2.03	0.76	0.200
18	24	333	163	2.04	0.70	0.186
20	22	370	180	2.06	0.65	0.175

Looking at Table 13.6, it is evident that liquefaction potential does not exist beyond a depth of 14 m. Here, N value is greater than 15 and (τ/σ') is less than 0.208.

Above 14 m depth, τ/σ'_o should be greater than 0.2 to prevent liquefaction. This will require an N value of 20. Hence, densification of the soil has to be done by vibrocompaction within this depth.

Example 13.2

During an earthquake, the maximum intensity of ground shaking at a site is 0.1g. The subsoil consists of a 20 m deposit of sand (fines content 15%) with SPT blow count of 10 blows/30 cm. Determine the zone of liquefaction (Take $\gamma = 18$ kN/m³ and water table at ground surface).

Solution

(a) Average shear stress in the soil

$$\frac{\tau_{av}}{\sigma'_o} = 0.65 \left(\frac{\alpha_{max}}{g} \right) \left(\frac{\sigma_o}{\sigma'_o} \right) r_d$$

The calculations give

Table 13.7 Calculation of cyclic stress ratio

Depth (m)	α_{max}/g	σ_o (kN/m²)	σ_o' (kN/m²)	r_d	τ/σ_o'
3	0.1	54	24	0.98	0.143
6	—	108	48	0.96	0.140
9	—	162	72	0.94	0.137
12	—	216	95	0.85	0.124
15	—	270	120	0.74	0.108
18	—	324	144	0.64	0.094
21	—	378	168	0.60	0.088

(b) Shear stress ratio causing liquefaction (see Fig. 13.10)

For $N = 10$ and 15% fines

$$\frac{\tau_{av}}{\sigma_o'} = 0.15$$

with $FS = 1.2$

$$\left(\frac{\tau_{av}}{\sigma_o'}\right)_{all} = \frac{0.15}{1.2} = 0.124$$

The data are plotted in Fig. 13.22. It is seen that liquefaction potential exists in the top 12 m of the deposit.

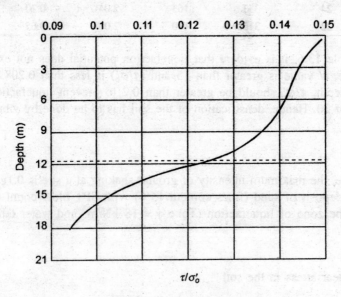

Fig. 13.22 Zone of liquefaction.

REFERENCES

Ambreseys, N. and S.K. Sarma (1969), *Liquefaction of Soils induced by Earthquakes,* Bulletin of the Seismological Society of America, Vol. 59, No. 2, pp. 651–664.

Byrne, P.M. (1976), An Evaluation of the Liquefaction Potential of the Frazer Delta, *Canadian Geotechnical Journal,* **15**, No. 1, pp. 32–46.

Grantz, A., G. Flafker, and R. Kachedoorian, (1964),—*Alaska's Good Friday Earthquake,* March 27, 1964, Geological Survey, US Dept. of Interior. Circular 491, Washington, DC

IS 1893 (1983), *Code of Practice for Design of Earthquake Resistant Structures,* Bureau of Indian Standards, New Delhi.

Ishihara, K. and S. Yasuda (1980), *Cyclic Strengths of Undisturbed Cohesive Soils of Western Tokyo,* Int. Symp. on Soils under Cyclic and Transient Loading Swansea, A.A. Balkema, pp. 57–66.

Kishida, H. (1969), Damage of Reinforced Concrete Buildings in Niigata City with special reference to Foundation Engineering, *Soil and Foundation Engineering,* **6**, No. 1, Tokyo, Japan.

Marsal, R.J. (1961), *Behaviour of a Sandy Uniform Soil during the Jaltipan Earthquake,* Mexico, Proceedings Fifth International Conference on Soil Mechanics and Foundation Engineering, Paris, France.

Matsui, T., H. Ohara, and T. Ito (1980), Cyclic Stress–Strain History and Shear Characteristics of Clay, *Journal of the Geotechnical Engineering Division,* ASCE, **106**, No. 10, pp. 1011–1020.

Newmann, N.M. (1965), Earthquake Effects on Dams and Embankments, *Geotechnique,* Vol. 15, No. 2, pp. 139–160.

Ohsaki, Yorihiko (1966), Niigata Earthquake 1964—Building Damage and Soil Condition, *Soils and Foundations,* VI, No. 2, pp. 14–37.

Oshaki, Y. (1970), Effects of Sand Compaction on Liquefaction during the Tokachioki Earthquake, *Soils and Foundations,* Vol. 10, No. 2, pp. 112–128.

Prakash, S. (1980), *Soil Dynamics and Machine Foundation,* Sarita Prakashan, Roorkee.

Rao, K.S. (2001), Magnitude Scales and Related Issues of Earthquakes, *IGS News,* Vol. 33, No. 3–4, Indian Geotechnical Society, New Delhi.

Richter, C.F. (1958), *Elementary Seismology,* W.H. Freeman, San Francisco.

Ross, G.A., H.B. Seed, and R.R. Migliaccio (1969), Bridge Foundation Behaviour in Alaska Earthquake, *Journal of the Soil Mechanics and Foundations Division,* ASCE, **95**, No. SM-4, pp. 1007–1036.

Seed, H.B. (1968), Landslides during Earthquake due to Soil Liquefaction, *Journal of the Soil Mechanics and Foundations Division,* ASCE, **94**, No. SM-5, pp. 1053–1122.

Seed, H.B., K. Tokimatsu, L.F. Harder and R.M. Chung (1984), *The Influence of SPT Procedures in Soil Liquefaction Resistance Evaluations,* Report No. UBC/EERC-84/15, Earthquake Engineering Research Centre, University of California, Berkeley, California.

Seed, R.B., S.E. Dickenson, M.F. Riemer, J.D. Bray, N. Sitar, J.K. Mitchell, I.M. Idriss, R.E. Kayen, A. Kropp, L.F. Harder and M.S. Power (1989), *Preliminary Report on the Principal Geotechnical Aspects of the October 17, 1989 Loma Prieta Earthquake,* Earthquake Engineering Research Centre, Report No. UCB/EERC-90/05, University of California.

Seed, H.B. and I.M. Idriss (1971), Simplified Procedure of Evaluating Soil Liquefaction Potential, *Journal of Soil Mechanics and Foundation Division,* ASCE, Vol. 97, No. SM 9, pp. 1249–1273

Seed, H.B. and I.M. Idriss (1982), *Ground Motions and Soil Liquefaction during Earthquakes,* Monograph Series, Earthquake Engineering Research Institute, Berkeley, Califorina.

Seed, H.B. and M.L. Silver (1972), Settlement of Dry Sands during Earthquakes, *Journal of the Soil Mechanics and Foundations Division,* ASCE, **98**, No. SM-4, pp. 381–397.

Seed, H.B. and K. Tokimatsu (1987), Evaluation of Settlements in Sands due to Earthquake Shaking, *Journal of Geotechnical Engineering,* **113**, No. 8, pp. 861–878.

Thiers, R.G. and H.B. Seed (1969), *Strength and Stress–Strain Characteristics of Clays subjected to Seismic Loading Conditions,* Vibration Effects of Earthquakes on Soils and Foundations, ASTM STP 450.

Togrol, E. and E. Guler (1984), Effect of Repeated Loading on the Strength of Clay, *Soil Dynamics and Earthquake Engineering,* **3**, No. 4, pp. 184–190.

UBC (1991), *Uniform Building Code,* USA.

Van Eekelen, H.A.M. and D.M. Potts (1989), The Behaviour of Drammen Clay under Cyclic Loading, *Geotechnique,* **28**, No. 2, pp. 173–196.

Zeevaert, L. (1989), Foundation Problems in Earthquake Region, Chapter 17 of *Foundation Engineering Handbook,* Ed. H.Y. Fang, CBS Publishers, New Delhi.

Construction Problems

14.1 INTRODUCTION

Foundation design is made on the basis of available knowledge about the structure to be built and the subsoil condition available at the given site. The structural design is done in a way to suit the facilities to be created for a given architectural layout. In general, the loads that are imposed on the soil can be evaluated fairly accurately. Foundation design requires evaluation of the safe bearing capacity and settlement, both immediate and long term. These factors require knowledge of the subsoil characteristics which are determined from an appropriate site investigation. However, soil being deposited at a site by natural geological processes over long periods of time, there are inherent variations which may not be fully reflected by even an elaborate subsoil investigation. Hence, simplifying assumptions are made about boundary conditions and average soil properties are to be assigned to the different strata for working out the detailed design. Also, the land use pattern of the area surrounding the site determines the vulnerability of existing buildings. Different degrees of precaution are to be taken to implement a given design without causing any distress to adjoining structures. The job of the foundation engineer does not, therefore, end at producing a design only. It is equally important to determine if any problems are to be anticipated during construction and work out proper construction procedure and remedial measures in time.

The construction problems vary widely and are often site specific. However, some general problems associated with foundation construction are discussed in this chapter.

14.2 COMMON CONSTRUCTION PROBLEMS

When a foundation design has been finalized, the job is given to a construction agency for doing the construction. Inevitably the work requires excavation of varying depth and magnitude. The problems multiply when excavation is to be made below the ground water table. The major construction problems, therefore, arise as a result of

1. stability of excavations,
2. dewatering, and
3. effect of adjoining structures.

14.3 STABILITY OF EXCAVATION

Excavation is done manually or by mechanical scrapers depending on the magnitude of earth work involved. Manual expansion is mostly adopted for foundation construction. Use of mechanical scrapers gives faster progress but they are not suitable for excavations in small areas since the ramp roads needed for the scrapers to move in and out of the cut need space. Mechanical scrapers are best suited to large area of shallow excavation.

Depending on the availability of space surrounding a construction site, an excavation with side slopes or braced cuts may be adopted. Sloped excavation only involves earth work for a stable slope designed from appropriate slope stability analysis. Figure 14.1 depicts excavation with side slopes. But a stable slope needs sufficient free space in the vicinity of the construction area. In particular, if there are existing buildings close to the area, excavation with side slopes does not become feasible. Therefore, in build up areas, braced cut is adopted.

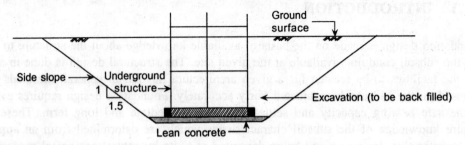

Fig. 14.1 Excavation with side slopes.

Braced cuts essentially consist of making vertical walls in the soil and suitably propping them by steel struts as the excavation is done. When the final excavation level is reached, the foundation is cast and backfilling done to restore the original ground surface as the struts are progressively removed, as presented in Fig. 14.2.

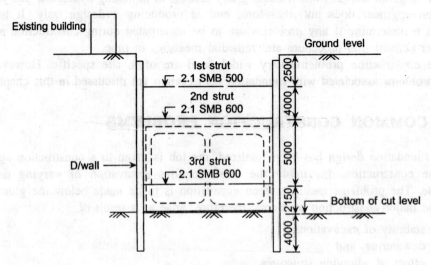

Fig. 14.2 Braced cut.

For shallow footings and raft foundations, the depth of excavation seldom exceeds 2 m and elaborate support is not required. However, if heterogeneous fill exists near the ground surface, side protection with timber planks and small struts should be sufficient.

Deep excavation for two or three basements (excavation depth: 6–10 m) would require adequate lateral support with diaphragm walls or contiguous bored piles and struts. Because of large excavation width, it may not be feasible to have horizontal struts which would tend to deflect under their own weight. In such cases, inclined props supported between the diaphragm wall and the already cast base raft at the centre may be adopted. Otherwise, a number of H-piles may be driven at close intervals and the struts be made to span among them to reduce the effective length. Figure 14.3 depicts inclined props to support diaphragm wall and Fig. 14.4 shows struts in wide cuts.

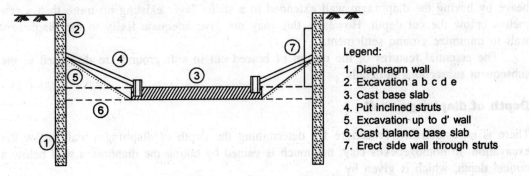

Legend:
1. Diaphragm wall
2. Excavation a b c d e
3. Cast base slab
4. Put inclined struts
5. Excavation up to d' wall
6. Cast balance base slab
7. Erect side wall through struts

Fig. 14.3 Inclined props to support diaphragm wall.

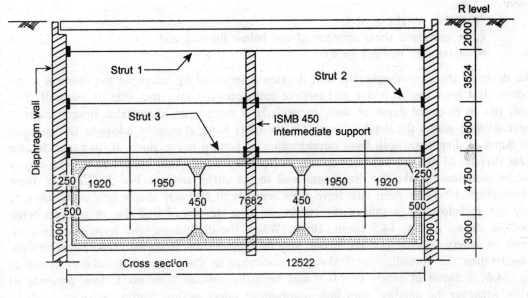

Fig. 14.4 Struts in wide cuts.

14.3.1 Design of Braced Cuts

The design of braced excavation involves two distinct yet interrelated features, namely

(a) stability of the excavation, ground movement, control of water into the excavation, effect on adjoining structures, and so on and

(b) design of structural elements, that is, diaphragm wall or sheet pile, struts or anchors and so forth.

Although the overall stability of braced cuts in soft ground does not depend to any great extent on the number and spacing of struts or anchors, they very much influence the pattern of ground movement expected in a given situation. The depth of diaphragm wall/ sheet pile determines both the stability of the system and the ground movement associated with it. Depending on the subsoil stratification, one may get adequate stability against bottom heave by having the diaphragm wall extended to a stiffer layer existing no more than a few metres below the cut depth. However, this may not give adequate fixity to the diaphragm wall to minimize ground settlement.

The essential features of the design of braced cut in soft ground are discussed in the subsequent subsections in details.

Depth of diaphragm wall

There is no established procedure for determining the depth of diaphragm wall below the excavation. In homogeneous clay, not much is gained by taking the diaphragm wall below a critical depth, which is given by

$$N_c = \frac{\gamma_H}{c_u} \tag{14.1}$$

where,

γ = unit weight of soil,

c_u = undrained shear strength of soil below the cut, and

N_c = stability number (≈ 6).

The depth of sheet pile/diaphragm wall is often determined by balancing the moment at the bottom strut level due to active and passive earth pressure on either side of the wall. This gives rise to extended depth of wall, particularly if there is no appreciable improvement of shear strength within the depth of wall. On the other hand, it may be adequate to determine the depth of diaphragm wall from consideration of bottom heave alone. If, in particular, the shear strength of the soil improves within shallow depth below the bottom of the cut, it may just be sufficient to take the diaphragm wall to the stiffer stratum. For the Calcutta metro construction, 10–14 m deep cuts have been made with 600 mm diaphragm walls taken to only 4–6 m below the bottom of the cut to rest in a stratum of stiff clay or medium/dense sand, as shown in Fig. 14.5 (Som, 1998). Where the diaphragm wall terminates in clay, factor of safety against bottom heave may be determined from Eq. (14.2), taking into consideration the contribution of shearing resistance at the soil–wall interface, shown in Fig. 14.6. A factor of safety of 2.0 would normally suffice. Needless to say, presence of struts, whatever the number, does not contribute to safety against bottom heave.

$$F = \frac{c_{u4}N_c + \gamma_3 D_2 + \gamma_4 D_f + \sum c_u (H + D_2 + D_f)/D_1}{\sum \gamma (H + D_2 + D_f)} \tag{14.2}$$

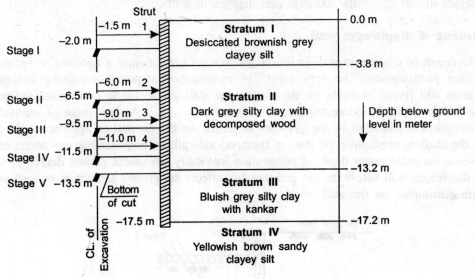

Fig. 14.5 Braced cut for Calcutta metro construction.

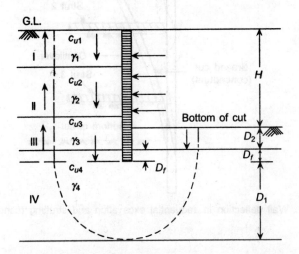

Fig. 14.6 Stability against bottom heave in stratified soil.

Number and spacing of struts

Struts are required to prevent failure of the diaphragm wall in flexure and to minimize lateral deflection of the wall. The diaphragm wall and the struts make up a rigid structural system which prevent excessive ground movement. Obviously, greater the number of struts, better is the rigidity of the system. On the other hand, too many struts create obstruction to the construction work. In general, a spacing of 3–4 m between struts may be adopted. Field measurements indicate that the settlement increases rapidly for unsupported cut depth of more than 4 m. Available case histories also suggest an optimum strut spacing of 3–4 m for deep excavations in soft clay with 500–600 mm diaphragm walls.

Movement of diaphragm wall

The movement of diaphragm wall in braced excavation gives rise to a complex soil-structure interaction phenomenon. The very nature of construction sequence involving successive excavation and fixing of struts on the diaphragm wall gives rise to conceptual deflection pattern which is highly indeterminate, as shown in Fig. 14.7. At any stage of excavation, wall movement is restricted by the presence of struts. As the excavation approaches the final depth, the shallow overburden on the cut (passive) side allows considerable movement of the wall below the cut. Greater depth of penetration inevitably introduces greater degree of fixity in the diaphragm wall below the cut and thereby, affects the ground settlement as well as the pressure distribution on the wall.

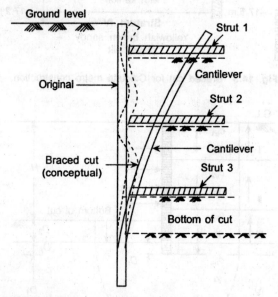

Fig. 14.7 Wall deflection in sequential excavation and strutting (conceptual).

Earth pressure

One of the most complex aspects of the behaviour of braced cuts in cohesive soil is the distribution of earth pressure on the diaphragm wall. This has important effect on the structural design of both the diaphragm wall and the struts. Obviously, the distribution of earth pressure on either side of the diaphragm wall corresponds to the K_o-stresses before the excavation commences (Som and Raju, 1989). The first stage of excavation allows the diaphragm wall to move freely as a cantilever although the rigidity of the wall may not allow the earth pressure to come down to the active value by the time the first strut is placed—normally, 2–3 m below ground level. Thereafter, the movement of diaphragm wall is restricted by the presence of the strut and stress concentration occurs in the vicinity of the strut as excavation is done below. Similar phenomenon occurs at the second and subsequent strut levels when excavation is continued below the respective struts. The earth pressure that develops is greatly dependent on the strut spacing and the rigidity of the diaphragm wall itself. However, at all stages, it is to be expected that the total earth pressure on the wall would be greater than the active pressure corresponding to that depth. Peck's apparent earth pressure diagram, Fig. 14.8, is generally used to determine the earth pressure distribution in braced cut (Peck, 1969).

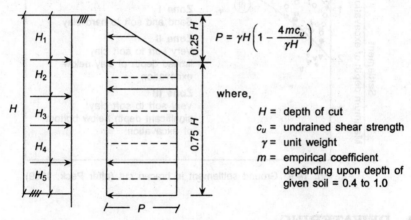

$$P = \gamma H \left(1 - \frac{4mc_u}{\gamma H}\right)$$

where,

H = depth of cut
c_u = undrained shear strength
γ = unit weight
m = empirical coefficient depending upon depth of given soil = 0.4 to 1.0

Fig. 14.8 Apparent earth pressure diagram in braced cut (after Peck, 1969).

Strut load

Estimation of strut load in braced excavation is a rather complex problem. The strut load depends primarily on the rigidity of the diaphragm wall, spacing of struts and, of course, the soil parameters at a given site. Obviously, the strut load would be a function of the lateral earth pressure that develops on the diaphragm wall at any stage of excavation. But the latter is a highly indeterminate phenomenon being primarily a function of the interface soil structure interaction. Based on extensive measurement of strut load in the Chicago subway, Peck proposed the apparent earth pressure distribution, mentioned above, for the estimation of strut load. However, the actual strut loads that are going to develop in a given situation may only be characteristic of the type of wall, construction methodology, speed of construction, manner of placing the struts, and so on. Nevertheless, an empirical procedure as proposed by Peck still appears to be the best way of estimating the strut load for initial design (refer Fig. 14.8).

Ground settlement

Ground settlement is the surface manifestation of the subsoil deformation in a braced cut. Both the magnitude of settlement and the zone of influence are of importance in determining whether adjacent buildings are going to be adversely affected by the construction. The variation of ground settlement for the final cut level may be obtained from Peck's normalized plot for braced cut with sheet pile supports in different types of soil, as depicted in Fig. 14.9. Different zones of settlement proposed by Peck are shown as zones I, II, and III for different soil conditions below the cut (Peck, 1969).

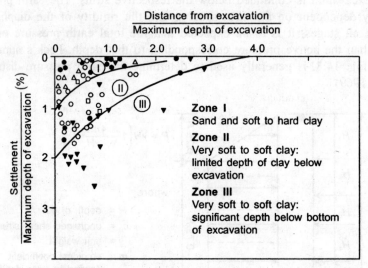

Fig. 14.9 Ground settlement in braced cut (after Peck, 1969).

14.4 DEWATERING

Dewatering is required for any deep excavation to facilitate construction of basements, power houses, pumping stations, and so on. Theoretically, any excavation below ground water table would necessitate some kind of dewatering. However, low permissibility soils, for example, clay and silty clay ($K < 10^{-6}$ cm/s) do not present much problem with seepage because the discharge is generally small and elaborate dewatering is not required. For medium to high degree of permeability ($K > 10^{-5}$ cm/s), suitable dewatering scheme has to be worked out.

The basic purpose of dewatering is to control seepage into the excavation either by pumping the water out of the excavation or by lowering the water table sufficiently below the bottom of the cut till the underground works are over. The extent of dewatering depends on the subsoil stratification, presence of water bearing stratum, aquifer parameters, and in-situ permeability.

14.4.1 Rate of Seepage

The first requirement in a dewatering job is to estimate the rate of seepage into the excavation for a certain degree of dewatering. For this, the excavation is considered as a large circular well from which water is pumped out to affect the desired ground water lowering. For a fully or partially penetrating well, as shown in Fig. 14.10, the rate of pumping is given by the well equation.

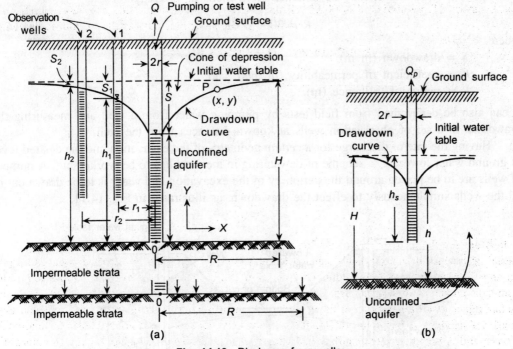

Fig. 14.10 Discharge from wells.

(a) Fully penetrating well [refer Fig. 14.10 (a)]

$$Q = \frac{\pi K(H^2 - h^2)}{\log_e (R/r)} \tag{14.3}$$

(b) Partially penetrating well [refer Fig. 14.10 (b)]

$$Q_p = Q\left[\frac{h_s}{H}\left(1 + 7\sqrt{\frac{r}{2h_s}}\ \cos\frac{\pi h_s}{2H}\right)\right] \tag{14.4}$$

where,

    Q = discharge from a fully penetrating well in an unconfined aquifer (m³/s),

    Q_p = discharge from a partially penetration well in an unconfined aquifer (m³/s),

    K = coefficient of permeability of soil (m/s),

    H = height of ground water above the top of aquifer before drawdown (m),

    h = height of water in the well above the top of aquifer after drawdown (m), that is, drawdown = $H - h$(m),

    r = radius of well or equivalent well (m),

    R = radius of influence (m), and

    h_s = penetrating of well below water table (m).

The radius of influence R may be obtained from Sichardt's expression

$$R = 3000 \, S\sqrt{K} \tag{14.5}$$

where,

S = drawdown (in m),

K = coefficient of permeability (in m/s), and

R = radius of influence (m).

R can also be determined from field tests by pumping water from a well and measuring the drawdown in a set of observation wells at known distances from the well.

Having the rate of discharge for a certain geometry of the excavation and the desired level of ground water lowering, a scheme of dewatering in the field has to be worked out. A number of wells are to be placed around the periphery of the excavation and water is to be drawn out of all the wells simultaneously to effect the drawdown, as illustrated in Fig. 14.11.

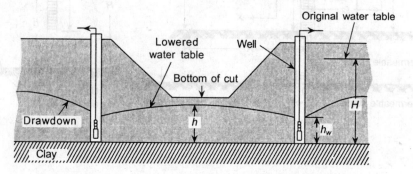

Fig. 14.11 Dewatering from an excavation.

14.4.2 Methods of Dewatering

The common dewatering methods for field application are

(a) Sump pumping

(b) Deep well pumping

(c) Well point dewatering

Sump pumping

If the soil consists of cohesive material having low permeability, shallow trenches are dug along the outer edge of the excavation to collect the water and conduct it to shallow sumps from which it is pumped out. Figure 14.12 is a schematic representation of sump pumping. This method is particularly suitable in clay and silty clay where the rate of seepage is low. Water is to be pumped out at the same rate at which seepage is occurring. The pump capacity is to be determined for the rate of discharge, the lift, and location of the discharge point.

This method is inexpensive and easy to operate. However, if ground water flows towards the excavation with high gradient, there is risk of soil loss and slope failure. The greatest depth to which water table may be lowered by this method is 5–6 m below the pump level. For greater depth, pumps are re-installed at intermediate depths.

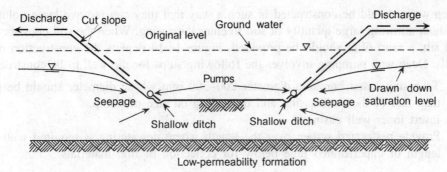

Fig. 14.12 Sump pumping.

Deep well pumping

Deep wells are installed along the sides of the excavation and water is continuously pumped out. This brings down the water table and a drawdown curve passing through the bottom of the cut is obtained. The diagrammatic representation of the same is given in Fig. 14.13. This method is particularly suitable for highly permeable soil such as sand and silty sand.

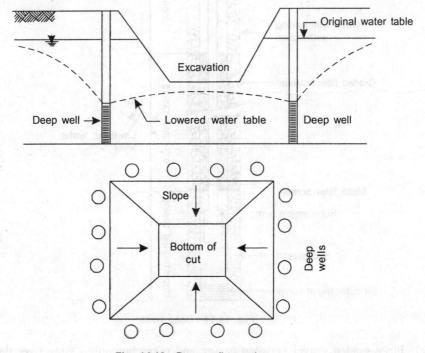

Fig. 14.13 Deep well pumping.

The method is sufficiently flexible. A group of wells is first installed and the effect of pumping is observed. If sufficient control is not achieved additional wells and pumps can be installed. Standby pumps and power supplies should be on hand in case of breakdown of equipment and drop of power supply.

Deep wells should be constructed in such a way that they can remove large volumes of water without allowing large quantity of soil to enter the casing. When the soil consists of fine sand and silt, a sand filter should be provided. Figure 14.14 depicts the construction of such deep wells. Deep well pumping involves the following steps for the well to be constructed.

(i) Take sink cased borehole diameter 200–300 mm whose diameter should be greater than that of the well casing and depending on size of pump.

(ii) Insert inner well casing.

(iii) Provide perforated screen over the length where dewatering is required with 3.5 m length of unperforated pipe below for collection of fine materials.

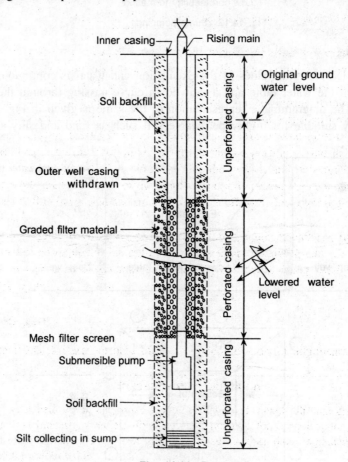

Fig. 14.14 Deep wells.

(iv) Place graded gravel between casing and outer borehole casing over the dewatering length.

(v) Fill the space above the screen by soil.

(vi) Water is then 'surged' by a boring tool to promote flow of water back and forth through the filter and fines collected at the bottom are bailed out.

(vii) Insert submersible pump.

The filter is designed on the conventional filter criteria suggested by Terzaghi which is

(a) $\dfrac{D_{15} \text{ (of filter)}}{D_{85} \text{ (of soil)}} < 4$ (b) $\dfrac{D_{15} \text{ (filter)}}{D_{15} \text{ (soil)}} > 5$

       (Piping criterion)   (Permeability criterion)

For slotted or perforated pipes,

(c) $\dfrac{D_{85} \text{(filter)}}{\text{width or diameter of hole}} > 1$

to prevent soil loss through the openings.

Well point dewatering

This consists of installation of a number of well points, usually 1 m long, around the excavation. They are connected by vertical riser pipes to a header pipe on the ground which, in turn, is connected to a pump. The ground water is drawn by the pump into the header pipe through the well points and discharged there. The well points are installed at 1–1.5 m spacing, as shown in Fig. 14.15.

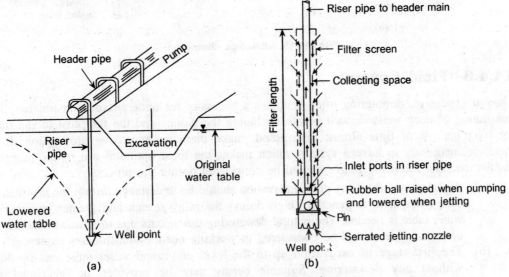

Fig. 14.15 Well point dewatering.

The filter wells or well points usually consist of a 1 m long screen, 60–75 mm diameter surrounding a central riser pipe. The well points are installed by pushing or jetting them into the ground.

Capacity of a single well point with 16 m riser pipe = 10 litres/min. (approx.)

The spacing of well points depends on the permeability of soil and the time available for affecting the drawdown. The spacings normally adopted for different soil types are as follows:

Fine to medium sand   0.75–1.0 m
Silty sand of low permeability   1.5 m

Well point dewatering is suitable in both fine and silty sands. In highly previous soil, the spacing required to handle the water may be too small and impractical. Well points are not generally suitable for clays because of slow water seepage.

Well points can lower a water table to a maximum of 6 m below the header pipe. For lowering water table to greater depth, multiple stage well point system may be adopted. Under average conditions, any number of stages can be used, each stage lowering the water table by about 5 m. A typical set-up for a two stage system is shown in Fig. 14.16. Multistage well point may be used for greater depth of dewatering but this requires additional header pipe, additional pumps, and larger excavation width for the provision of berms.

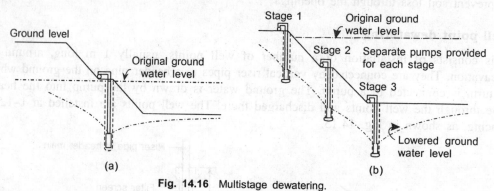

Fig. 14.16 Multistage dewatering.

14.4.3 Field Control

Though expensive, dewatering often becomes a necessity for underground construction. The installation of deep wells or well points, including the pumps and the fuel cost to run them for long periods of time almost interrupted, make them an expensive proposition. It is, therefore, necessary to have a system which makes the most optimum use of the pumping installation. The following points should be considered during the process:

(a) As far as possible, any deep excavation should be undertaken during the dry season. The water table is generally high during the rainy season and greater lowering of water table is needed. The natural dewatering that occurs due to seasonal fluctuation of water table should be considered in working out the construction sequence.

(b) The first stage of excavation up to the level of ground water table can be done without any dewatering. Suitable berms may be provided at this level for installation of deep wells/well points before commencing further excavation.

(c) The rate of discharge obtained by using well equation gives only an indication of the rate of discharge to be expected for a given drawdown. There is no need to put all the deep wells/well points into operation right at the beginning.

(d) Initially a limited number of wells may be commissioned and depending on field observation, further wells may by added taking into consideration the gradual excavation process. The number of wells required for a certain depth of cut may be installed initially. Further, additions may be done as the depth of cut is increased.

(e) Adequate number of standby pumps should be available at site particularly when concreting is in progress so that any pump breakdown may be compensated without loss of time.

(f) The effect of ground subsidence on adjacent structures due to dewatering should be evaluated and adequate precautions taken to prevent major damage.

14.5 LAND FILLING

In many construction projects, low lying areas are to be reclaimed to raise the formation level of the ground. Even in an apparently high ground, often there are ponds and depressions which had been filled up in the past. In general, except for large scale land reclamation, not much attention is given to such landfilling. All kinds of materials, including garbage and rubbish are used to fill up these areas. A few years after filling when grass has grown on the land, there is no apparent indication of filling and problems arise only during construction. Any load bearing fill should be done with care and all data pertaining to the filling, namely type of soil, method of compaction, date of filling, and so on, should be properly recorded so that the relevant data pertaining to the foundation design are available during design.

14.5.1 Cohesive Fill

Cohesive soil may be used for land filling in dry areas. Compaction is done in layers, 250–300 mm thick, with a number of roller passes to achieve the desired degree of compaction. Prior to placement, the physical properties of the soil, including grain-size, consistency limits, water content and the like should be determined. Soils with high organic content and high expansive potential should be avoided in filling work.

The compaction characteristics of cohesive soil are determined in the laboratory from the standard or modified Proctor Compaction Test. Representative soils from the borrow area are tested to obtain the optimum moisture content (OMC) and the maximum dry density (MDD).

Most compaction specifications provide for field compaction at 90–95% of MDD as determined from laboratory compaction tests. Figure 14.17 gives the typical compaction curve for clayey soil. Higher compaction in the modified proctor test gives higher MDD and lower OMC than the standard proctor test. The curve of dry density versus moisture condition for zero air voids is superimposed on the proctor test data for the purpose of field control. It is generally a good practice to have the field moisture content slightly on the wet side of the OMC to ensure that all void spaces of the compacted soil are filled with water.

It may be noted here that not all soils yield the same type of compaction curve as shown in Fig. 14.17. Lee and Suedkamp (1972) reported the test results of 35 samples and suggested four types compact ion curves that are shown in Fig. 14.18. They also gave some guidelines for predicting the nature of compaction curve based on the liquid limit of the soil (Table 14.1).

Table 14.1 Type of compaction centre

Liquid limit of soil	Nature of compaction curve (Fig. 14.18)
30–70	Type I
< 30	Type II and III
> 70	Type III and IV

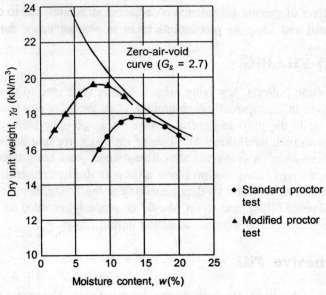

Fig. 14.17 Dry density versus moisture content relationship from Proctor test.

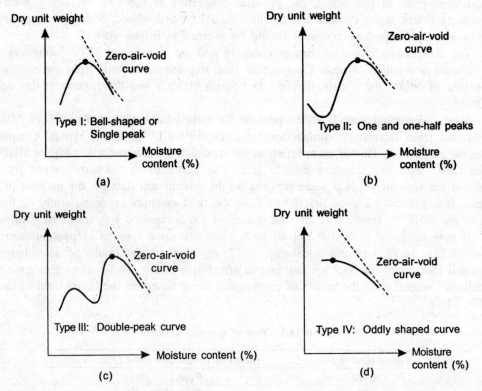

Fig. 14.18 Typical moisture density relationships of cohesive soil (after Lee and Suedkamp, 1972).

It may however be understood that irrespective of the type of compaction curve, the highest value of MDD as marked by point A should be taken for specifying the field compaction parameters in cohesive soil.

14.5.2 Granular Fill

Granular soil, predominantly sand, happens to be the most suitable material for land filling. Well graded sand ($D_{60}/D_{10} > 4$) gives good compaction when saturated and vibrated. These give high bearing capacity and low settlement potential and sand appears to be best suited for under water filling.

The compaction of granular soil is determined by the relative density, defined as

$$R_p = \frac{e_{max} - e}{e_{max} - e_{min}} \times 100\% \tag{14.6}$$

where,

e_{max} = maximum void ratio under loose condition,

e_{min} = minimum void ratio under dense condition, and

e = void ratio achieved at site.

In terms of dry density, the relative density can be expressed as

$$R_p = \left(\frac{\gamma_d - (\gamma_d)_{min}}{(\gamma_d)_{max} - (\gamma_d)_{min}} \right) \left(\frac{(\gamma_d)_{max}}{\gamma_d} \right) \tag{14.7}$$

where,

γ_d = dry density achieved in the field,

$(\gamma_d)_{max}$ = maximum dry density as determined in the laboratory, and

$(\gamma_d)_{min}$ = minimum dry density as determined in the laboratory.

For field compaction, a relative density of at least 80% should be specified. This can be easily achieved by placing the loose sand in 300 mm layers, flooding it with water and then compacting it with vibratory rollers to layers 250 mm thick.

Numerous buildings have suffered damages due to inadequate attention given to the ground filling in the construction area. Som and Sahu (1993) report the collapse of a 4-storey building soon after construction near Calcutta. The building was supported on a central raft with two-way interconnected strip placed 5.3 m below the ground which had indicated a bowl-shaped profile sloping across the building. A differential excavation was made to locate the foundation below the lowest point. Subsequent filling on the foundation put an overburden pressure of varying magnitude across the building area. This caused a non-uniform foundation pressure varying from 16 t/m^2 on western side to 11 t/m^2 on the eastern side of the foundation, as shown in Fig. 14.19. The factor of safety against bearing capacity failure was only 1.5. This is believed to have led to significant yielding of the soil and the building tilted towards the western face due to heavier stress concentration. Subsequent consolidation settlement (the building collapsed almost 3 years after start of construction) appears to have led to further tilting of the building and an estimated angular distortion of 1/86 occurred towards the western side as against a permissible angular distortion of 1/300 for conventional RCC framed structures. Analysis of the building frame and the nondestructive tests on concrete had shown that the beams and columns were not adequately designed to withstand the additional stresses due to excessive angular distortion. The failure of individual structural members, one by one, appears to have led to the ultimate collapse of the building.

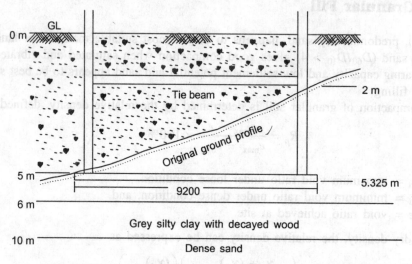

Fig. 14.19 Foundation on differential filling (Som and Sahu, 1993).

Som (2000) reported the severe tilt in a 4-storey residential building built on filled up soil. The subsoil consisted of 2 m top soil followed by 3 m of rice mill waste which was used to fill up the low lying ditches in the area. No soil tests were done before construction. The building was provided with RCC raft foundation with a net bearing pressure of 50 kN/m^2, as shown in Fig. 14.20.

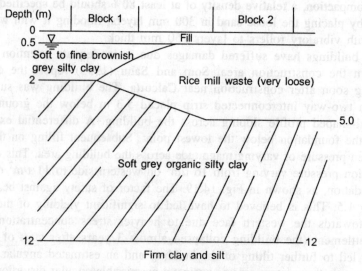

Fig. 14.20 Building tilt on filled up soil.

The building started to tilt soon after construction and in spite of cement grouting in the surrounding area, it went on to tilt by as much as 800 mm within one year after construction due to the compression of the fill. Subsequently, counterbalancing weight on the heaved up portion and removal of soil from underneath by pumping water through perforated pipes inserted below the building gradually brought the building back to verticality, although the ground floor ended up 800 mm below the surrounding ground.

Major land reclamation is done in submerged ground or even on the sea bed by sand filling. Sand compacts well under water and provides good foundation bed for structures to be built subsequently. Salt Lake City near Calcutta was reclaimed in the early sixties by pumping dredged silty sand from the river Hooghly on soft clay 2–3 m depth of sand was sedimented on the low lying land to raise the ground surface and also to provide good foundation for one or two storeyed residential buildings (refer Fig. 14.19). In recent years, Kansai international airport in Osaka, Japan has been built on an airport island created on the sea bed by reclamation with 150 million cu m of sand fill (Soda, 2001). The sea bed soil consists of 20 m thick alluvial clay with further depth of alluvial clay and sand, gravel extending to several hundred metres below the sea bed, the depth of reclaimed soil being 30 m. Sand drains were installed in the sea bed prior to land filling to force the compressible clay to undergo consolidation during the reclamation itself. A pilot construction area was set up before the full scale reclamation. Settlement studies were made to evaluate the settlement potential of the reclaimed ground. Such observational method helped to build and monitor the foundations of the airport structures.

14.6 EFFECT ON ADJOINING STRUCTURES

Underground construction at any location affects surrounding areas by way of ground movement, vibration and the like. Every construction has a zone of influence and facilities existing without this zone are liable to get affected. Whether these will be any damage to a building will depend on the severity of such influence and the condition of the building.

Excavation during construction may cause damage to existing structures due to the effects of ground movement in the vicinity of the excavation. It is to be remembered that an existing building has, in all probability, undergone settlement under its own weight and has probably reached the limit of permissible settlement already. Any additional settlement would, therefore, take the building beyond permissible limits and damages may occur. Proper bracing system has, therefore, to be adopted to protect the sides of the excavation.

As a general rule, the ground settlement due to braced excavation extends to a distance of three times the depth of the cut, as depicted in Fig. 14.21. All buildings within this zone should be thoroughly surveyed. Wherever necessary, photographs should be taken to assess the condition of the building. Settlement points should be established on the buildings plinth and measurements taken with reference to suitable benchmarks to monitor the movement during and after construction. These will provide valuable evidence if claims of damage are to be faced subsequently.

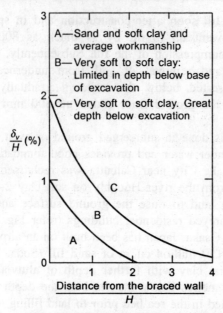

A—Sand and soft clay and average workmanship

B—Very soft to soft clay: Limited in depth below base of excavation

C—Very soft to soft clay. Great depth below excavation

$\frac{\delta_v}{H}$ (%)

Distance from the braced wall
H

Fig. 14.21 Influence zone in braced cut.

Time is the essence of underground construction. Faster the underground works are completed, lesser will be the problems of settlement. Deep excavation should never be kept open for long periods of time as this invites additional settlement and damage to adjoining structures. Figure 14.22 shows the effect of Calcutta metro construction on a building close

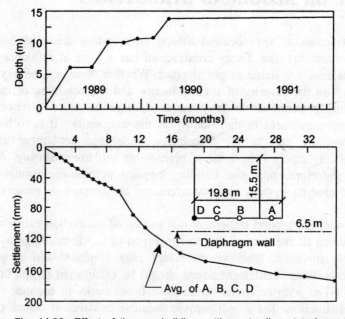

Fig. 14.22 Effect of time on building settlement adjacent to braced cut.

to the diaphragm wall. For various reasons, the metro work was stopped repeatedly at this section and the construction took the best part of three years. The excavation was kept open f : 90 days at 8 m depth and thereafter for almost two years at 14 m depth. It may be noticed that out of a total settlement of 172 mm, no less than 93 mm had occurred during stoppages of construction. The data clearly show the effect of time on ground settlement adjacent to a braced cut.

14.6.1 Effect of Vibration

Vibrations caused by construction operations such as blasting, pile-driving, compaction by drop hammer, compaction piles, and so on can damage existing structures and equipment if not designed to resist such motions. The peak particle velocity near a building due to a source of single impact is shown in Fig. 14.23 (Mayne et al., 1982).

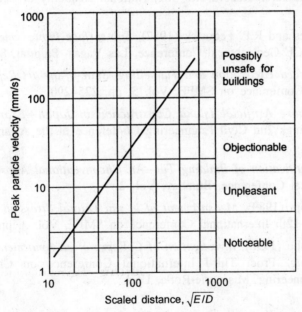

Fig. 14.23 Particle velocity versus distance due to a single impact (after Mayne et al., 1982).

A safe upper limit for small buildings, is given as

$$v = 70 \left(\frac{E}{10 D^2} \right)^{0.56} \tag{14.8}$$

where,

E = energy per blow in KJ, and
D = distance between the source and the structure under consideration.

A maximum peak particle velocity of 50 mm/s is normally considered safe for conventional buildings.

If the vibration level is intolerable, the soil foundation system may be modified to minimize the effect of vibration. This may be done by providing vibration barriers. Trenches may be made around the vibration source for *active isolation* or may be made near the object to form *passive isolation*. For active isolation, the depth of trench should be 0.6 times the distance from the structure. This would reduce the ground motion to one quarter of the amplitude of 'no trench' condition. For passive isolation, a semicircular plan area behind the trench at a distance L_R with the depth of trench equal to $1.33L_R$ would give effective isolation.

REFERENCES

Lee and R.J. Suedkamp (1972), *Characteristics of Irregular Shaped Compaction Curves of Soils,* Highway Research Record No. 381, National Academy of Seeing, Washington DC, pp. 1–9.

Mayne, P.W., J.S. Jones and R.P. Fedosick (1982), *Sub-surface Improvement due to Impact Densification,* ASCE Geotechnical Conference, Las Vagas, Prepriat, Session B 2.

Peck, R.B. (1969), *Braced Excavation and Tunnelling, State-of-the-art Report,* Proceedings 5th International Conference on SMFE, Vol. 3, pp. 225–290.

Soda, S. (2001), *Offshore Artificial Island Construction in Japan—Example of Offshore Airports,* Proceedings 2nd Civil Engineering Conference in the Asian Region, Tokyo. p. 77.

Som, N.N. (2000), *Rectification of Building Tilt—An Unconventional Approach,* Proceeding Indian Geotechnical Conference, Bombay, Vol. 2.

Som, N. and I.R.K. Raju (1989), *Measurement of in-situ Lateral Stress in Normal Calcutta Soil,* Proceedings 12th International Conference on SMFE, Vol. 3, pp. 1519–1522.

Som, N.N. and R.B. Sahu (1993), *Investigation of Collapse of an Apartment Building due to Differential Filling,* Proc. Third International Conference on Case Histories in Geotechnical Engineering, Missoorie-Rolla, USA.

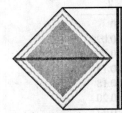

Appendix

(a) Vertical stress influence coefficient for uniform circular load

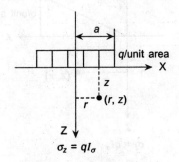

$$\sigma_z = qI_\sigma$$

Table A.1   Values of I_σ

z/a \ r/a	0	0.5	1.0
0	1.000	1.000	1.000
0.5	0.910	0.790	0.417
1.0	0.646	0.560	0.332
1.5	0.424	0.375	0.256
2	0.286	0.258	0.196
4	0.087	0.083	0.075

(b) Vertical stress influence coefficient for uniformly distributed strip load

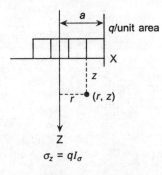

$$\sigma_z = qI_\sigma$$

Table A.2 Values of I_σ

z/a\\x/a	0	0.5	1.0	1.5	2
0	1.00	1.00	0.50	0	0
0.5	0.96	0.90	0.49	0.09	0.02
1.0	0.82	0.73	0.38	0.25	0.08
1.5	0.67	0.61	0.45	0.27	0.15
2	0.55	0.51	0.41	0.29	0.18
3	0.39	0.40	0.30	0.28	0.20
4	0.31	0.30	0.25	0.26	0.22

(c) Vertical stress influence coefficient for triangular load

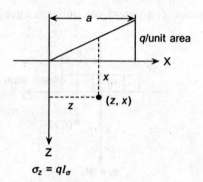

$$\sigma_z = qI_\sigma$$

Table A.3 Values of I_σ

z/a\\x/a	0	0.5	1	2
0	0	0.5	0.5	0
0.5	0.13	0.41	0.35	0.03
1	0.16	0.28	0.25	0.07
2	0.13	0.16	0.15	0.08

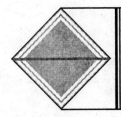

Index